Physiological and Molecular Biology Research on Ornamental Flower

Physiological and Molecular Biology Research on Ornamental Flower

Editors

Caiyun Wang
Tuo Zeng

MDPI

Basel • Beijing • Wuhan • Barcelona • Belgrade • Novi Sad • Cluj • Manchester

Editors
Caiyun Wang
Huazhong Agricultural University
Wuhan
China

Tuo Zeng
Guizhou Normal University
Guiyang
China

Editorial Office
MDPI AG
Grosspeteranlage 5
4052 Basel, Switzerland

This is a reprint of articles from the Special Issue published online in the open access journal *Horticulturae* (ISSN 2311-7524) (available at: https://www.mdpi.com/journal/horticulturae/special_issues/B51E576FB0).

For citation purposes, cite each article independently as indicated on the article page online and as indicated below:

Lastname, A.A.; Lastname, B.B. Article Title. *Journal Name* **Year**, *Volume Number*, Page Range.

ISBN 978-3-7258-1491-6 (Hbk)
ISBN 978-3-7258-1492-3 (PDF)
doi.org/10.3390/books978-3-7258-1492-3

Contents

About the Editors

Caiyun Wang

Dr. Caiyun Wang is a professor at the National Key Laboratory for Germplasm Innovation and Utilization of Horticultural Crops, College of Horticulture and Forestry Sciences, Huazhong Agricultural University. She was promoted to professor in 2002 and received her Ph.D. from Beijing Forestry University in 2005. From 2003 to 2004, she was a visiting scholar at Wageningen University in The Netherlands. Dr. Wang's research, spanning over two decades, primarily focuses on the physiological and molecular biology of pyrethrum and chrysanthemums, emphasizing both ornamental and functional floral qualities. Her team has advanced the selection and green cultivation techniques of ornamental and insect-resistant chrysanthemums, and actively promoted these methods. Additionally, her team has made significant contributions to the study of osmanthus germplasm resources, particularly in enhancing the quality and regulation of their fragrance and color. Dr. Wang has published over 200 peer-reviewed articles as a corresponding author.

Tuo Zeng

Dr. Tuo Zeng is an associate professor at the College of Life Sciences, Guizhou Normal University. He holds a bachelor's degree in Applied Chemistry and a master's degree in Biochemistry and Molecular Biology. He obtained his Ph.D. in Ornamental Horticulture from the College of Horticulture and Forestry Sciences at the Huazhong Agricultural University in 2022. His research primarily focuses on the regulation of gene expression and molecular biology in pyrethrums and chrysanthemums, especially the varieties valued for their ornamental and insecticidal properties. Since joining Guizhou Normal University, Dr. Tuo has continued his earlier research and expanded into new areas, such as crop stress resistance, genomics, and the exploration and utilization of horticultural germplasm resources. He has published nearly 30 scientific articles and currently serves as a guest editor for this Special Issue of the journal *Horticulturae*.

horticulturae

MDPI

Editorial

Ornamental Plant Physiology and Molecular Biology

Tuo Zeng [1,*] and Caiyun Wang [2,*]

[1] School of Life Sciences, Guizhou Normal University, Guiyang 550025, China
[2] National Key Laboratory for Germplasm Innovation & Utilization of Horticultural Crops, College of Horticulture & Forestry Sciences, Huazhong Agricultural University, Wuhan 430070, China
* Correspondence: zengtuo@gnzu.edu.cn (T.Z.); wangcy@mail.hzau.edu.cn (C.W.)

1. Introduction

Ornamental plants not only beautify our environment, thanks to their vivid colors and diverse forms, but also play a vital role in agriculture, industry, and medicine, reflecting significant scientific and economic value. However, the genetic and physiological complexities that contribute to their unique characteristics are often underexplored.

This Special Issue, titled "Ornamental Plant Physiology and Molecular Biology", seeks to elucidate the underlying mechanisms governing the growth, development, and stress responses of ornamental plants. It presents a carefully selected collection of research papers that spotlight cutting-edge advances in the field. The topics covered include novel insights into alterations in flower color and scent, the development and utilization of ornamental flowers, and environmental adaptation regulation. Our aim in publishing this Special Issue is to emphasize this collection's importance and potential to enhance our understanding of ornamental flower physiology and molecular biology, alongside their applications in horticulture.

2. Floral Color in Ornamental Plants

Flower color is a key trait that attracts pollinators in nature and captivates human interest, significantly influencing the ornamental plant industry. The diverse palette of colors displayed by flowers is not merely a visual delight; it signifies a complex interaction of genetics, environmental factors, and evolutionary processes [1,2]. Advances in biotechnology have deepened our understanding of the molecular mechanisms underlying flower color. Through the manipulation of genetic and biochemical pathways, scientists can now introduce novel hues and patterns into ornamental plants, enhancing their appeal and marketability. Techniques such as mutagenesis, induced by gamma rays and X-rays, have enabled the creation of new chrysanthemum varieties with minimal impact on plant health, promoting sustainable horticultural practices (contribution 1).

Transcriptomic and metabolomic integration play critical roles in exploring flower color, offering valuable clues for enhancing color modification breeding programs [3]. By examining the expression and metabolic profiles of different *Rhododendron liliiflorum* petal colors, researchers can identify the genes and compounds responsible for these variations, which are crucial for developing new plant varieties with desirable traits (contributions 2 and 3).

Furthermore, research on leaf color variation in *Primulina serrulata* provides insights into different aspects of plant pigmentation. Proteomic analyses have shown how variations in leaf coloration, especially the species' distinctive white markings, are linked to alterations in chloroplast biosynthesis and photosynthesis. These findings highlight the potential for targeted genetic and cultivation strategies to enhance leaf variegation, adding another layer of value to ornamental plants (contribution 4).

Collectively, these research endeavors significantly enrich our understanding of ornamental plant pigmentation, guiding the breeding of new esthetically pleasing varieties.

Citation: Zeng, T.; Wang, C. Ornamental Plant Physiology and Molecular Biology. *Horticulturae* **2024**, *10*, 532. https://doi.org/10.3390/horticulturae10060532

Received: 8 May 2024
Accepted: 16 May 2024
Published: 21 May 2024

3. Floral Scent in Ornamental Plants

Floral scent has long been integral to human culture and commerce, serving roles beyond mere esthetic appeal. Floral fragrances facilitate vital interactions between plants and pollinators [4,5], while also being exploited across various industries, including perfumery and pharmaceuticals, highlighting their significance in both natural and man-made environments.

Research on species like *Iris uniflora*, *Iris typhifolia*, and *Iris sanguinea* has identified specific aromatic compounds, such as terpenes and sulfur-containing molecules, that contribute to their alluring fragrances and medicinal properties, thus enhancing their economic value. These discoveries underscore these compounds' potential to aid in improving food flavor profiles, developing new pharmaceutical applications, and crafting sophisticated fragrances (contribution 5).

In studies on ornamental and medicinal plants like *Dendrobium*, genetic pathways critical for scent production, particularly those involved in terpene synthesis, have been mapped. Genes encoding linalool synthase play a vital role in producing rich, complex scents. Understanding these genetic pathways facilitates the breeding of new plant varieties with customized aromatic profiles, presenting exciting opportunities for agricultural, horticultural, and commercial applications (contribution 6).

4. Nutritional Development and Utilization of Ornamental Plants

Ornamental plants are renowned for their esthetic value, enhancing landscapes and enriching our living spaces with their vibrant colors and scents. However, their significance extends beyond visual beauty, as many of these plants also possess substantial nutritional and medicinal properties that have been utilized in various cultures for centuries [6]. In ornamental plants, flower development is not only a key aspect of their reproductive cycle but also a critical phase during which significant biochemical transformations take place. These transformations often result in accumulating valuable phytochemicals, such as anthocyanins, which have well-documented health benefits. Understanding these processes can lead to a more effective use of ornamental plants in the food and pharmaceutical industries [7].

For instance, *Cercis chinensis* is valued not only for its striking flowers but also for its role in traditional medicine. These flowers are rich in anthocyanins, serving as a natural source of red pigments widely used in the food industry. Ongoing research focuses on the nutritional and antioxidative changes occurring during the flower development of *C. chinensis* to optimize harvest times and maximize the extraction of these functional components for broader applications in pharmaceuticals and food products (contribution 7).

Similarly, *Michelia maudiae 'Rubicunda'*, known for its esthetic and aromatic qualities, contains a rich array of nutrients crucial for flower development. Research into hormonal balances such as ABA, GA3, and CTK during the flowering stages of *M. maudiae* provides valuable insights into flower bud differentiation and physiological mechanisms. This knowledge not only deepens our understanding of plant biology but also aids in breeding new varieties with optimized characteristics for both decorative and practical uses (contribution 8).

5. Adaptive Strategies of Ornamental Plants against Environmental Stress

Ornamental plants are widely appreciated for their visual appeal, yet beneath this beauty lies a complex array of adaptive strategies that enable them to thrive under environmental stresses. These adaptations are crucial not only for the survival of the plants but also for preserving their ornamental value in landscapes facing diverse climatic challenges. Understanding these strategies is essential for optimizing horticultural practices and enhancing plant resilience in natural and cultivated environments. For instance, in response to drought, plants can adapt through physical modifications, biochemical adaptions, hormonal regulation, and transcriptional reprogramming [8,9]. Recent research has illuminated the genetic and physiological mechanisms that ornamental plants utilize to manage adverse conditions. For example, studies on *Heimia myrtifolia* have identified specific genes responsible for drought tolerance, providing insights into the molecular foundations that support this sun-loving shrub in conditions of water scarcity (contribution 9). Similarly, research on *Impatiens uliginosa* has shown that environmental stressors like elevated copper levels can induce significant changes in petal coloration by altering pigment biosynthesis genes (contribution 10).

Additionally, research on Crassulacean Acid Metabolism (CAM) in plants highlights a significant evolutionary adaptation to arid conditions. CAM plants, including many ornamental varieties, demonstrate an impressive ability to switch between C3 and CAM photosynthesis, facilitating efficient water usage and improved drought tolerance. This photosynthetic flexibility not only aids plant survival during water deficits but also preserves their esthetic qualities, making them especially valuable in drought-prone areas (contribution 11).

The exploration of these adaptive strategies enriches our understanding of plant resilience and assists in the development of robust ornamental plants that can maintain their beauty and functional roles in landscapes despite environmental challenges.

6. Conclusions

This Special Issue on "Ornamental Plant Physiology and Molecular Biology", collating exciting new research on aspects ranging from flower color variations, scent profiles, nutritional content, and resilience to environmental stresses, provides fundamental insights into the genetic and molecular mechanisms driving these traits. The advancements discussed herein not only deepen our understanding of plant biology but also offer practical applications that can lead to the development of more robust and visually appealing plant varieties fulfilling both commercial and ecological demands.

Author Contributions: Writing—original draft preparation, Writing—review and editing, T.Z. and C.W. All authors have read and agreed to the published version of the manuscript.

Funding: This research received no external funding.

Conflicts of Interest: The authors declare no conflicts of interest.

List of Contributions

1. Haspolat, G. Variations in Flower Color of Mutant Chrysanthemums. *Horticulturae* **2024**, *10*, 385.
2. Zhang, H.; Chen, M.; Wang, X.; Dai, J.; Zhang, X.; Zhang, Z.; Zhang, X.; Tang, M.; Tang, J.; Gong, J.; et al. Transcriptome Analysis of *Rhododendron liliiflorum* H. Lév. Flower Colour Differences. *Horticulturae* **2023**, *9*, 82.
3. Dai, J.; Wang, X.; Meng, X.; Zhang, X.; Zhou, Q.; Zhang, Z.; Zhang, X.; Yi, Y.; Liu, L.; Shen, T. UPLC–MS/MS and Gene Expression Research to Distinguish the Colour Differences of *Rhododendron liliiflorum* H. Lév. *Horticulturae* **2023**, *9*, 1351.
4. Dou, Q.-L.; Xie, D.-J.; Deng, T.; Chen, M.-F.; Qian, Z.-M.; Wang, S.-S.; Zhang, R.-B. Comparative Proteomics Analysis of *Primulina serrulata* Leaves Reveals New Insight into the Formation of White Veins. *Horticulturae* **2024**, *10*, 19.
5. Cai, K.; Ban, Z.; Xu, H.; Chen, W.; Jia, W.; Zhu, Y.; Chen, H. Analysis of Floral Scent Component of Three *Iris* Species at Different Stages. *Horticulturae* **2024**, *10*, 153.
6. Yang, Y.; Xia, K.; Wu, Q.; Lu, X.; Lu, S.; Zhao, Z.; Qiu, S. Combined Analysis of Volatile Compounds and Extraction of Floral Fragrance Genes in Two *Dendrobium* Species. *Horticulturae* **2023**, *9*, 745.
7. Ren, H.-Y.; Qian, W.-Z.; Yi, L.; Ye, Y.-L.; Gu, T.; Gao, S.; Cao, G.-X. Nutrient Composition and Antioxidant Activity of *Cercis chinensis* Flower in Response to Different Development Stages. *Horticulturae* **2023**, *9*, 961.
8. Yu, T.; Yang, Y.; Wang, H.; Qian, W.; Hu, Y.; Gao, S.; Liao, H. The Variations of C/N/P Stoichiometry, Endogenous Hormones, and Non-Structural Carbohydrate Contents

in Micheliamaudiae 'Rubicunda' Flower at Five Development Stages. *Horticulturae* **2023**, *9*, 1198.

9. Zhang, G.; Gu, C.; Ye, Y.; Zhao, Y.; Shang, L.; Shao, W.; Hong, S.; Ma, J. Characterization, Evolutionary Analysis, and Expression Pattern Analysis of the Heat Shock Transcription Factors and Drought Stress Response in *Heimia myrtifolia*. *Horticulturae* **2023**, *9*, 588.
10. Tan, Y.; Zhang, X.; Li, Q.; Li, X.; Luo, L.; He, H.; Liang, G.; Huang, H.; Huang, M. Transcriptomic Analysis of Flower Color Changes in *Impatiens uliginosa* in Response to Copper Stress. *Horticulturae* **2024**, *10*, 412.
11. Qiu, S.; Xia, K.; Yang, Y.; Wu, Q.; Zhao, Z. Mechanisms Underlying the C3–CAM Photosynthetic Shift in Facultative CAM Plants. *Horticulturae* **2023**, *9*, 398.

References

1. Sapir, Y.; Gallagher, M.K.; Senden, E. What maintains flower colour variation within populations? *Trends Ecol. Evol.* **2021**, *36*, 507–519. [CrossRef] [PubMed]
2. Dalrymple, R.L.; Kemp, D.J.; Flores-Moreno, H.; Laffan, S.W.; White, T.E.; Hemmings, F.A.; Moles, A.T. Macroecological patterns in flower colour are shaped by both biotic and abiotic factors. *New Phytol.* **2020**, *228*, 1972–1985. [CrossRef] [PubMed]
3. Ahmad, S.; Yuan, C.; Cong, T.; Yang, Q.; Yang, Y.; Zhang, Q. Transcriptome and chemical analyses identify candidate genes associated with flower color shift in a natural mutant of *Chrysanthemum* × *morifolium*. *Ornam. Plant Res.* **2022**, *2*, 19. [CrossRef]
4. Shen, Y.; Rao, Y.; Ma, M.; Li, Y.; He, Y.; Wang, Z.; Liang, M.; Ning, G. Coordination among flower pigments, scents and pollinators in ornamental plants. *Hortic. Adv.* **2024**, *2*, 6. [CrossRef]
5. Pichersky, E. Biochemistry and genetics of floral scent: A historical perspective. *Plant J.* **2023**, *115*, 18–36. [CrossRef] [PubMed]
6. Takahashi, J.A.; Rezende, F.A.G.G.; Moura, M.A.F.; Dominguete, L.C.B.; Sande, D. Edible flowers: Bioactive profile and its potential to be used in food development. *Food Res. Int.* **2020**, *129*, 108868. [CrossRef] [PubMed]
7. Teixeira, M.; Tao, W.; Fernandes, A.; Faria, A.; Ferreira, I.M.; He, J.; de Freitas, V.; Mateus, N.; Oliveira, H. Anthocyanin-rich edible flowers, current understanding of a potential new trend in dietary patterns. *Trends Food Sci. Technol.* **2023**, *138*, 708–725. [CrossRef]
8. Zhang, Y.; Gu, J.; Xia, X.; Zeng, J.; Sun, H.; Chen, F.; Fang, W.; Jiang, J.; Chen, S. Contrasting responses to drought stress between *Chrysanthemum japonense* and *C. nankingense*. *Ornam. Plant Res.* **2022**, *2*, 16.
9. Li, T.; Zhou, T.; Liang, J.; Zhang, D.; Teng, N.; Wu, Z. Overexpression of lily LlWRKY22 enhances multiple abiotic stress tolerances in transgenic Arabidopsis. *Ornam. Plant Res.* **2022**, *2*, 17. [CrossRef]

horticulturae

MDPI

Article

Transcriptomic Analysis of Flower Color Changes in *Impatiens uliginosa* in Response to Copper Stress

Yi Tan [†], Xiaoli Zhang [†], Qinmei Li, Xinyi Li, Liang Luo, Haihao He, Guangrong Liang, Haiquan Huang * and Meijuan Huang *

College of Landscape Architecture and Horticulture Sciences, Southwest Research Center for Engineering Technology of Landscape Architecture (State Forestry and Grassland Administration), Yunnan Engineering Research Center for Functional Flower Resources and Industrialization, Research and Development Center of Landscape Plants and Horticulture Flowers, Southwest Forestry University, Kunming 650224, China

* Correspondence: haiquanl@163.com (H.H.); xmhhq2001@163.com (M.H.)

† These authors contributed equally to this work.

Abstract: *Impatiens uliginosa* is a native and potential water body-restoring ornamental plant. In this study, RNA-Seq technology was used to analyze the transcriptome of its floral organs. Candidate genes related to flower color changes under copper stress were investigated through transcriptome screening, and the intrinsic mechanism of the effects of different concentrations of copper on *I. uliginosa* was revealed at the molecular level. The main findings were as follows: (1) Transcriptome sequencing analysis was performed on the flower organs of *I. uliginosa* treated with different concentrations of copper (0 $mg{\cdot}L^{-1}$, 10 $mg{\cdot}L^{-1}$, and 20 $mg{\cdot}L^{-1}$). A total of 70,319 transcripts and 39,949 unigenes were obtained. An analysis of differentially expressed genes revealed structural genes including *GT, ANS, CHI,* and *PAL*; transcription factors including MYB and WD40; and transport factors including *GST* and *ABC*. (2) The gene expression levels of flower color changed in the flowering period of *I. uliginosa* at different copper concentrations. The expression levels of *IuGT* and *IuGST* genes in *I. uliginosa* were significantly different under different concentrations of copper treatments. Their expression levels were the highest at a copper concentration of 0 $mg{\cdot}L^{-1}$ and the lowest at 20 $mg{\cdot}L^{-1}$. In summary, the low expression of *IuGT* and *IuGST* genes was more conducive to the formation of white flowers of *I. uliginosa*.

Keywords: *Impatiens uliginosa*; copper stress; transcriptome analysis; flower pigmentation changes

Citation: Tan, Y.; Zhang, X.; Li, Q.; Li, X.; Luo, L.; He, H.; Liang, G.; Huang, H.; Huang, M. Transcriptomic Analysis of Flower Color Changes in *Impatiens uliginosa* in Response to Copper Stress. *Horticulturae* **2024**, *10*, 412. https://doi.org/10.3390/horticulturae10040412

Academic Editor: Jiafu Jiang

Received: 3 March 2024
Revised: 14 April 2024
Accepted: 14 April 2024
Published: 19 April 2024

1. Introduction

The rapid development of industry and agriculture, as well as the unreasonable discharge of domestic sewage and the burning of fossil fuels, has caused urban and rural water bodies in my country to be polluted by heavy metals, hindering the utilization of water resources. Cu is one of the common heavy metal pollutants in polluted water bodies. It is widely distributed, difficult to degrade, and extremely biologically toxic. It can threaten the health of humans and other aquatic organisms through food chain accumulation and biomagnification. In southwest China, Dianchi Lake, as the largest freshwater body, carries important ecological functions. However, since the 1980s, the water quality of Dianchi Lake has begun to be seriously polluted, especially due to heavy metal pollutants, such as zinc, copper, cadmium, chromium, lead, etc. The copper content in the water is as high as 126.36 $\mu g{\cdot}L^{-1}$. However, there is no normal pollution. The average copper concentration in polluted natural water bodies is 2 $\mu g{\cdot}L^{-1}$, indicating that the current copper pollution level in Dianchi Lake water body is quite serious [1,2]. Compared with physical and chemical methods, phytoremediation technology is more popular due to its advantages of small environmental interference, easy control of secondary pollution, and low cost. It is considered to be one of the current research hotspots in the field of environmental remediation of heavy metal pollution [3]. The selection of tolerant plant species is one of the basic steps for green plant remediation of Cu-polluted water bodies. Therefore, in-depth study of the molecular mechanism of plant defense against Cu stress can provide theoretical reference for the phytoremediation of Cu water pollution.

Research on plant response to adverse stress can help reveal the response mechanism of plants to adverse environments so that preventive measures can be taken to reduce the damage to plants caused by adverse stress. Transcriptomics is a discipline that studies gene transcription and transcriptional regulation in cells at an overall level and can reveal the molecular mechanisms underlying specific biological processes and diseases. Revealing the expression at the entire genome level under stress from the entire transcription level is of great significance to increase the understanding of complex regulatory networks related to stress adaptation and tolerance and to construct the stress genome transcription regulatory network [4].

Plant glycosyltransferases (GTs) are a multi-member gene superfamily that specifically catalyzes glycosylation reactions and are responsible for transferring sugar moieties to various small molecules and controlling numerous metabolic processes. Glycosylation is one of the key modification steps in plants to produce a broad spectrum of flavonoids with different structures and colors. This process allows anthocyanins to be easily transferred from the production site in the cytoplasm to the vacuole, which shows that glycosylation plays a key role in the formation of flower color [5,6]. It also plays a crucial role in enhancing the stability and solubility of plant anthocyanins. At present, glycosyltransferases have been successfully cloned in model plants, ornamental plants, and horticultural crops such as *Arabidopsis thaliana* [7], *Nelumbo nucifera* [8], and *Prunus persica* [9].

Studies have shown that the glutathione transferase (GST) family is involved in the transport of anthocyanins [10,11]. At present, *GST* genes have been found to be crucial for anthocyanin transport in a variety of plants, such as *Chrysanthemum morifolium*, lily, and *Pericallis* × *hybrida*. For example, in *C. morifolium*, members of the GST and MATE families are involved in anthocyanin accumulation, thereby affecting the coloration of floral organs [12]; *LhGST* plays a key role in the transport and accumulation of anthocyanins in the perianth [13]; Jin [14] found that *PhGST3* in *Pericallis* × *hybrida* may be a candidate gene related to anthocyanin transport. Therefore, the study of plant GSTs in flavonoid accumulation is of great significance.

Impatiens uliginosa is an annual or perennial plant characterized by its unique flower shape, long flowering period, wide adaptability, and high ornamental value [15]. It is widely distributed in Yunnan, Guizhou, Guangxi, and other places. It often grows in forests, moist places in ditches, or by streams, at an altitude of 1500–2600 m. Research on *I. uliginosa* mainly focuses on the cloning of flower color and flower development regulatory genes [16,17], determination of metal elements [18], and the spur development [19,20]. The fact is that discussions on changes in *I. uliginosa* flower color under abiotic stress are rare, especially heavy metal stress. *I. uliginosa* is an important local plant distributed in wetlands around Dianchi Lake, serving as an aquatic flower and water purification plant. Flower color is also a significant factor in determining the ornamental value of impatiens. Based on the previous studies of our research group, it was found that the color of *I. uliginosa* flowers changed significantly under different concentrations of copper. It was observed that the flower color gradually became lighter as the copper concentration increased. However, there are no relevant studies that have reported on the impact of the flower color change so far.

In view of this, this study conducted transcriptome sequencing of *I. uliginosa* petals under Cu stress, and further explored genes related to color changes in petals under copper stress. It lays a theoretical foundation for future in-depth study of the molecular mechanism of *I. uliginosa* in defense against heavy metal toxicity, and also provides a theoretical basis for the practical application of *I. uliginosa* in restoring heavy metal-polluted water bodies.

2. Materials and Methods

2.1. Plant Materials and Treatments

I. uliginosa seeds were collected from Laoyuhe Wetland Park in Kunming, Yunnan Province. They were sown and propagated by cuttings in the Arboretum of Southwest Forestry University. The growth conditions of these plants are an average temperature of about 25 °C, 14 h of sunshine, and a relative humidity of 60% to 75%.

I. uliginosa plants with consistent traits and robust growth were selected. The plant height was about 30 cm. The root soil was rinsed with deionized water and then hydroponically cultivated for a week. Then, we used 1/2 Hoagland (without $CuSO_4$) to acclimate and culture them. Different concentrations of $CuSO_4{\cdot}5H_2O$ were then applied, and a total of 5 copper concentration gradients were set: 0 mg·L^{-1}, 5 mg·L^{-1}, 10 mg·L^{-1}, 15 mg·L^{-1}, and 20 mg·L^{-1}. The nutrient solution and treatment solution were all prepared with deionized water. In addition, in order to ensure the consistency of the experimental results and the copper stress suffered by the plants during the experiment, all treatments were replaced with newly prepared nutrient solutions containing corresponding copper concentrations every 6 days, with 10 plants in each treatment. After being treated with copper at different concentrations for 30 days, full petals at the peak flowering stage were collected for subsequent related experiments.

2.2. Experimental Methods

2.2.1. Transcriptome Sequencing and Analysis

The petals of *I. uliginosa* were stored in liquid nitrogen, and samples treated with copper concentrations of 0mg·L^{-1}, 10mg·L^{-1}, and 20mg·L^{-1} were selected, because we found that under these three concentrations, the color change in *I. uliginosa* flowers was the most significant (Figure 1), and sent to the company for sequencing and subsequent analysis.

Total RNA was extracted from the *I. uliginosa* tissue using Plant RNA Purification Reagent for plant tissue according to the manufacturer's instructions (Invitrogen, Carlsbard, CA, USA) and genomic DNA was removed using DNase I (TaKara TaKaRa, Tokyo, Japan). Then, the integrity and purity of the total RNA quality was determined by 2100 Bioanalyser (Agilent Technologies, Inc., Santa Clara, CA, USA) and quantified using the ND-2000 (NanoDrop Thermo Scientific, Wilmington, DE, USA). Only high-quality RNA sample (OD260/280 = 1.8~2.2, OD260/230 $\geq$ 2.0, RIN $\geq$ 8.0, 28S:18S $\geq$ 1.0, >1 μg) was used to construct sequencing library.

RNA purification, reverse transcription, library construction, and sequencing were performed at Shanghai Majorbio Bio-pharm Biotechnology Co., Ltd. (Shanghai, China) according to the manufacturer's instructions (Illumina, San Diego, CA, USA). The *I. uliginosa* RNA-seq transcriptome libraries were prepared using Illumina TruSeqTM RNA sample preparation kit (San Diego, CA, USA). Poly(A) mRNA was purified from total RNA using oligo-dT-attached magnetic beads and then fragmented by fragmentation buffer. Taking these short fragments as templates, double-stranded cDNA was synthesized using a SuperScript double-stranded cDNA synthesis kit (Invitrogen, CA, USA) with random hexamer primers (Illumina). Then, the synthesized cDNA was subjected to end-repair, phosphorylation, and 'A' base addition according to Illumina's library construction protocol. Libraries were size selected for cDNA target fragments of 200–300 bp on 2% Low Range Ultra Agarose followed by PCR amplified using Phusion DNA polymerase (New England Biolabs, Boston, MA, USA) for 15 PCR cycles. After being quantified by TBS380, two RNAseq libraries were sequenced in a single lane on an Illumina Hiseq xten/NovaSeq 6000 sequencer (Illumina, San Diego, CA) for 2 × 150 bp paired-end reads.

The raw paired-end reads were trimmed and quality controlled by SeqPrep (https://github.com/jstjohn/SeqPrep, accessed on 15 January 2022) and Sickle (https://github.com/najoshi/sickle, accessed on 17 January 2022) with default parameters. Then, clean data from the samples (*I. uliginosa*) were used to perform de novo assembly with Trinity (http://trinityrnaseq.sourceforge.net/, accessed on 18 January 2022). All the assembled transcripts were searched against the NCBI protein nonredundant (NR), COG, and KEGG databases using BLASTX to identify the proteins that had the highest sequence similarity with the given transcripts to retrieve their function annotations and a typical cut-off E-value less than 1.0×10^{-5} was set. BLAST2GO (http://www.blast2go.com/b2ghome, accessed on 22 January 2022) program was used to obtain GO annotations of unique assembled transcripts for describing biological processes, molecular functions, and cellular components. Metabolic pathway analysis was performed using the Kyoto Encyclopedia of Genes and Genomes (KEGG, http://www.genome.jp/kegg/, accessed on 24 January 2022).

To identify DEGs (differential expression genes) between two different samples, the expression level of each transcript was calculated according to the transcripts per million reads (TPM) method. RSEM (http://deweylab.biostat.wisc.edu/rsem/, accessed on 27 January 2022) was used to quantify gene abundances. Essentially, differential expression analysis was performed using the DESeq2/DEGseq/EdgeR with Q value ≤ 0.05; DEGs with $|\log2FC| > 1$ and Q value ≤ 0.05 (DESeq2 or EdgeR)/Q value ≤ 0.001 (DEGseq) were considered to be significantly different expressed genes. In addition, functional enrichment analysis including GO and KEGG were performed to identify which DEGs were significantly enriched in GO terms and metabolic pathways at Bonferroni-corrected p-value ≤ 0.05 compared with the whole-transcriptome background. GO functional enrichment and KEGG pathway analysis were carried out by Goatools (https://github.com/tanghaibao/Goatools, accessed on 29 January 2022) and KOBAS (http://kobas.cbi.pku.edu.cn/home.do, accessed on 31 January 2022).

Figure 1. Variation in flower color of *I. uliginosa* treated with different concentrations of copper. Note: Copper concentration: (**A**). 0 mg·L^{-1}; (**B**). 5 mg·L^{-1}; (**C**). 10 mg·L^{-1}; (**D**). 15 mg·L^{-1}; (**E**). 20 mg·L^{-1}.

2.2.2. *IuGT* and *IuGST* Gene Cloning and Sequence Analysis

The total RNA of *I. uliginosa* was extracted using Plant RNA Kit (Plant RNA kit, Omega, Norcross, GA, USA), and the first strand of *I. uliginosa* cDNA was transcribed and synthesized using a reverse transcription kit (TransGen, Beijing, China) and stored at −80 °C. The design of gene cloning primers is detailed in Table 1. PCR reaction system (20 µL): 2× *Taq* PCR StarMix (Dye) 10 µL, ddH_2O 7 µL, forward and reverse primers 1 µL each, template cDNA 1 µL. The specific program is as follows: pre-denaturation at 94 °C for 2 min, denaturation at 94 °C for 30 s, annealing at 55 °C for 30 s, extending at 72 °C for 2 min, and final extending at 72 °C for 10 min, 35 cycles. Perform gel recovery on PCR amplification products using the gel recovery kit (Biomed, Beijing, China). Connect the recovered product to the pMDTM19-T Vector carrier (TaKaRa, Tokyo, Japan) according to the instructions, and connect the carrier to the product in DH5α in receptive cells. Use PCR to screen positive clones and send them to the company for sequencing (Sangon Biotech (Shanghai) Co., Ltd., Shanghai, China). Analyze the physicochemical properties of proteins using online software ExPasy ProtParam (https://web.expasy.org/protparam/, accessed on 31 January 2022). Construct a phylogenetic tree of *IuGST* and *IuGT* using MAGA7.0.

Table 1. Primers of *IuGT* and *IuGST* genes of *I. uliginosa.*

Primer Name	Primer Sequence (5′-3′)
IuGTF	GATCCAGTGCCACTCTGAAACAAAGAAAG
IuGTR	TTAGACATGAGACAGTAAGACACTTATGAGC
IuGSTF	ATGGCAGGAAGAGATTCTTCCGTC

Table 1. *Cont.*

Primer Name	Primer Sequence (5′-3′)
IuGSTR	TTAGTTCTTAGCTTGAGATTCAACCATGGC
qIuGTF	GTCACCTTCGTCAACACCGA
qIuGTR	AGTCCATCGGGAATGGCTTG
qIuGSTF	CACTTAATCCTCCCCTCGG
qIuGSTR	TCAACCACCCCAAGAAGCAT
IuActinF	TGAATGTCCCTGCTGTTTG
IuActinR	ACCTTCCGCATAACTTTACC

2.2.3. Related Gene Expression and Analysis

Primers were designed based on the sequences of *IuGT* and *IuGST*, and *IuActin* was used as the internal reference gene. qRT-PCR was performed on samples treated with different concentrations of copper during the flowering period of *I. uliginosa* to analyze the expression differences of *IuGT* and *IuGST*. Roche: LightCycler®480 II fluorescence quantitative PCR instrument (Roche, Basel, Switzerland) was used for amplification, and qRT-PCR reaction system (20 μL): qPCR SYBR Green Master Mix 10 μL, ddH_2O 7 μL, forward and reverse primers 1 μL each, and template cDNA 1 μL. The specific program is as follows: pre-denaturation at 95 °C for 15 s, denaturation at 60 °C for 30 s, and annealing at 72 °C for 30 s, 40 cycles. Primer information is shown in Table 1. The $2^{-\Delta\Delta Ct}$ calculation method was used to analyze the relative expression of genes.

2.2.4. Data Processing

Microsoft Excel 2019 and SPSS 27.0 software were used for data analysis, and Origin 26.0 software was used for graphing.

3. Results

3.1. Transcriptome Sequencing and Assembly from De Novo

The transcriptome analysis of floral organs was conducted across various copper concentrations designated as Cu0 (0 mg·L^{-1}), Cu10 (10 mg·L^{-1}), and Cu20 (20 mg·L^{-1}). The sequencing process generated a substantial dataset, comprising 169,755,424 raw reads and 25.63 GB of raw data. After rigorous quality control measures, a refined dataset of 24.86 GB of high-quality clean data was obtained, with approximately 56 million clean reads per sample. Notably, the Q30 base quality score ranged from 94.5 to 94.58%, indicating excellent sequence quality, and the GC content exceeded 44.55%. Furthermore, the mapping ratio of all three samples exceeded 83.04%, as detailed in Table 2, demonstrating the reliability and accuracy of our transcriptome analysis.

Table 2. Summary of sequencing data of *I. uliginosa* transcriptome.

Attributes	Cu0	Cu10	Cu20
Raw reads	55,657,964	58,102,838	55,994,622
Clean reads	55,287,682	57,653,418	55,597,734
Clean bases	8,188,113,484	8,477,755,077	8,199,632,407
Q30 (%)	94.57	94.58	94.5
GC content (%)	44.55	44.69	44.82
Total mapped	22,954,371 (83.04%)	24,066,941 (83.49%)	23,599,843 (84.89%)

A de novo assembly approach was applied to the high-quality reads, resulting in the generation of 70,319 transcripts, encompassing a total of 84.6 Mb of sequence data after rigorous optimization and filtering processes. The transcripts exhibited a diverse range of lengths, with the longest transcript measuring 15,701 bp and the shortest 201 bp. The average transcript length was 1203.37 bp, and the N50 value was 1810 bp, indicating a high-quality assembly. Furthermore, a total of 40,610 unigenes were identified from the assembled transcripts, encompassing 44.2 Mb of sequence data. The length distribution of these unigenes was similar to that of the transcripts, with the maximum and minimum lengths matching those of the transcripts. The average unigene length was 1088.24 bp, and the N50 value was 1812 bp, reflecting the consistency and reliability of the assembly process (Table 3). In terms of unigene size distribution, 83% of the unigenes fell within the range of 200 to 2000 bp, while 15% ranged from 2000 to 4000 bp. Notably, 827 unigenes exceeded 4000 bp in length, representing approximately 2% of the total unigenes identified. This comprehensive length distribution analysis provides valuable insights into the transcriptome complexity and gene expression patterns of the studied floral organs (Figure S1).

Table 3. Statistics of transcriptome assembly.

Attributes	Unigenes	Transcripts
Total number	40,610	70,319
Total base	44,193,278	84,620,119
Largest length (bp)	15,701	15,701
Smallest length (bp)	201	201
Average length (bp)	1088.24	1203.37
N50 length (bp)	1812	1810

To annotate the transcriptome, BLAST searches were conducted to align the assembled sequences with various databases, including NR (NCBI Non-Redundant Protein Sequence Database), Swiss-Prot, Pfam (Protein families), COG (Clusters of Orthologous Groups of proteins), GO (Gene Ontology), and KEGG (Kyoto Encyclopedia of Genes and Genomes). This comprehensive analysis allowed us to obtain valuable annotation information for the transcriptome, enhancing our understanding of the gene expression patterns and functional characteristics of the studied floral organs. Of all the 40,610 unigenes, 98.37% (39,949 unigenes) were successfully annotated in the six databases (Table 4). *Camellia sinensis* (5711, 23.49%), *Actinidia chinensis* (4171, 17.15%), *Nyssa sinensis* (3332, 13.70%), *Vitis vinifera* (631, 2.6%), and *Rhododendron williamsianum* (538, 2.21%) were the top five species that showed similarity with the unigenes of *I. uliginosa* (Figure 2). In total, 19,465 and 19,695 unigenes were assigned to Swiss-Prot and Pfam databases, respectively.

Table 4. Functional annotation of *I. uliginosa* unigenes.

	Unigene Number	Unigene Percent
GO	20,926	52.38%
KEGG	10,945	27.40%
COG	22,343	55.93%
NR	24,238	60.67%
Swiss-Prot	19,456	48.70%
Pfam	19,695	49.30%
Total	39,949	100%

A total of 22,343 unigenes were categorized into 23 distinct COG functional groups. Notably, 12,437 of these unigenes exhibited ambiguous characteristics and were associated with unknown functional roles. Among the effectively annotated unigenes, the largest category was devoted to 'Transcription', closely followed by 'Posttranslational modification, protein turnover, chaperones', and 'Signal transduction mechanisms' (Figure 3).

A total of 20,926 unigenes were assigned to at least one GO term, providing a comprehensive overview of their functional roles. Within the biological process (BP) category, 'Cellular process' and 'Metabolic process' emerged as the most prevalent subcategories. In terms of cellular component (CC), a significant proportion of unigenes were grouped under 'Cell part'. Finally, within the molecular function (MF) category, 'Binding' and 'Catalytic activity' stood out as the most significant subcategories (Figure 4).

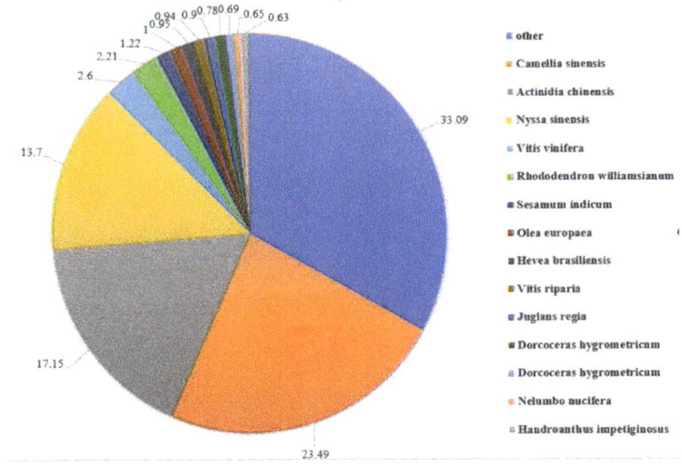

Figure 2. Distribution of NR-annotated species.

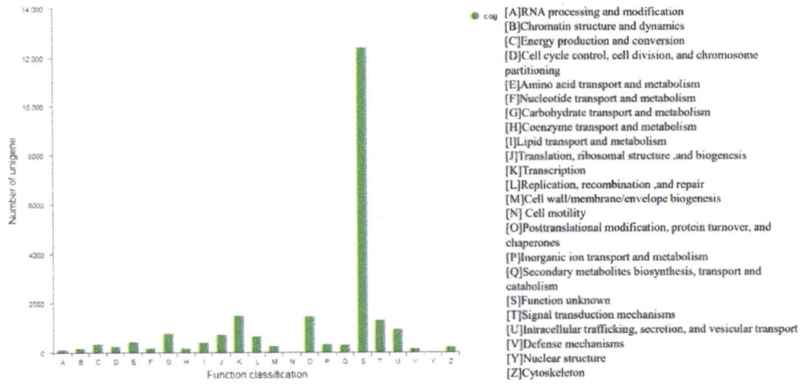

Figure 3. COG classification of unigenes in *I. uliginosa*.

A total of 10,945 unigenes were successfully annotated to KEGG and categorized into 20 distinct pathways across six primary categories. Notably, the majority of these genes were associated with the categories of 'Metabolism' and 'Genetic Information Processing'. Among the various pathways, 'Translation', 'Carbohydrate Metabolism', and 'Folding, Sorting, and Degradation' emerged as the most prominent representatives. This comprehensive analysis, as depicted in Figure 5, offers valuable insights into the function of the transcriptome and highlights the key biological processes and metabolic pathways involved.

In order to explore the genes that may be involved in regulating the flower color changes in *I. uliginosa* under copper stress, the differential expression at different copper concentrations was analyzed. The results showed that a total of 15,610 genes were differentially expressed among three groups. There were 4974 differentially expressed genes between the Cu0 group and the Cu10 group, including 3290 up-regulated genes and 1684 down-regulated genes; there were 4017 differentially expressed genes between the Cu0 group and the Cu20 group, including 1699 up-regulated genes and 2318 down-regulated genes. There were 6619 differentially expressed genes between the Cu10 group and the Cu20 group, including 1846 up-regulated genes and 4773 down-regulated genes (Table 5). There were 760 DEGs shared among these three groups (Figure 6).

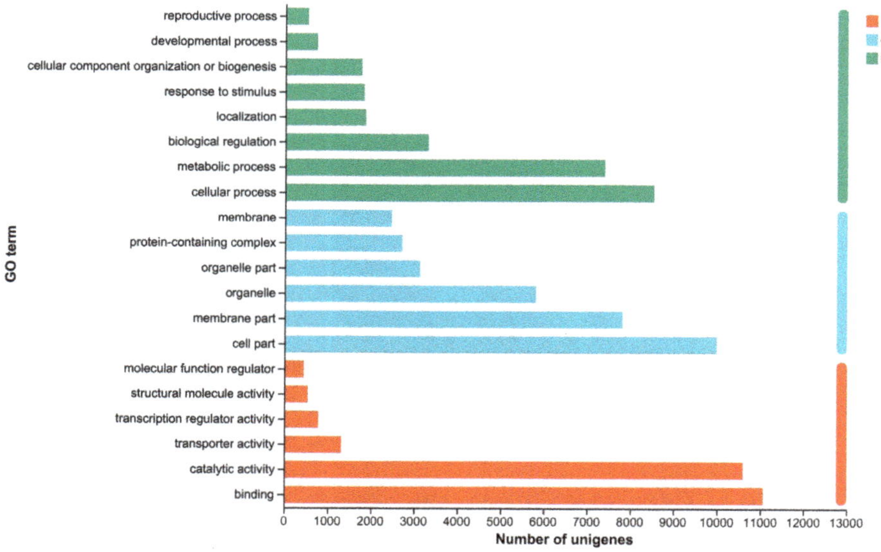

Figure 4. Main GO categories of unigenes in *I. uliginosa* transcriptome.

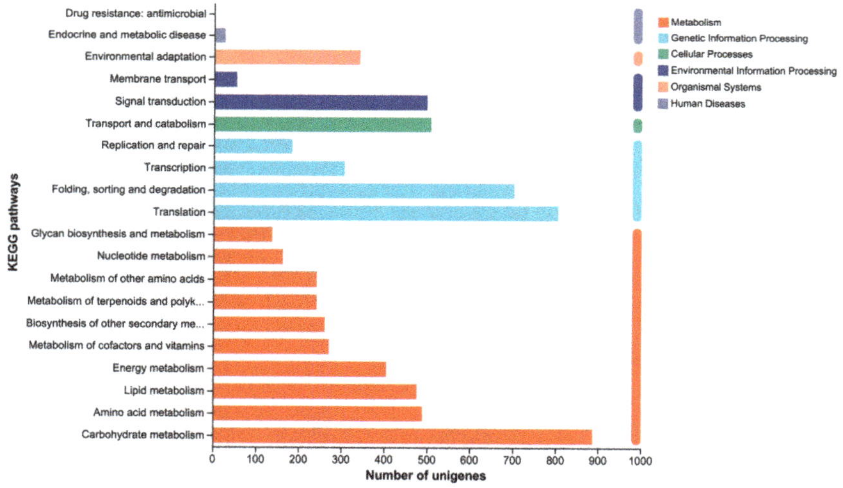

Figure 5. KEGG metabolic pathway of unigenes in *I. uliginosa.*

GO classification was performed on the differentially expressed genes of *I. uliginosa* under different concentrations of copper treatment. The GO classification results showed that there was a total of 20 secondary function entries. In BP, the number of genes annotated to secondary functional items such as cellular process, metabolic process, and biological regulation was relatively large, 197, 162, and 68, respectively. In MF, the number of genes classified into the catalytic activity and binding secondary function entries was relatively large, 303 and 269, respectively. The most significantly enriched Cu0 vs. Cu10 vs. Cu20 are clathrin coat assembly (GO:0048268), 1-phosphatidylinositol binding (GO:0005545), and clathrin-coated vesicle (GO:0030136) (Figure S1).

KEGG classification was performed on the differentially expressed genes of *I. uliginosa* under different concentrations of copper treatment. The annotated DEGs were distributed in 62 metabolic pathways and belong to five categories. Among them, the metabolism category contains the largest number of genes and the largest number of secondary pathways. This category includes Carbohydrate metabolism, Lipid metabolism, Energy metabolism and biosynthesis of other secondary metabolites, with the number of genes being 38, 12, 10, and 10, respectively. KEGG enrichment pathway analysis was performed on the differentially expressed genes of *I. uliginosa* under different concentrations of copper treatment. A total of six significant metabolic pathways were enriched, among which the most significantly enriched were Pentose and glucuronate interconversions (map00040), nitrogen metabolism (map00910) and Phenylpropanoid biosynthesis (map00940), with 22, 6, and 10, respectively (Figure S1).

Table 5. Statistical table of the number of differential genes.

Diff_Group	Total DEG	Up	Down
Cu0 vs. Cu10	4974	3290	1684
Cu0 vs. Cu20	4017	1699	2318
Cu10 vs. Cu20	6619	1846	4773

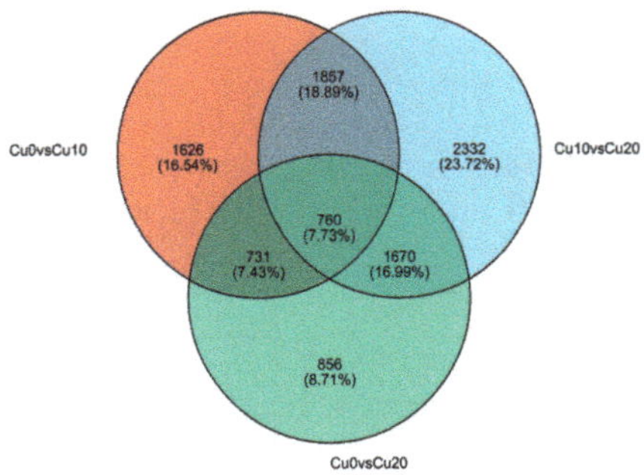

Figure 6. Venn diagram of DEGs.

In order to further study the differentially expressed genes related to the flower color of *I. uliginosa* under different concentrations of copper treatment, the flavonoid biosynthetic pathway related to flower color regulation was analyzed, the DEGs were compared with the metabolic pathways, and the metabolic pathway diagram of DEGs was obtained (Figure S2).

In order to further analyze the role of structural genes in anthocyanin accumulation under different concentration treatments, we can see from Figure S3 that a total of 359 *ANS* genes, 138 *CHI* genes, 4 *PAL* genes, and 3 *GT* genes were screened. From Figure S3, we can see that only the expression pattern of *GT* genes in Cu0 vs. Cu10 vs. Cu20 is down-regulated; that is, as the copper concentration increases, the expression level gradually decreases. The expression level was the highest when the copper concentration was 0 $mg{\cdot}L^{-1}$, and the expression level was low at high concentrations (20 $mg{\cdot}L^{-1}$), which is consistent with the flower color phenotype and anthocyanin content under different concentrations of copper treatment.

Based on GO and KEGG functional classification and enrichment analysis of differentially expressed genes, factors related to flower color transcription were further analyzed. From Figure S3, we can see that a total of nine MYB genes and three WD40 protein genes were screened from DEGs. From Figure S3, we can see that only the expression pattern of the MYB (TRINITY_DN7677_c0_g2) gene in Cu0 vs. Cu10 vs. Cu20 is down-regulated; that is, as the copper concentration increases, the expression level gradually decreases. The expression level was the highest when the copper concentration was 0 $mg{\cdot}L^{-1}$, and the expression level was low at high concentrations (20 $mg{\cdot}L^{-1}$), which is consistent with the flower color phenotype and anthocyanin content under different concentrations of copper treatment.

In order to further analyze the role of transport genes in anthocyanin accumulation under different concentration treatments, we can see from Figure S3 that a total of four *ABC* genes and three *GST* genes were screened. As can be seen from Figure S3, only the *ABC* (TRINITY_DN14115_c0_g1) gene and the *GST* (TRINITY_DN5047_c0_g1) gene have a down-regulated expression pattern in Cu0 vs. Cu10 vs. Cu20; that is, as the copper concentration increases, the expression level gradually decreases. The expression level was the highest when the copper concentration was 0 mg·L^{-1}, and the expression level was low at high concentrations (20 mg·L^{-1}), which is consistent with the flower color phenotype and anthocyanin content under different concentrations of copper treatment.

3.2. *Physical and Chemical Properties and Physiological Analysis of Candidate Genes*

The *IuGT* gene sequence of *I. uliginosa* has a length of 1482 bp, encoding 493 amino acids with a molecular formula of $C_{2469}H_{3808}N_{654}O_{718}S_{21}$. The theoretical isoelectric point PI value is 5.41; the instability index is 44.99, indicating that the protein is an unstable protein. The total average hydrophilicity index is −0.151, indicating that the protein belongs to hydrophilic protein. The total length of the cDNA of the *IuGST* gene in *I. uliginosa* is 696 bp, encoding 231 amino acids with the molecular formula $C_{1202}H_{1859}N_{293}O_{335}S_6$. The theoretical isoelectric point PI value is 5.42, and the instability index is 40.86. Therefore, it is speculated that the protein is an unstable protein with a total average hydrophilicity index of 0.015, indicating that it belongs to hydrophobic proteins (Table 6). To explore the evolutionary relationships of candidate genes, a phylogenetic tree was constructed using the amino acid sequences of IuGT and IuGST, respectively. As shown in the figure, in the evolutionary tree, IuGT and *Vitis riparia* (XP:034674887.1) are clustered together with the closest genetic relationship; IuGST is closely related to *Jatropha curcas* (KDP21251.1) (Figure 7).

Table 6. The basic physical and chemical properties of *IuGT* and *IuGST*.

Basic Physicochemical Properties	*IuGT*	*IuGST*
Molecular formula	C2469H3808N654O718S21	C1202H1859N293O335S6
Theoretical isoelectric point	5.41	5.42
Molecular weight	54,814.61 Da	25,967.10 Da
Total number of atoms	7670	3695
Negatively charged residue (Asp + Glu)	60	31
Positively charged residue (Arg + Lys)	44	25
Instability index	44.99	40.86
Total average hydrophilicity index	−0.151	0.015

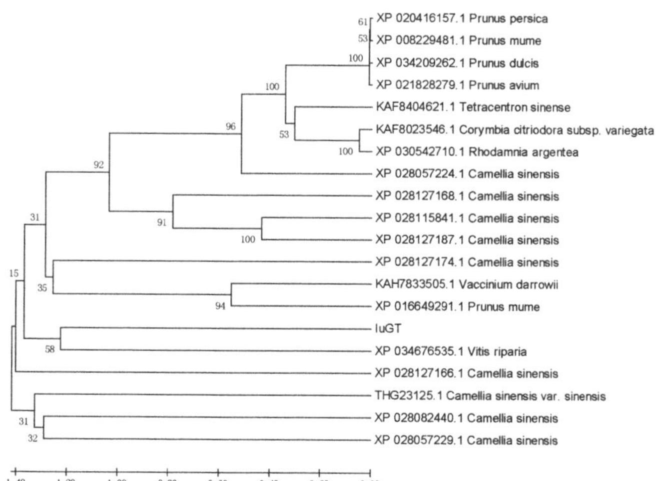

Figure 7. *Cont.*

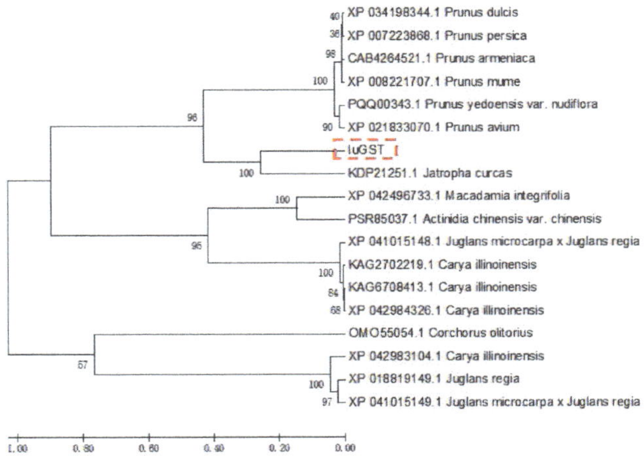

Figure 7. Phylogenetic tree of the amino acid sequence of IuGT and IuGST.

3.3. *qRT-PCR Validation of the Candidate Genes*

Two candidate genes were selected for qRT-PCR to verify the accuracy of transcriptome data (Figure 8). The expression levels of *IuGT* and *IuGST* genes in *I. uliginosa* were significantly different under different concentrations of copper. As the copper concentration increased, the expression levels of *IuGT* and *IuGST* genes showed a gradually decreasing trend. When the copper concentration was 0 mg·L^{-1}, the expression level was the highest, and when the copper concentration was 20 mg·L^{-1}, the expression level was the lowest. The expression levels of *IuGT* and *IuGST* genes had decreased by more than 10 times.

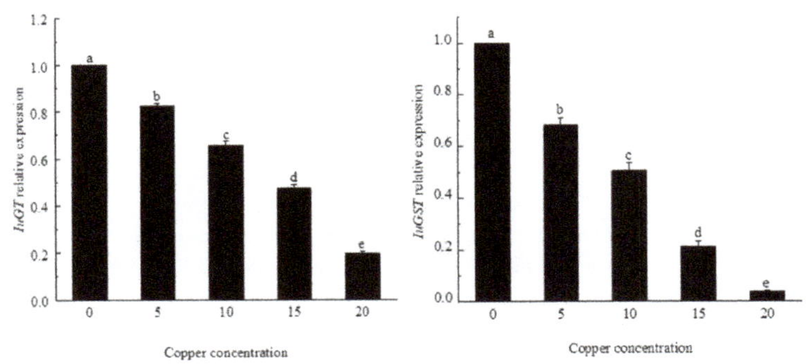

Figure 8. The relative expression levels of *IuGT* and *IuGST* genes under different copper concentrations. Different letters represent statistically significant differences ($\alpha = 0.05$).

4. Discussion

Different concentrations of different metal ions have different effects on plants. For example, Li [18] found that the Impatiens hawkeri color of red was affected by the concentration of Cu^{2+} and Al^{3+}. When the Cu^{2+} concentration was 3.2×10^{-7} mol·L^{-1}, the flowers were the darkest in color; as the Al^{3+} concentration increased, the flower color first became lighter and then gradually deepened [21]. In previous studies, we measured the chromaticity value, anthocyanins, total flavonoids, carotenoids, copper ion content in petals, and related physiological and biochemical indicators of *I. uliginosa* under copper stress. The color of the *I. uliginosa* organs changed under different concentrations of copper treatment. The flower color gradually became lighter as the copper concentration increased. It was also found that as the copper concentration increased, the color of *I. uliginosa* petals became lighter, and the total flavonoid content and anthocyanin content showed a downward trend [22,23]. When the copper concentration was 20 mg·L^{-1}, the copper ion content of *I. uliginosa* was 2363.90 mg·kg^{-1}, which was 39 times that of CK.

This study used copper-treated (0 mg·L^{-1}, 10 mg·L^{-1}, and 20 mg·L^{-1}) flower organs of *I. uliginosa* during the blooming period as research materials and conducted transcriptome sequencing so as to explore the molecular mechanism of flower color fading as copper concentration increases. After sequencing, a total of 40,610 unigenes were obtained. In order to obtain the functional information of *I. uliginosa* unigene, these genes were annotated on NR, NT, Swissprot, KEGG, KOG, Pfam, and GO databases, and a total of 39,949 unigenes were assembled. Analysis of differentially expressed genes found that 1626, 856, and 2332 genes only existed in Cu0 vs. Cu10, Cu0 vs. Cu20, and Cu10 vs. Cu20, respectively. Meanwhile, only 760 differentially expressed genes appeared in all comparison groups. This study analyzed differentially expressed genes and screened out structural genes that play a key role in the accumulation of flavonoids in *I. uliginosa*, including *ANS*, *CHI*, *PAL*, and *GT*. Among them, the expression pattern of *IuGT* (TRINITY_DN3511_c0_g1) is consistent with the flower color phenotype and anthocyanin content under different concentrations of copper, namely, the transcription level of the *IuGT* gradually decreases as the copper concentration increases. *GTs* have been identified in many plants, such as *Fagopyrum esculentum* [24], *Rosa rugosa* [25], and *Arabidopsis thaliana* [26]. The above confirmed that glycosylation is very important for flower color formation. In addition, phytochrome transport factors have also become important factors affecting plant color, such as the GST protein family. Studies have shown that the glutathione transferase (*GST*) family is the key factor of the accumulation of anthocyanins in the vacuole, and it can directly bind to anthocyanins and thus become a carrier for anthocyanin transport [8]. A total of three *GST* genes were screened out in this study. The expression pattern of *IuGST* (TRINITY_DN5047_c0_g1) in Cu0 vs. Cu10 vs. Cu20 is down-regulated, namely, as the copper concentration increases, the expression level gradually decreases. The expression level is the highest when the copper concentration is 0 mg·L^{-1}, and the expression level is low at high concentrations (20 mg·L^{-1}), which is consistent with the flower color phenotype and anthocyanin content under different concentrations of copper treatment.

Different genes directly affect the direction and metabolic process of flavonoids according to their different expression levels, producing different metabolites, which in turn affect the color of plants. Hassani et al. [27] studied the regulation of peach flower color and discovered that the *GT* gene expression level was the highest in red peach flowers and almost not expressed in white flowers. This study demonstrated that as the copper concentration increased, the expression level of *IuGT* in the organs of *I. uliginosa* showed a gradually decreasing trend. When the copper concentration was 20 mg·L^{-1}, the expression level was the lowest, and the flower color showed a change from pink to white. Studies have shown that *GST* family genes are involved in responses to stress in many plants. Tasaki et al. [28] found that *GST1* is not only a gene responsible for anthocyanin transport in Japanese gentian under sugar-induced stress conditions but also essential for the accumulation of gentisone in flowers. This study found that as the copper concentration increased, the expression of *IuGST* in the organs of *I. uliginosa* showed a gradually decreasing trend. When the copper concentration was 20 mg·L^{-1}, the expression level was the lowest. Meanwhile, as the copper concentration increased, the anthocyanin content continued to decrease, which was consistent with the expression pattern of *IuGST*.

5. Conclusions

This study used *I. uliginosa* as the material and treated it with different concentrations of copper (0 mg·L^{-1}, 5 mg·L^{-1}, 10 mg·L^{-1}, 15 mg·L^{-1}, and 20 mg·L^{-1}). After 30 days, the petals of *I. uliginosa* were taken to explore the effect of different concentrations of copper on *I. uliginosa* flower color from a molecular biology perspective, revealing the mechanism of the effect of different concentrations of copper on Dianshui Jinfeng flower color. It lays a theoretical foundation for future in-depth study of the molecular mechanism of *I. uliginosa* in defense against heavy metal toxicity, and also provides a theoretical basis for the practical application of *I. uliginosa* in restoring heavy metal-polluted water bodies.

Supplementary Materials: The following supporting information can be downloaded at: https://www.mdpi.com/article/10.3390/horticulturae10040412/s1, Figure S1. (**a**) Length distribution of unigenes; (**b**) Bubble chart of go enrichment of differential genes; (**c**) KEGG Pathway enrichment of differentially expressed genes. Figure S2. Key metabolic pathways involved in color changes of I. uliginosa under copper stress. Figure S3. (**a**) Venn diagram of structural genes; (**b**) Differential expression of GT Unigenes; (**c**) Venn diagram of transcription factors; (**d**) Differential expression of MYB Unigenes; (**e**) Venn diagram of transport factors; (**f**) Differential expression of ABC and GST Unigenes.

Author Contributions: Y.T. and H.H. (Haiquan Huang) were responsible for the experimental design. Y.T. and X.Z. carried out sample collection, experiments, data analysis, and article writing. Q.L., X.L., L.L., H.H. (Haihao He) and G.L. participated in the experiment. M.H. supervised the research and revised the manuscript. All authors have read and agreed to the published version of the manuscript.

Funding: This work was supported by Joint Special Key Project for Agriculture Basic Research in Yunnan Province (202101BD070001-018), the National Natural Science Foundation of China (32060364, 32060366), and the Doctoral Supervisor Team Project of Landscape Plants Genetic Improvement and Efficient Breeding in Yunnan Province.

Data Availability Statement: All data generated in this article are included within the article and its additional flies.

Acknowledgments: We thank Shanghai Majorbio Bio-pharm Technology Co., Ltd. (Shanghai, China) for its help in sequencing. The data were analyzed through the free online platform of Majorbio Cloud Platform (www.majorbio.com, accessed on 15 February 2022).

Conflicts of Interest: The authors declare no conflicts of interest.

References

1. Li, B.; Dao, J.R.; Zhu, R.Y.; He, H.; Meng, X.Q.; Han, F.X. Distribution, accumulation and risk assessment of heavy metal pollution in Dianchi Lake. *Environ. Chem.* **2021**, *40*, 1808–1818.
2. Zhang, G.H.; Xie, Q.; Yan, K.; Li, Z.Y.; Su, T.; Yang, Y.L. Pollution Characteristics and Ecological Risk Assessment of Heavy Metals in Sediments of Dianchi Outer Lake During the '13th Five-Year Plan Period'. *J. Soil Water Conserv.* **2023**, *37*, 240–247.
3. Peng, G.Y.; Hu, L.; Huang, C.; Yang, K.; Wan, W.; Huang, C.G. Transcriptome Analysis of Response to Heavy Metal Copper Stress in *Setcreasea purpurea* Root Tissue. *Biotechnol. Bull.* **2022**, *38*, 83–94.
4. Zhang, C.; Tang, C.C.; Wang, J.Y.; Guo, L.; Wang, L.; Li, W. Advances on transcriptome of plants under stresses. *J. Biol.* **2017**, *34*, 86–90.
5. García-Vico, L.; Sánchez, R.; Fernández, G.; Sanz, C.; Pérez, A.G. Study of the olive β-glucosidase gene family putatively involved in the synthesis of phenolic compounds of virgin olive oil. *J. Sci. Food Agric.* **2021**, *101*, 5409–5418. [CrossRef]
6. Gachon, C.M.; Langlois-Meurinne, M.; Saindrenan, P. Plant secondary metabolism glycosyl transferases: The emerging functional analysis. *Trends Plant Sci.* **2005**, *10*, 542–549. [CrossRef]
7. Li, P.; Li, Y.J.; Zhang, F.J.; Zhang, G.Z.; Jiang, X.Y.; Yu, H.M.; Hou, B.K. The Arabidopsis UDP-glycosyltransferases UGT79B2 and UGT79B3, contribute to cold, salt and drought stress tolerance via modulating anthocyanin accumulation. *Plant J.* **2017**, *89*, 85–103. [CrossRef]
8. Hu, Z.; He, J.; Chen, K.; Wang, Z.; Liu, J.; Qiao, X.; Ye, M. Molecular cloning and biochemical characterization of a new flavonoid glycosyltransferase from the aquatic plant lotus. *Biochem. Biophys. Res. Commun.* **2019**, *510*, 315–321. [CrossRef]
9. Cheng, J.; Wei, G.; Zhou, H.; Gu, C.; Vimolmangkang, S.; Liao, L.; Han, Y. Unraveling the mechanism underlying the glycosylation and methylation of anthocyanins in peach. *Plant Physiol.* **2014**, *166*, 1044–1058. [CrossRef]
10. Young, L.; Akhov, L.; Kulkarni, M.; You, F.; Booker, H. Fine-mapping of a putative glutathione S-transferase (GST) gene responsible for yellow seed colour in flax (*Linum usitatissimum*). *BMC Res. Notes* **2022**, *15*, 72. [CrossRef]
11. Lu, S.W.; Zheng, X.A.; Wang, J.Y.; Fang, J.G. Research Progress on the Metabolism of Flavonoids in Grape. *Acta Hortic. Sin.* **2021**, *48*, 2506–2524.
12. Kim, S.H.; Sung, S.Y.; Kim, Y.S.; Jo, Y.D.; Kang, S.Y.; Kim, J.B.; Ahn, J.W.; Ha, B.K.; Kim, D.S. Isolation and characterization of differentially expressed genes in petals of Chrysanthemum mutant cultivars developed by irradiation. *Sci. Hortic.* **2015**, *189*, 132–138. [CrossRef]
13. Cao, Y.; Xu, L.; Xu, H.; Yang, P.; He, G.; Tang, Y.; Qi, X.; Song, M.; Ming, J. LhGST is an anthocyanin-related glutathione S-transferase gene in Asiatic hybrid lilies (*Lilium* spp.). *Plant Cell Rep.* **2021**, *40*, 85–95. [CrossRef]

14. Jin, X.H.; Hong, Y.; Huang, H.; Dai, S.L.; Zhu, Y. Isolation and Expression Analysis of GST Gene Encoding Glutathione S-transferase from *Senecio cruentus*. *Acta Hortic. Sin.* **2013**, *40*, 1129–1138.
15. Yu, S.X. *Balsaminaceae of China*; Peking University Press: Beijing, China, 2012; p. 147.
16. Li, Y.; Huang, W.L.; Lin, Y.F.; Luo, C.; Li, X.Y.; Liu, Y.L.; Zhu, J.P.; Feng, Z.X.; Huang, M.J.; Huang, H.Q. Cloning and Expression Analysis of CHI Genes in Impatiens uliginosa. *Acta Agric. Univ. Jiangxiensis* **2020**, *42*, 468–474.
17. Liu, Y.L.; Feng, Z.X.; Zhu, J.P.; Luo, C.; Liang, Y.; Li, Y.; Tong, Z.K.; Huang, H.Q.; Huang, M.J. Cloning and Expression Analysis of AP3/DEF Homologous Genes Associated with Floral Development in *Impatiens uliginosa*. *Mol. Plant Breed.* **2020**, *18*, 6626–6632.
18. Guo, J.W.; Huang, Q.; Ma, M.L.; Sun, Z.H.; Wen, Y.H.; Huang, H.Q.; Huang, M.J.; Wang, Q. Determination of Metal Elements in the Petals of Yunnan Wild Impatiens in Different Colors. *J. Anhui Agric. Sci.* **2019**, *47*, 179–181.
19. Li, Y.; Wei, C.M.; Li, X.Y.; Meng, D.C.; Gu, Z.J.; Qu, S.P.; Huang, M.J.; Huang, H.Q. De novo transcriptome sequencing of *Impatiens uliginosa* and the analysis of candidate genes related to spur development. *BMC Plant Biol.* **2022**, *22*, 553. [CrossRef]
20. Li, Y.; Huang, W.L.; Li, X.Y.; Zhang, Y.D.; Meng, D.C.; Wei, C.M.; Huang, M.J.; Huang, H.Q. The cellular and molecular basis of the spur development in *Impatiens uliginosa*. *Hortic. Res.* **2024**, *11*, uhae015. [CrossRef]
21. Li, R.H. Effects of Copper and Aluminum on Growth and Flower Color of *Impatiens hawkeri*. Master's Thesis, Hebei Agricultural University, Baoding, China, 2005.
22. Li, Q.M.; Li, W.X.; Cao, M.H.; Liu, S.; Zhang, T.Y.; Wang, Q.; Huang, M.J.; Huang, H.Q. Physiological and Biochemical Correlations between Color of *Impatiens uliginosa* Flower and Nutrient Supply on Copper. *Fujian J. Agric. Sci.* **2021**, *36*, 1323–1329.
23. Li, Q.M.; Li, W.X.; Li, X.Y.; Li, Y.; Qu, S.P.; Huang, M.J.; Huang, H.Q. Effects of Copper Stress on Flower Development and Physiological and Biochemical Characteristics of *Impatiens uliginosa*. *Shandong Agric. Sci.* **2022**, *54*, 74–79.
24. Matsui, K.; Tomatsu, T.; Kinouchi, S.; Suzuki, T.; Sato, T. Identification of a gene encoding glutathione S-transferase that is related to anthocyanin accumulation in buckwheat (*Fagopyrum esculentum*). *J. Plant Physiol.* **2018**, *231*, 291–296. [CrossRef] [PubMed]
25. Sui, X.; Zhao, M.; Han, X.; Zhao, L.; Xu, Z. RrGT1, a key gene associated with anthocyanin biosynthesis, was isolated from *Rosa rugosa* and identified via overexpression and VIGS. *Plant Physiol. Biochem.* **2019**, *135*, 19–29. [CrossRef]
26. Yonekura-Sakakibara, K.; Fukushima, A.; Nakabayashi, R.; Hanada, K.; Matsuda, F.; Sugawara, S.; Inoue, E.; Kuromori, T.; Ito, T.; Shinozaki, K.; et al. Two glycosyltransferases involved in anthocyanin modification delineated by transcriptome independent component analysis in *Arabidopsis thaliana*. *Plant J.* **2012**, *69*, 154–167. [CrossRef] [PubMed]
27. Hassani, D.; Liu, H.L.; Chen, Y.N.; Wan, Z.B.; Zhuge, Q.; Li, S.X. Analysis of biochemical compounds and differentially expressed genes of the anthocyanin biosynthetic pathway in variegated peach flowers. *Genet. Mol. Res.* **2015**, *14*, 13425–13436. [CrossRef]
28. Tasaki, K.; Yoshida, M.; Nakajima, M.; Higuchi, A.; Watanabe, A.; Nishihara, M. Molecular characterization of an anthocyanin-related glutathione S-transferase gene in Japanese gentian with the CRISPR/Cas9 system. *BMC Plant Biol.* **2020**, *20*, 370. [CrossRef]

horticulturae

MDPI

Article

Variations in Flower Color of Mutant Chrysanthemums

Gulden Haspolat

Aegean Agricultural Research Institute, Cumhuriyet Street, Izmir 35661, Turkey;
gulden.haspolat@tarimorman.gov.tr

Abstract: The induction of variation in chrysanthemums using gamma radiation under in vitro conditions is an effective technique in ornamental plants. The purpose of this study is to obtain new mutants by isolating desirable properties from the three-colored single chrysanthemum mutant using in vitro cultures. Bud explants were cultured four times, the plantlets were acclimatized, and 520 plants were planted in outdoor conditions. Observations of the mutants were collected during flowering time, and 97 of the mutants were compared to the control group. Plants with pink, white, and chimeric flowers were obtained. Mutant plants with white flowers constituted the majority of the population followed by plants with variegated flower colors. The population is divided into six clusters, based on the plant height, plant diameter, flower number, flower diameter, number of flower colors, ray flowers' number, leaves' number, stem weight, and lengths and widths of leaves. The population decreased in plant height, flower number, and stem weight, while other features increased compared to the control group. Thus, a new population with similar characteristics to the parent plant was obtained from a single mutant. Chrysanthemum plants exposed to mutagens showed major changes in flower parts as well as other parts of the plant.

Keywords: chrysanthemum; mutant; in vitro; color; variation

1. Introduction

Chrysanthemums originate from China and are considered one of the most significant cut flowers in the world [1,2]. In addition to their ornamental uses, chrysanthemum flowers offer nutritional and medicinal value due to their high antioxidant content [3–6]. *Chrysanthemum morifolium* Ramat (*Dendranthema* × *grandiflora* Tzvelv = *Chrysanthemum* × *grandiflorum* Ramat) is the most commercially significant ornamental species with most cultivars being autohexaploid and self-incompatible, which presents challenges for modern breeding [7–9].

Individuals in the genus Chrysanthemum have a wide range of flowers which are used in floriculture. They have a unique capitulum that consists of disc and ray florets, and clear distinctions can be observed between the two florets. The visual color of ray flowers is referred to as flower color and is a crucial feature of these plants, and they are available in an abundance of color and form [10–14].

Improving and innovating chrysanthemums by modifying their ornamental attributes such as floral color, shape, plant type, flowering time, and vase life, while also enhancing their ability to tolerate biotic and abiotic stress, is the main purpose of chrysanthemum breeding [15]. Breeders have placed a strong emphasis on manipulating floral color with a wide range of colors in the ray florets, as it is a major factor that influences customer selection [16,17]. Older colors, including yellow, pink, and white, are attributed to the presence of carotenoids and anthocyanins or the absence of these pigments. However, modern flowers have been bred to display a wider range of colors such as purplish red, orange, scarlet, and deep red by increasing the pigment content or by combining different pigments [18–20].

Mutation breeding in plants has proven to be a successful breeding method as mutants can also be produced directly [21–23]. Using novel breeding methods such as chimera

Citation: Haspolat, G. Variations in Flower Color of Mutant Chrysanthemums. *Horticulturae* **2024**, *10*, 385. https://doi.org/10.3390/horticulturae10040385

Academic Editors: Caiyun Wang and Tuo Zeng

Received: 13 March 2024
Revised: 4 April 2024
Accepted: 5 April 2024
Published: 11 April 2024

management and in vitro mutation results in targeted modifications which have enhanced and changed the procedures as needed [24]. Numerous new and promising plant species, including chrysanthemums, have benefited from mutation procedures using ionizing radiation and other mutagens [25]. Chrysanthemums are the plants with the most mutant variations developed by mutation breeding among vegetatively propagated crops [26]. Chemical agents, gamma and X-rays, and other radiation sources are used as mutagens in studies carried out in chrysanthemum mutation breeding [8]. As a result of increased exposure to irradiation, more mutations in flower color have occurred in chrysanthemums, while the vegetative growth of all cultivars has only been minimally altered [27,28]. Although polyploids with significant genetic heterogeneity make up the majority of cultivated chrysanthemum varieties, mutants with related flower shapes, floral sizes, and colors are commonly observed. Radiation can easily activate associated floral colors in chimeric tissues and isolate them using in vitro techniques [29].

Mutation breeding is useful for developing new ornamental plants with visible changes in color, shape, and size. By exposure to a mutagen dose, plant height variations and significant variations in floral parts and leaves were observed on chrysanthemums. Additionally, gamma radiation is an effective mutagen when applied to in vitro bud explants for creating new types. The results of in vitro treatments with mutagenic gamma radiation show that the method of induced homogeneous mutation is an effective approach for chrysanthemum breeding. Selected mutants can be propagated vegetatively to obtain new plants with desirable traits that will be well received. Numerous experiments using tissue culture techniques for mutation breeding resulted in similar results, and diversity has been achieved with novel genotypes possessing the desired characteristics [30].

The present study aimed to create a new population from a single mutant plant with different flower types and three different colors. These three flower colors were isolated from the single mutant using tissue culture methods. The mutant had more ray flowers and was obtained by irradiating the in vitro bud explants of a white chrysanthemum variety with gamma rays. This aimed to isolate distinct colors from the single mutant and create a new population from the clones of this parent mutant plant. With this goal, rapid propagation techniques of tissue culturing were used to create mutant plants from this single mutant. In this way, morphological observations were made and compared to the control plants focused on the possibilities of obtaining new candidate varieties by selecting individuals with superior characteristics among the mutant population.

2. Materials and Methods

Materials from the single mutant plant of *Chrysanthemum morifolium* (Ramat.) (*Dendranthema* × *grandiflora* Tzelev.), variety 'Bacardi', were propagated from the bud explants by tissue culture (Figure 1). The in vitro plantlets of the control plant were irradiated with 20 Gy (Gray) of cobalt 60 (Co^{60}) gamma source at the Energy, Nuclear and Mining Research Association, Nuclear Energy Research Institute. Then, they were subcultured four times till the M_1V_4 (Mutation 1: Vegetation 4) growing period and then planted in outdoor conditions after acclimatization. This part was the study's initial section, which was published in a different journal [30].

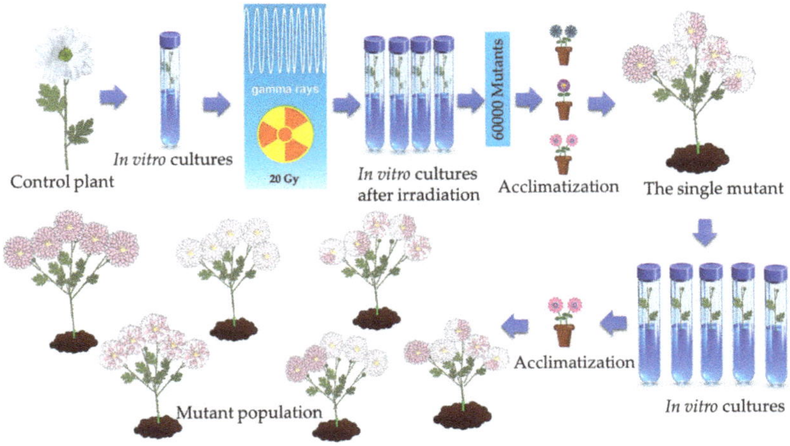

Figure 1. Summary of the study.

The research was conducted from 2022 to 2023. One of the mutants among the 60,000 mutants showed an increase in the number of ray flower rows (daisy-eyed double flower head type) and had pink, white, and pink and white variegated (multicolored chimeric) flowers (Figure 2). This mutant was selected and propagated by tissue culture. After four subcultures were complete, rooted plantlets were transferred to outdoor conditions by acclimatization and 520 mutants were planted in an open field (Figure 1). The 97 mutants with desirable features that bloomed earlier than others in the mutant population were marked in November 2022. Morphological observations of the mutants confirming their stable status as well as control plants were recorded in November 2023 when they were in full bloom. Plant height, plant diameter, flower number, number of ray flowers, number of flower colors, number of leaves, stem weight, and lengths and widths of leaves are among the attributes that were observed.

Figure 2. The single mutant with daisy-eyed double flower heads. (**a**,**b**) General appearance of plant; (**c**) white-colored flower; (**d**) variegated flower; (**e**) pink flowers of the single mutant (bars: 1 cm).

2.1. Tissue Culture

To prepare explants, the lateral buds were washed with running tap water with a few drops of dishwashing detergent for 60 min. The explants were surface sterilized in 70 percent (v/v) ethanol (EtOH) for one minute after three rounds of rinsing with distilled water. Next, three rinses were performed with a 25% hydrogen peroxide (H_2O_2) solution for the next 10 min, followed by autoclaved distilled water for five minutes each [30].

2.2. Planting In Vitro Explants on Nutrient Medium

The nodes of the sterilized shoot explants were planted on MS (Murashige and Skoog) [31] with 3% sucrose, 0.7% agar, and 1 mg/L BAP (benzylaminopurine) in glass tubes. The culture media with the pH adjusted to 5.8 was autoclaved at 121 °C for 15 min. The cultured explants were grown in the climate room under a 16 h photoperiod (30 mol/m^2 s^1). The nodes of the shoot explants were subcultured after one month [30].

2.3. Rooting and Acclimatization Stages

After four subcultures, the shoots longer than 1 cm were transferred to the plant growth regulator-free MS medium for rooting. The roots appeared 4 weeks later, and the rooted 3 cm long shoots were moved to the acclimatization stage. Rooted plantlets were removed from glass tubes and their roots were cleaned of the agar-based nutrient medium to acclimate them to the outside environment. The plants were transplanted into plastic vials containing a 3:1 mixture of peat and perlite placed in boxes and covered with transparent wrapping paper. The vials were kept for one week at 22 °C under cool white light with a 16 h photoperiod (30 mol/ m^2 s^1) in the climate room. Tiny holes were pierced daily on the transparent wrapping paper, as they hardened during the week. The plantlets that had hardened off were placed in the greenhouse for one week and then planted in soil in an open field in June 2022 [30].

2.4. Observations of Mutants

The observations of the mutants were collected by taking into account the plant height (cm; from the soil surface to the uppermost part of the plant), plant diameter (cm; the widest aboveground plant width) [32], number of leaves, leaf length (cm; the longest part of leaves), leaf width (cm; the widest part of leaves), number of flowers per plant, flower width (cm; the widest part of the flower head), number of flower colors each plant, number of ray florets, and stem weight in early November 2023 during anthesis. Somatic mutations of the flower color variegations chimeras were recorded as variegated or multicolored. The color variations from white to pink of ray florets were determined according to the Royal Horticultural Society (RHS) Color Chart cards [33].

2.5. Statistical Analysis

Morphological measurements were performed on 97 mutants and a control group (mean of 10 control plants). Based on these measurements, changing ratios and clusters were created according to the plant height, plant diameter, flower diameter, number of flowers, number of ray flowers, number of flower colors, flower color, number of leaves, leaf length, leaf width, and stem weight. Changing ratios of mutant genotypes according to the control group were calculated with Formula (1):

$$\%\ Change = \frac{Genotype - Control}{Control} \times 100 \quad (1)$$

The genetic closeness in terms of morphological features was determined using the MINITAB 20 software program. A hierarchical cluster dendrogram was created to analyze the distance and similarity between mutant populations. The core algorithm was used to calculate the similarity matrix between the mutants.

3. Results

While the flower head type of the control plants was semi-double, the mutant's flower head type was daisy-eyed double due to the increased number of ray flower rows. In the mutant group, there were six ray floret rows, compared to three in the control group, resulting in a change in flower head type. In addition to the change in flower head type, we also observed some changes in the variation in flower colors.

3.1. Color Distribution of the Mutant Population

The mutant population presented a distribution of pink, white, and pink and white variegated (multicolored) flowers despite maintaining the same type of flower heads. Ray floret color variations from white to pink were identified using the RHS Color Chart cards. While the pink color was 69A, the white one was NN155A [33]. The majority of the population consisted of white-colored mutant plants, followed by plants with variegated colored flowers. In some mutants, variegated flowers as well as pink or white flowers have been observed on the same plant. There was only one mutant plant in the population which had white, pink, and variegated flowers like the parent mutant plant (Figure 3). The number of flower colors on a plant varied from 1 to 3 and increased by 23% according to the control group (Figure 4).

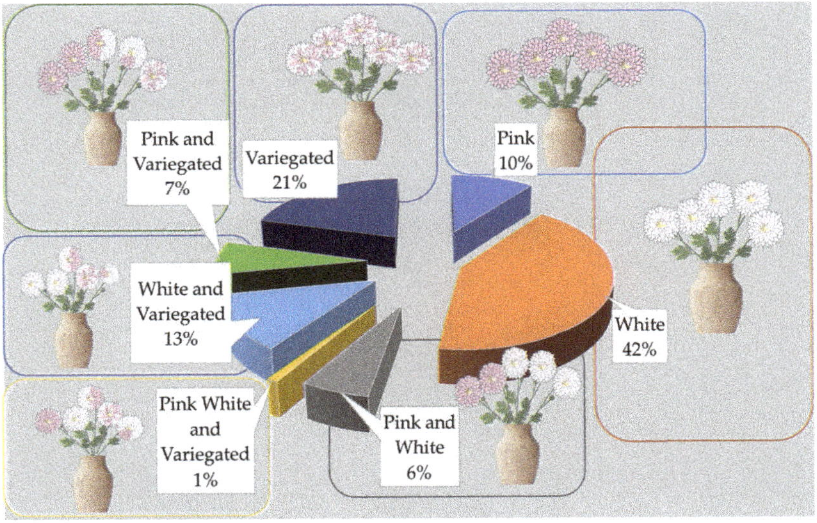

Figure 3. Color distribution in flowers of mutant genotypes.

3.2. The Comparison between the Control Group and the Mutant Population

The population showed increasing and decreasing properties compared to the control plants in terms of plant height, plant diameter, flower diameter, number of flowers, number of ray flowers, number of flower colors, number of leaves, leaf length, leaf width, and stem weight. All characteristics of the mutant population increased except for the plant height, the number of flowers, and stem weight, which decreased compared to the control group (Figure 4).

According to the control plants, the plant height decreased by 3% (Figure 4). The tallest mutant plant was coded M31 mutant, and the shortest one was M38 (Figure 5). Plant diameter increased by 69% (Figure 4). The widest plant was M1 followed by the control group, and M32 was the narrowest plant in the population (Figure 6). Flower numbers differed from 5 to 40 flowers (Figure 7), and the flower number decreased by 23% compared to the control plants (Figure 4).

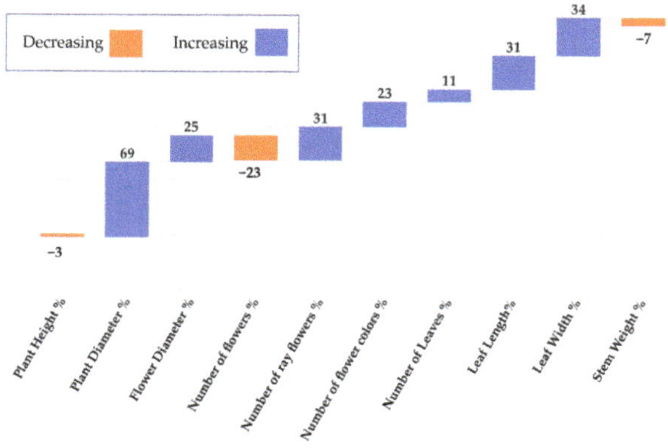

Figure 4. Changing ratios of the population compared to the control plants.

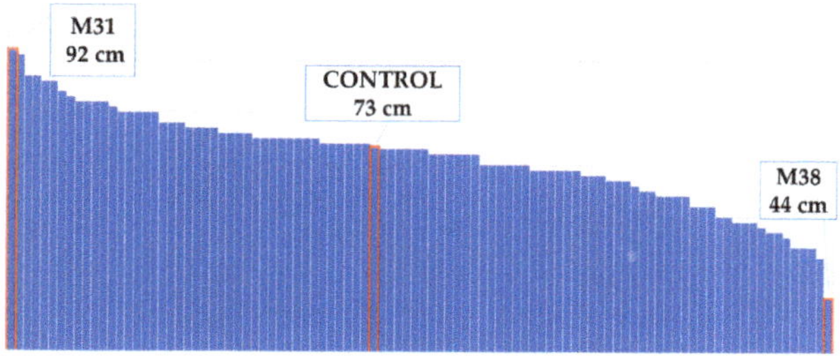

Figure 5. Maximum and minimum plant heights of the population.

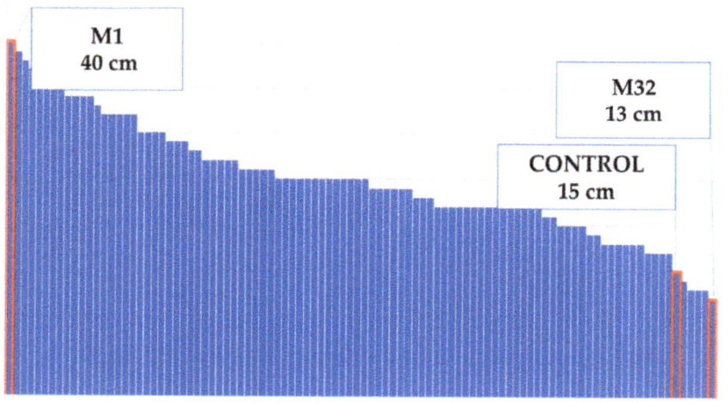

Figure 6. Maximum and minimum plant diameters of the population.

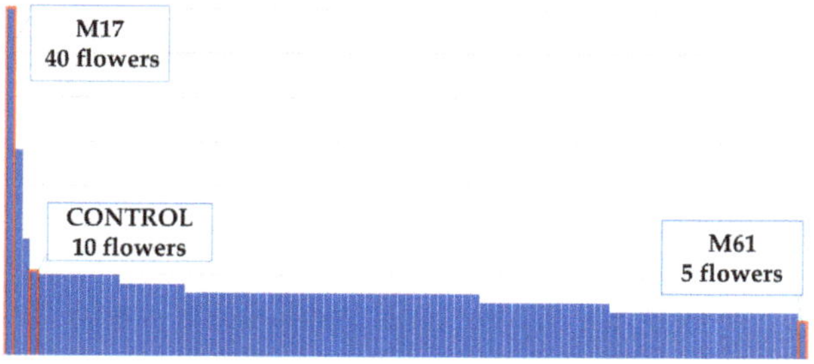

Figure 7. Maximum and minimum flower numbers of the population.

Flower diameter increased by 25% compared to the control plants (Figure 4). The widest flower was observed on the M52 mutant; the control plants and the mutants M25–M41 had the same flower diameters (Figure 8). The M38 mutant formed the greatest amount of ray flowers, whereas the control group formed the fewest (Figure 9), and the number of ray florets increased by 31% compared to the control group (Figure 4). The number of leaves increased by 11% compared to the control group (Figure 4). M24 was the mutant with the maximum number of leaves, whereas M11 had the minimum leaf number (Figure 10). Leaf length increased by 31% compared to the control group (Figure 4). While the longest leaves were observed in the mutant plant coded M37, the mutant plant numbered M84 was the plant with the shortest leaves (Figure 11).

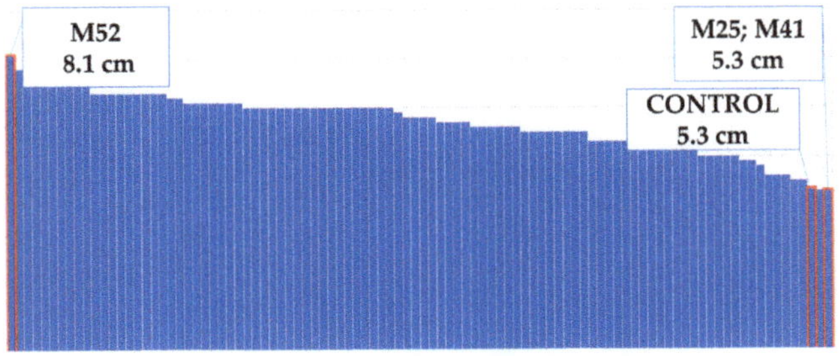

Figure 8. Maximum and minimum flower diameters of the population.

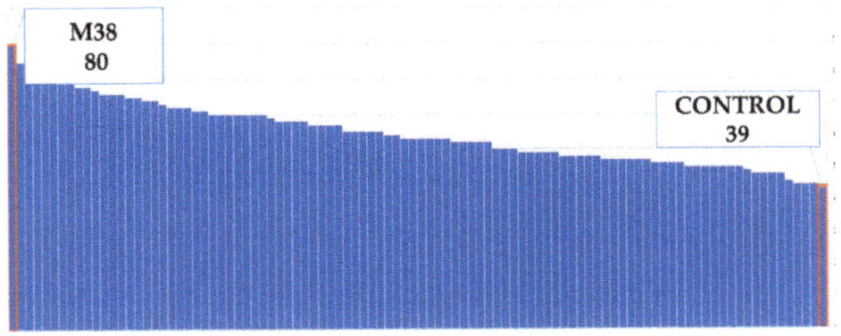

Figure 9. Maximum and minimum ray flower numbers of the population.

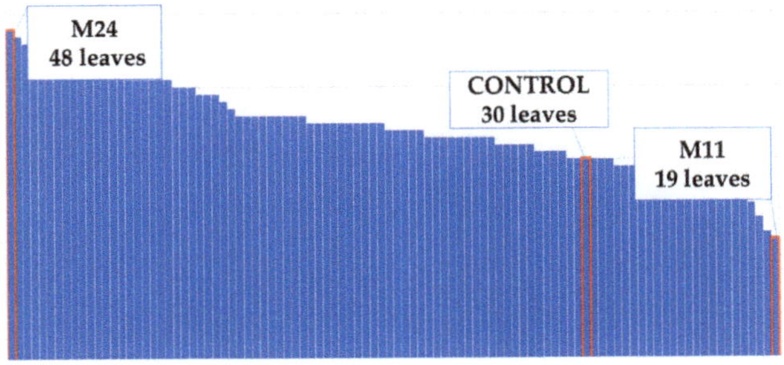

Figure 10. Maximum and minimum leaf numbers of the population.

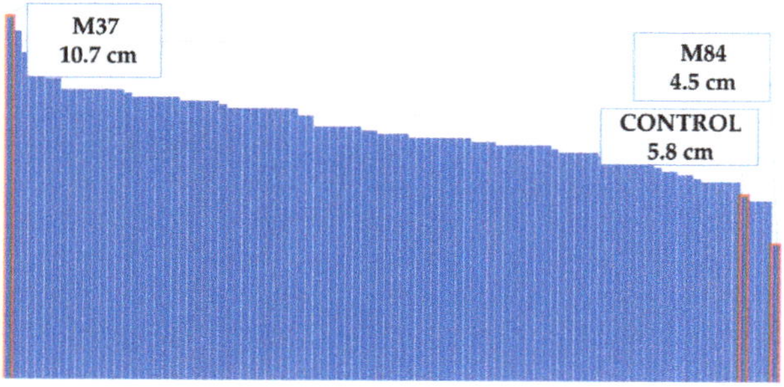

Figure 11. Maximum and minimum leaf lengths of the population.

In comparison to the control group, the mutant plant coded M39 had the widest leaves, whereas the mutant plant coded M84 had narrowest leaves (Figure 12). The difference in leaf diameter was 34% according to the control group (Figure 4).

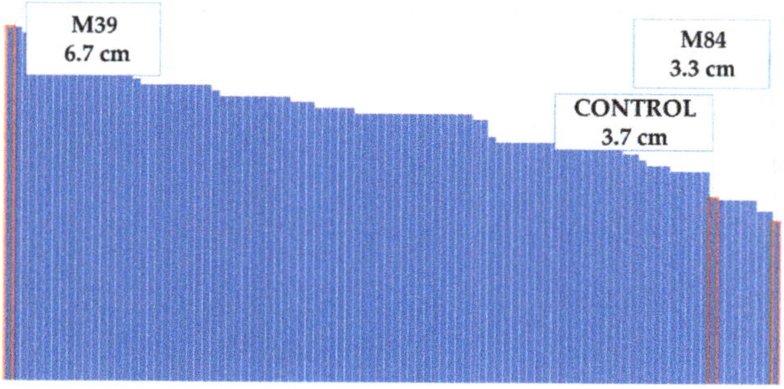

Figure 12. Maximum and minimum leaf widths of the population.

The M74-coded mutant plant had the lightest stem weight, while the M17-coded mutant plant had a heavier stem than the control group (Figure 13). Compared to the mutant population in the control group, stem weights decreased by 7% (Figure 4).

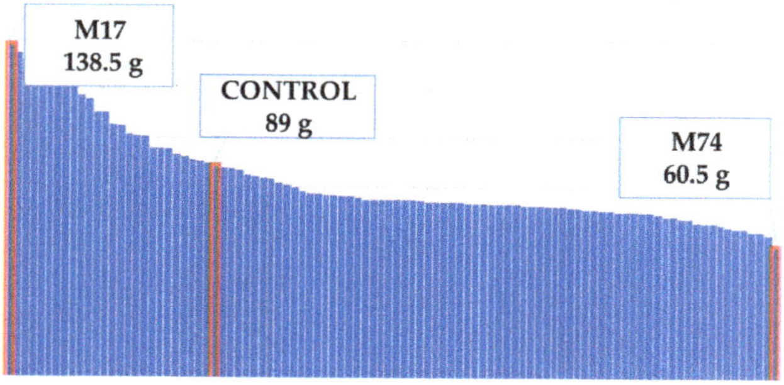

Figure 13. Maximum and minimum stem weights of the population.

3.3. Cluster of the Mutant Population

Cluster analysis was performed on plant height, plant diameter, flower diameter, number of flowers, number of ray flowers, flower colors, number of flower colors, number of leaves, leaf length, leaf width, and stem weight. We obtained six groups; control plants and the only mutant with three colors (pink, white, and variegated) were located in the first group among 43 genotypes. There were 6 genotypes in the second group and 33 genotypes in the third group. The shortest plant (M38) was the only mutant in the fourth group, 11 genotypes were located in the fifth group, and the multi-flowered plants (M7 and M11) formed the sixth group. Both pink-flowered mutant genotypes and the pink and variegated flowered plants were observed in the first and third groups (Figures 14 and 15). According to the correlation matrix, the similarity ratio was 97.2% between the closest mutants M53 and M72 (Table 1).

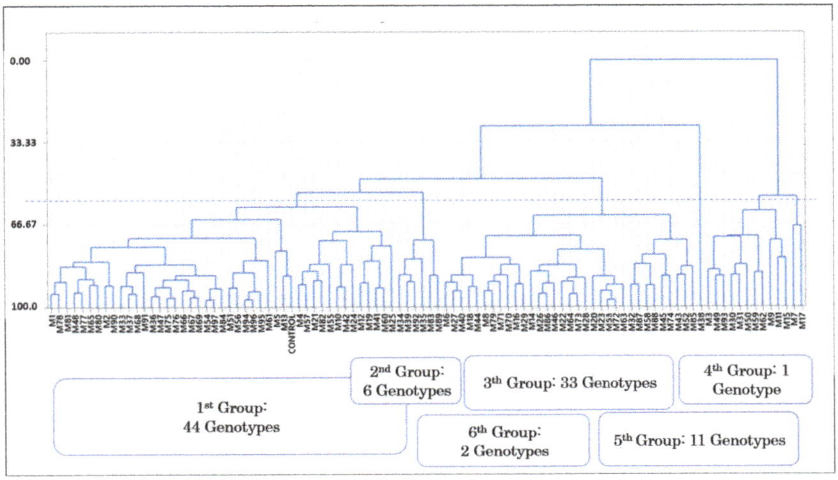

Figure 14. Dendrogram of mutant genotypes.

(a) (b) (c)

Figure 15. The mutant plants: (**a**) white-colored mutant—M38; (**b**) the mutant with pink, white, and variegated flowers; (**c**) white and pink–white variegated flowers (bars: 1 cm).

Table 1. Distance between the genotypes according to the cluster groups.

Genotype 1	Genotype 2	Distance
M38	M1	73.18
M3	M7	44.67
M4	M12	30.45
M14	M20	23.01
M11	M15	19.71
M4	M21	14.51
M36	M54	12.83
M13	CONTROL	12.58
M10	M42	8.09
M37	M68	5.45
M53	M72	2.83

In the principal component analysis, eigenvalues are shown in the scree plot. As can be seen from the figure, there are four components with eigenvalues greater than 1 (Figure 16).

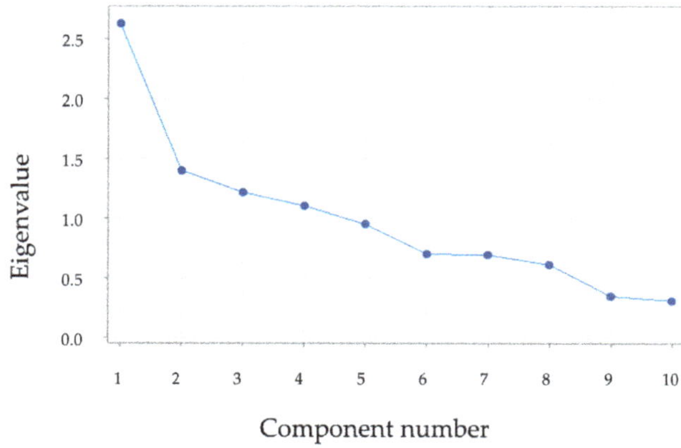

Figure 16. Eigenvalues in the scree plot.

The first four principal components were determined to be responsible for 0.636 of the total variance in the principal component analysis (Table 2). The first group constitutes 0.263 of the total variation, and plant height, number of leaves, leaf length, and width were more effective in the formation of this group in a positive direction (Table 3).

Table 2. Eigenvalue, variation percentage, and cumulative variation percentage values obtained as a result of principal component analysis in observed mutant genotypes.

	Eigenvalue	Proportion	Cumulative
PC 1	2.6279	0.263	0.263
PC 2	1.4048	0.140	0.403
PC 3	1.2211	0.122	0.525
PC 4	1.1082	0.111	0.636
PC 5	0.9547	0.095	0.732
PC 6	0.7058	0.071	0.802
PC 7	0.6974	0.070	0.872
PC 8	0.6162	0.062	0.934
PC 9	0.3508	0.035	0.969
PC 10	0.3132	0.031	1

In the second main component, while plant height, plant diameter, number of leaves, and stem weight had a positive effect, the number of flowers, number of flower colors, number of ray flowers, flower diameter, leaf length, and leaf width had a negative effect. Although the number of flowers, leaf length, and stem weight had a positive effect on the formation of the third group, the other traits had the opposite impact on it. The creation of the fourth group was positively impacted by flower diameter, the number of flower colors, the number of leaves, leaf length, and leaf width, whereas the other factors had the opposite effect (Table 3).

As a result of the analysis, plant height, leaf number, leaf length, and stem weight were determined as traits with high factor coefficients on the first PC component. It was determined that flower diameter and the number of ray flowers were important in the second PC component. The prominent features of the third PC component were the number of flowers and the number of flower colors. The features of the fourth PC component were the plant diameter and leaf width (Table 3).

Table 3. PC component and factor coefficients obtained as a result of principal component analysis in the observed mutant genotypes.

	PC 1	PC 2	PC 3	PC 4	PC 5
Plant Height (cm)	0.403	0.329	−0.182	−0.127	0.163
Plant Diameter (cm)	0.200	0.163	−0.391	−0.626	0.100
Flower Diameter (cm)	0.064	−0.629	−0.283	0.029	−0.138
Number of Flowers	0.276	−0.121	0.443	−0.221	−0.610
Number of Ray Flowers	0.083	−0.553	−0.053	−0.476	0.165
Number of Flower Colors	−0.100	−0.015	−0.615	0.195	−0.571
Number of Leaves	0.364	0.222	−0.303	0.094	−0.079
Leaf Length (cm)	0.408	−0.237	0.038	0.233	0.338
Leaf Width (cm)	0.433	−0.183	−0.065	0.458	0.110
Stem Weight (g)	0.463	0.103	0.244	−0.071	−0.293

4. Discussion

The market for ornamental plants is always in need of new commercial varieties and mutation breeding is frequently utilized to quickly achieve alterations that are significant to the market. Through this work, flowers from the semi-double white control group were used to create new daisy-eyed double white and pink mutants with more ray flowers. By using tissue culture methods, the mother mutant's various colored flowers were separated and propagated rapidly.

Studies on the effects of mutation breeding in chrysanthemums have revealed that the results provide comparable changes in plants. The calli of chrysanthemum varieties were exposed to cobalt-60 gamma rays to modify the morphology of the original varieties. Advantageous mutant lines were obtained with the lower stem, shorter growth process, smaller leaves, changes in flower color, and larger blooms [34]. In the present study, in addition to the color changes in flowers, an increase in flower diameter and decreases in plant height were observed, while leaf size increased compared to the control group.

It was indicated that when treated with gamma rays, differences were generated in the form and color of the flowers like the yellow–white chimera, the deeper color, the pink–white color, and the dark purple color. The flower variations were the changes in the flower head diameter, ray flower number, flowering time, flower type, and flower color [35]. Our study's findings showed a rise in the diameter of the flower head and the number of ray flowers, and we also obtained pink and white chimeric flowers which were compatible with the findings of the mentioned study. While the transformation from white to yellow was observed more depending on the carotenoid content, the transformation from white to pink was also recorded due to anthocyanin formation. Additionally, it was discovered that plants exposed to varying levels of gamma radiation showed changes in flower color, form, and shape [36]. Comparably to this research, the present research's wide spectrum of mutants with floral patterns was observed when we compared the control plants. We have seen distinct variations in plant height, plant width, leaf number, leaf size, number of flower colors, and flower head size. With the increase in the number of ray flowers, the flower form turned into a daisy-eyed double. Among other traits, the highest increase was recorded in plant diameter compared to control group plants.

However, other types of mutagens, such as microwave radiation or synchrotron light irradiation, may also be helpful for the chrysanthemum variants. After being exposed to microwave radiation (MW), the explants of the chrysanthemum variety showed longer stems with bigger flowers [8]. It was discovered that synchrotron light irradiation caused changes in mutants that displayed a range of color patterns, including a paler and darker shade of colors than the original flowers [37]. We attained similar results in the present study such as alterations in the plant height and flower characteristics of the plants. Even if the mutagen changes, distinct variations in the flowers and other characteristics of the plant can be used in the creation of novel types. The benefit of utilizing tissue culture methods on mutants is the ability to quickly and easily develop new plants with the desired characteristics.

It was stated that treatment of gamma irradiation caused a reduction in floral size and an increase in flower yield in small-sized plants [38]. In the present study, while plant height was shortened, plant diameter and flower sizes increased compared to the control group. When we look at the findings of a different study in which a white chrysanthemum variety was irradiated by gamma radiation, the irradiated plants differed from control plants in plant height, number of leaves, leaf length and width, number of flowers, and flower diameter. The first of the selected three mutants had tubular ray florets and the others both had yellow flowers, while one of them had spoon ray florets and the other one had flat-shaped ray florets [39]. In this study, based on the control group, changes were detected in the previously mentioned characteristics such as plant height, number of leaves, leaf length and width, number of flowers, and flower diameter. In contrast to the described study, the flowers in the present study displayed pink and white variegation instead of yellow, and the ray florets did not exhibit any changes in morphology.

Consistent with the findings of previous research, it was discovered that the flower colors of mutants exposed to gamma radiation differed from the original varieties. While the original variety had pink–orange flowers, the mutants were discovered to have entirely pink ray flowers. The second mutant had fewer yellow shade ray florets than the original pink–orange variety, while the third mutant had more yellow ray florets in the center of the flower than the original pink–yellow variety. The fourth mutant had a mutation in flower morphology that resulted in multiple ray florets [40]. In addition to modifications in other parts of the mutants, we also noticed differences in color and ray floret number between the mutant group and the control group. Further support for future research is provided by the fact that the control and mutant plants produced results that were equivalent in terms of flower color, plant heights, flower head size, number of ray florets, number of leaves, leaf size, and stem weight.

These kinds of variances can be used to create distinctive and desirable properties that will appeal to producers and customers both. The reason why height and number of flowers in plants and stem weight decreased while other traits were increased compared to the control group was due to the effect of applied gamma rays and somatic variations in tissue cultures. In addition, too many flowers and extreme flower height values that are too short or too tall are already undesirable in cut flower cultivation. Considering that up to five stems are placed in bouquets, developing more flowering plants is not suitable for commercial use. On the other hand, short–tall and multi-flowered plants come to the fore in outdoor and potted use. It is an appropriate and desired feature to evaluate such plants in terms of their usage.

5. Conclusions

Our research points to great potential for expanding the understanding of breeding mutant chrysanthemums and creating new possibilities for the creation of new distinct genotypes with improved functional qualities as well as attractiveness. We can optimize the potential of these mutants and significantly impact the floriculture market by carefully choosing or breeding them in addition to continuing research on mutants and utilizing them as new varieties. These findings provide opportunities for further research into the fundamental mechanisms generating mutant populations that give rise to these variations. Results from in vitro treatments using mutagenic gamma radiation show that this method is effective in observing mutations in chrysanthemums.

Funding: This research was funded by the Republic of Türkiye Ministry of Agriculture and Forestry, General Directorate of Agricultural Research and Policies, Aegean Agricultural Research Institute.

Data Availability Statement: Data are contained within the article.

Acknowledgments: This study is a part of the project in collaborations with the Turkey Energy, Nuclear and Mining Research Association, Nuclear Energy Research Institute and Bademler Village Agricultural Development Cooperative. I would like to thank all my colleagues for their assistance with this paper. My special thanks to Yaprak Kantoglu, Burak Kunter, and Alameddin Bayav for their devoted help.

Conflicts of Interest: The author declares no conflicts of interest. The funders had no role in the design of the study; in the collection, analysis, or interpretation of data; in the writing of the manuscript; or in the decision to publish the results.

References

1. Anderson, N.O. Chrysanthemum. In *Flower Breeding and Genetics*; Anderson, N.O., Ed.; Springer: Dordrecht, The Netherlands, 2007; pp. 389–437. [CrossRef]
2. Miler, N.; Jędrzejczyk, I.; Trafara, P.; Winiecki, J. Effect of high-energy ionizing radiation on the DNA content and genetic variation in chrysanthemum plants regenerated from irradiated ovaries. *Acta Sci. Pol. Hortorum Cultus* **2023**, *22*, 117–134. [CrossRef]
3. Eisa, E.A.; Tilly-Mándy, A.; Honfi, P.; Shala, A.Y.; Gururani, M.A. Chrysanthemum: A comprehensive review on recent developments on in vitro regeneration. *Biology* **2022**, *11*, 1774. [CrossRef] [PubMed]
4. Welch, C.W. Crysanthemums. In *Breeding New Plants and Flowers*; The Crowood Press: Wiltshire, UK, 2002; p. 144.
5. Xu, Y.; Liao, B.; Ostevik, K.L.; Zhou, H.; Wang, F.; Wang, B.; Xia, H. The maternal donor of chrysanthemum cultivars revealed by comparative analysis of the chloroplast genome. *Front. Plant Sci.* **2022**, *13*, 923442. [CrossRef] [PubMed]
6. Yu, Y.P.; Lin, K.H.; Shih, M.C.; Chen, C.L.; Lu, C.P. Optimization of aqueous extraction of antioxidants from chrysanthemum (*C. morifolium* Ramat and *C. indicum* L.) flowers and evaluation of their protection from glycoxidation damage on human αA-crystallins. *Exp. Eye Res.* **2023**, *235*, 109629. [CrossRef] [PubMed]
7. Luo, C.; Dongliang Chen, D.; Cheng, X.; Hua Liu, H.; Li, Y.; Huang, C. SSR analysis of genetic relationship and classification in chrysanthemum germplasm collection. *Hortic. Plant J.* **2018**, *4*, 73–82. [CrossRef]
8. Miler, N.; Kulus, D. Microwave treatment can induce chrysanthemum phenotypic and genetic changes. *Sci. Hortic.* **2018**, *227*, 223–233. [CrossRef]
9. Nakano, M.; Taniguchi, K.; Masuda, Y.; Kozuka, T.; Aruga, Y.; Han, J.; Motohara, K.; Nakata, M.; Sumitomo, K.; Hisamatsu, T.; et al. A Pure line derived from a self-compatible chrysanthemum seticuspe mutant as a model strain in the genus chrysanthemum. *Plant Sci.* **2019**, *287*, 110174. [CrossRef] [PubMed]
10. Li, M.; Wen, Z.; Meng, J.; Cheng, T.; Zhang, Q.; Sun, l. The genomics of ornamental plants: Current status and opportunities. *Ornam. Plant Res.* **2022**, *2*, 1–18. [CrossRef]
11. Li, Y.; Yang, P.; Luo, Y.; Gao, B.; Sun, J.; Lu, W.; Liu, J.; Chen, P.; Zhang, Y.; Yu, L.L. Chemical compositions of chrysanthemum teas and their anti-inflammatory and antioxidant properties. *Food Chem.* **2019**, *286*, 8–16. [CrossRef] [PubMed]
12. Qi, S.; Yang, L.; Wen, X.; Hong, Y.; Song, X.; Zhang, M.; Dai, S. Reference gene selection for RT-qPCR analysis of flower development in *Chrysanthemum morifolium* and *Chrysanthemum lavandulifolium*. *Front. Plant Sci.* **2016**, *7*, 186000. [CrossRef]
13. Ryu, J.; Nam, B.; Kim, B.R.; Kim, S.H.; Jo, Y.D.; Ahn, J.W.; Kim, J.B.; Jin, C.H.; Han, A.R. Comparative Analysis of Phytochemical Composition of Gamma-Irradiated Mutant Cultivars of Chrysanthemum morifolium. *Molecules* **2019**, *24*, 3003. [CrossRef] [PubMed]
14. Song, X.; Tian, Y.; Gao, K.; Li, J.; Li, Y.; Wang, J.; Deng, C.; Zhang, F.; Kong, K.; Fan, G.; et al. Genetic and QTL analysis of flower color and pigments in small-flowered chrysanthemum based on high-density genetic map. *Ornam. Plant Res.* **2023**, *3*, 17. [CrossRef]
15. Mekapogu, M.; Kwon, O.K.; Song, H.Y.; Jung, J.A. Towards the improvement of ornamental attributes in chrysanthemum: Recent progress in biotechnological advances. *Int. J. Mol. Sci.* **2022**, *23*, 12284. [CrossRef] [PubMed]
16. Din, A.; Qadri, Z.A.; Wani, M.A.; Rather, Z.A.; Iqbal, S.; Malik, S.A.; Hussain, P.R.; Rafiq, S.; Nazki, I.T. Congenial In Vitro γ-ray-Induced Mutagenesis Underlying the Diverse Array of Petal Colours in Chrysanthemum (*Dendranthema grandiflorum kitam*) cv. "Candid". *Biol. Life Sci. Forum* **2021**, *4*, 21. [CrossRef]
17. Mekapogu, M.; Vasamsetti, B.M.K.; Kwon, O.K.; Ahn, M.S.; Lim, S.H.; Jung, J.A. Anthocyanins in floral colors: Biosynthesis and regulation in chrysanthemum flowers. *Int. J. Mol. Sci.* **2020**, *21*, 6537. [CrossRef]
18. Nakano, M.; Hirakawa, H.; Fukai, E.; Toyoda, A.; Kajitani, R.; Minakuchi, Y.; Itoh, T.; Higuchi, Y.; Kozuka, T.; Bono, H.; et al. A chromosome-level genome sequence of chrysanthemum seticuspe, a model species for hexaploid cultivated chrysanthemum. *Commun. Biol.* **2021**, *4*, 1167. [CrossRef] [PubMed]
19. Lu, C.; Li, Y.; Wang, J.; Qu, J.; Chen, Y.; Chen, X.; Huang, H.; Dai, S. Flower color classification and correlation between color space values with pigments in potted multiflora chrysanthemum. *Sci. Hortic.* **2021**, *283*, 110082. [CrossRef]
20. Ohmiya, A. Molecular mechanisms underlying the diverse array of petal colors in chrysanthemum flowers. *Breed. Sci.* **2018**, *68*, 119–127. [CrossRef] [PubMed]
21. Bado, S.; Forster, B.P.; Maghuly, F. Physical and chemicals mutagenesis in plant breeding. In *Mutation Breeding for Sustainable Food Production and Climate Resilience*; Penna, S., Jain, S.M., Eds.; Springer: Singapore, 2023; p. 97. [CrossRef]

22. Udage, A. Introduction to plant mutation breeding: Different approaches and mutagenic agents. *J. Agric. Sci. Sri Lanka* **2021**, *16*, 466. [CrossRef]
23. Bezie, Y.; Tilahun, T.; Atnaf, M.; Taye, M. The potential applications of site-directed mutagenesis for crop improvement: A review. *J. Crop Sci. Biotechnol.* **2021**, *24*, 229–244. [CrossRef]
24. Datta, S.K. Breeding of ornamentals: Success and technological status. *Nucleus* **2021**, *65*, 107–128. [CrossRef]
25. Datta, S.K. Introduction/Review. In *Induced Mutation Breeding*; Springer: Singapore, 2023; p. 71. [CrossRef]
26. Melsen, K.; Van De Wouw, M.; Contreras, R. Mutation Breeding in Ornamentals. *Am. Soc. Hortic. Sci. Mutat. Breed. Ornam.* **2021**, *56*, 1154–1165. [CrossRef]
27. Sawada, Y.; Sato, M.; Okamoto, M.; Masuda, J.; Yamaki, S.; Tamari, M.; Tanokashira, Y.; Kishimoto, S.; Ohmiya, A.; Abe, T.; et al. Metabolome-based discrimination of chrysanthemum cultivars for the efficient generation of flower color variations in mutation breeding. *Metabolomics* **2019**, *15*, 1–15. [CrossRef] [PubMed]
28. Anne, S.; Hee Lim, J. Variability of chrysanthemum cultivars induced by gamma irradiation. *Hortic. Sci. Technol.* **2021**, *39*, 660–672. [CrossRef]
29. Din, A.; Qadri, Z.A.; Wani, M.; Iqbal, S.; Malik, S.; Bhat, Z.A.; Banday, N. Developing an efficient in vitro callusing and regeneration protocol in *Dendranthema* × *grandiflorum* Kitam. *J. Crop Sci. Biotechnol.* **2022**, *25*, 393–405. [CrossRef]
30. Haspolat, G.; Kunter, B.; Kantoglu, Y. Determination of mutagenic-sensitivity and induced variability in the mutant populations of 'Bacardi' chrysanthemum cultivar. *Genetika* **2022**, *54*, 161–172. [CrossRef]
31. Murashige, T.; Skoog, F. A revised medium for rapid growth and bioassay with tobacco tissue culture. *Physiol. Plant.* **1962**, *15*, 473–479. [CrossRef]
32. Anderson, N.O.; Ascher, P.T. Inheritance of seed set, germination and day neutrality/heat delay insensitivity of garden chrysanthemums (*Dendranthema* × *grandiflora*) under glasshouse and field conditions. *J. Am. Soc. Hortic. Sci.* **2004**, *129*, 509–516. [CrossRef]
33. The Royal Horticultural Society, London, in Association with the Flower Council of Holland. R.H.S. Colour Chart. London: The Society. 1986. Available online: https://www.rhs.org.uk/ (accessed on 5 March 2024).
34. Thao, L.; Dung, N.; Tham, N. Study on Chrysanthemum breeding by gamma (co60) irradiation on callus of 4 exotic varieties. *Int. J. Agric. Technol.* **2015**, *11*, 1813–1822.
35. Wu, J.; Zhang, J.; Lan, F.; Fan, W.; Li, W. Morphological, cytological, and molecular variations induced by gamma rays in ground-grown chrysanthemum 'Pinkling'. *Can. J. Plant Sci.* **2019**, *100*, 68–77. [CrossRef]
36. Chowdhury, J.; Hoque, M.I.; Sarker, R.H. Evaluation of the Effect of Different Doses of Gamma Radiation to Induce Variation in *in vitro* Raised Plants of Chrysanthemum. *Plant Tissue Cult. Biotech.* **2023**, *33*, 155–165. [CrossRef]
37. Sakamoto, K.; Nishi, M.; Ishiji, K.; Takatori, Y.; Chiwata, R. Induction of flower-colour mutation by synchrotron-light irradiation in spray chrysanthemum. *Acta Hortic.* **2019**, *1237*, 73–78. [CrossRef]
38. Anitha, G.; Shiragur, M.; Patil, B.C.; Nishani, S.; Seetharamu, G.K.; Ramanagouda, S.H.; Naika, M.B. Mutation studies in chrysanthemum cultivar Poornima white. *J. Pharmacogn. Phytochem.* **2021**, *10*, 1235–1239.
39. Soliman, T.M.; Lv, S.; Yang, H.; Hong, B.; Ma, N.; Zhao, L. Isolation of flower color and shape mutations by gamma radiation of Chrysanthemum morifolium Ramat cv. Youka. *Euphytica* **2014**, *199*, 317–324. [CrossRef]
40. Puripunyavanich, V.; Piriyaphattarakit, A.; Chanchula, N.; Taychasinpitak, T. Mutation induction of in vitro Chrysanthemum by gamma irradiation. *Chiang Mai J. Sci.* **2019**, *46*, 609–617.

horticulturae

MDPI

Article

Analysis of Floral Scent Component of Three Iris Species at Different Stages

Keyu Cai [1,†], Zhengjie Ban [1,†], Haowen Xu [2], Wanlin Chen [1], Wenxu Jia [3], Ying Zhu [4] and Hongwu Chen [1,*]

1 College of Horticulture, Northwest A&F University, Xianyang 712100, China; caikeyu@nwafu.edu.cn (K.C.); banzhengjie@nwafu.edu.cn (Z.B.); cwl@nwafu.edu.cn (W.C.)
2 College of Forestry, Northwest A&F University, Xianyang 712100, China; xhw@nwafu.edu.cn
3 College of Information Engineering, Northwest A&F University, Xianyang 712100, China; jiawenxu@nwafu.edu.cn
4 Beijing Botanical Garden Management Office, Beijing 100093, China; zhuying@chnbg.cn
* Correspondence: ylchhw1971@nwafu.edu.cn
† These authors contributed equally to this study.

Abstract: The research investigates the variations in floral scent composition among different species and developmental stages of Iris plants: *Iris uniflora*, *Iris typhifolia*, and *Iris sanguinea*. The study analyzes the fragrance components by utilizing electronic nose technology in tandem with headspace solid-phase microextraction-gas chromatography-mass spectrometry (HS-SPME-GC-MS). Principal component analysis (PCA), linear discriminant analysis (LDA), and loading analysis are applied to discern whether floral scents of the same Iris species at distinct stages could be differentiated. The results show that the electronic nose significantly distinguishes the aromas from different stages and that there are differences in aroma composition. Gas Chromatography-Mass Spectrometry confirms significant differences in volatile components regarding the three Iris species, with common compounds like alcohols, aromatics, and aldehydes present throughout stages. Notably, nonyl aldehyde, capric aldehyde, 2,4-di-tert-butylphenol, and n-heptadecane are consistently found. Cluster analysis reveals a grouping of decay stage samples of *Iris typhifolia* and *Iris sanguinea* due to terpene and ester abundance. Nonyl aldehyde significantly contributes to the aroma profiles of all species, owing to its high odor activity value. The significant content of volatile compounds in these Iris varieties suggests economic and medicinal potential beyond ornamental value, providing references for the development of Iris-scented products, aromatherapy, and the extraction of pharmacologically active substances from Iris.

Keywords: Iris; floral fragrance; GC-MS; OAV

Citation: Cai, K.; Ban, Z.; Xu, H.; Chen, W.; Jia, W.; Zhu, Y.; Chen, H. Analysis of Floral Scent Component of Three Iris Species at Different Stages. *Horticulturae* **2024**, *10*, 153. https://doi.org/10.3390/horticulturae10020153

Academic Editor: Jiri Gruz

Received: 9 January 2024
Revised: 24 January 2024
Accepted: 5 February 2024
Published: 6 February 2024

1. Introduction

Floral fragrances are mainly composed of small-molecule volatile compounds, which can be extracted, collected, or biosynthesized, and are widely used in various fields, such as daily chemicals, medicine, food, and others [1]. Floral fragrances are crucial characteristics of plants, having several important functions. Plants release volatile components through flower organs to attract insects for pollination and fruit development. Additionally, floral fragrances aid in the defense of plants by attracting natural enemies and deterring potential damage. Moreover, floral fragrances enhance the ornamental value of plants. As people's understanding continues to deepen and the application of flower fragrances keeps being promoted, breeding fragrant flowers has become an important direction in flower cultivation.

To date, more than 1700 natural aroma substances have been discovered and identified in plants. Although thousands of plant floral aroma compounds have been identified, only 12 are found in more than 50% of seed plant families, such as limonene (71%), (E)-Ocimene

(71%), Myrcene (70%), and so on [2]. Observing many studied plants, it is evident that there are usually between 20 and 60 kinds of floral fragrance substances [3].

During the entire flowering process, the flower itself regulates the synthesis and volatilization of volatile components. Different plants share similar regulatory mechanisms. Generally, in the early stage of flowering, the volatilization of floral aroma components gradually increases, reaching its peak when flowers are ready for pollination [4].

In a study on the floral fragrance of the German bearded Iris [5], 219 species in 10 categories are identified, including 42 terpenes, 19 alkanes, 11 aromatic compounds, 52 esters, 41 alcohols, 17 ketones, 21 aldehydes, 9 acids, and 4 phenols. The volatile components of different species and cultivars show significant variations.

Furthermore, it should be noted that, during the flowering process, there is a certain time lag between the peak expression of the gene encoding the floral aroma synthase and the volatilization of the corresponding floral aroma components, sometimes even lasting 1 to 2 days [6,7]. These findings also indicate that the expression of structural genes that are responsible for producing floral aroma substances ultimately affects the volatilization of floral aroma compounds.

There are approximately 260 to 300 species of Iris plants worldwide, mainly distributed in the temperate regions of the northern hemisphere [8]. However, in my country's northwest region, there are 18 species of Iris, many of which remain under-explored. Some research teams have made progress in studying seed dormancy, germination characteristics, and interspecific hybridization barriers of *Iris sativa* [9].

As a traditional folk medicine, Iris plants have a long history of medicinal use and are mainly employed to treat various diseases, such as cancer, inflammation, and bacterial and viral infections [10]. Extracts from Iris plants are also utilized in treating atherosclerosis and osteoporosis [11]. Some Iris plants' roots, stems, and leaves contain precious essence, which can be ground into powder to obtain high-quality fragrance powder.

Iris uniflora stands out for its unique sweet and fragrant rhizome extract, making it a valuable raw material for perfume [12]. The floral fragrance characteristics of *Iris typhifolia* are primarily provided by its pistil, giving rise to a distinctive fresh aroma. [13]. The stems and leaves of *Iris sanguinea* can be used to make paper, while the flowers, with their large and colorful appearance, hold high ornamental value, and their rhizomes are used medicinally.

For this study, three types of Iris flowers in the bud stage, bloom stage, and decay stage are selected. It has been observed that the aroma of each flower stage can be significantly distinguished using an electronic nose, and there are notable differences in aroma components [14]. To determine the types and contents of various floral aroma substances in different flowering stages, electronic nose measurement, and headspace solid-phase microextraction gas chromatography-mass spectrometry, are employed [15–17]. The experimental results reveal the different flowering periods of the Iris and the main floral components, as well as the pattern of fragrance release.

2. Methods and Materials

2.1. *Overview of the Test Field and Test Materials*

In this study, flowers at various stages (bud, bloom, and decay) from three different Iris varieties, namely, *Iris uniflora*, *Iris typhifolia*, and *Iris sanguinea*, were selected as the sample subjects. The wild species of Iris used in the experiment were collected at the Caoxinzhuang Experimental Farm of Northwest A&F University in Yangling, Shaanxi Province, China. This region experiences a temperate continental monsoon climate, with an average altitude of 530 m, an average annual temperature of 12.9 °C, four distinct seasons, and sufficient sunshine conditions. *Iris uniflora*, *Iris typhifolia*, and *Iris sanguinea* in three stages—flower bud, bloom, and decay—were chosen as samples to determine the composition of floral aroma substances.

Sampling Method: Under sunny conditions, samples were collected between 8:00 a.m. and 10:00 a.m. Experimenters wore nitrile gloves, cut the inflorescences with scissors, and placed them into pre-cooled 50 mL sterile transparent plastic centrifuge tubes using liquid nitrogen. The tubes were sealed with tin foil, the bottle caps were tightened, and they were marked before being immersed in liquid nitrogen. Samples were temporarily stored in nitrogen and moved to the refrigerator as soon as possible, where they were stored at −80 °C.

Test Equipment: The ISQ&TRACE ISQ gas chromatography-mass spectrometry (GC-MS) coupled instrument (Type: QP-2010) from Shimadzu in Kyoto, Japan, PEN3 electronic nose from Airsense in Schwerin, Germany, analytical balance, manual headspace sampler, chromatographic column Rtx-1MS (30 m × 0.25 mm × 0.25 µm, Shimadzu, Kyoto, Japan), 15 mL headspace vials, foil, centrifuge tubes, etc., were used.

Before conducting the test, the laboratory air was ensured to be clean and free of odors, with minimal human presence. Before sample injection, the air was washed until the response values of all 10 sensors reached 1, indicating that the radar chart was approximately circular. Each centrifuge tube containing Iris in each flowering stage was taken out and allowed to stand for 30 min to allow the volatile matter in the headspace of the sample to reach an equilibrium state. The electronic nose sampling needle and the air supply needle were then inserted into the headspace bottle through the tin foil simultaneously for headspace sampling and detection. Each test included gas washing, zeroing, and sample injection.

2.2. *Research Methodology*

2.2.1. Electronic Nose

Electronic nose measurement conditions: gas flow rate 400 mL·min^{-1}, gas washing time 60 s, zeroing time 5 s, preparation time 5 s, measurement time 120 s, taking the response value of 115~117 s for data analysis. The injection flow rate is 150 mL·min^{-1}. Clean air is used as the carrier gas during cleaning, and both the flow rates of carrier gas and sample injection are 200 mL·min^{-1}. The responses of electronic nose sensors to different types of compounds are shown in Table 1. The built-in software WinMuster (Version 1.6.2) on the PEN3 was used for data collection and analysis.

Table 1. Response characteristics of each sensor of PEN3 electronic nose.

Array Serial Number	Sensor Name	Performance Description
1	W1C	aromatic
2	W5S	broadrange
3	W3C	aromatic
4	W6S	hydrogen
5	W5C	arom-aliph
6	W1S	broad-methane
7	W1W	sulphur-organic
8	W2S	broad-alcohol
9	W2W	sulph-chlor
10	W3S	methane-aliph

2.2.2. HS-SPME-GC-MS Analysis

Headspace solid-phase microextraction-gas chromatography-mass spectrometry (HS-SPME-GC-MS, Shimadzu, Kyoto, Japan) was employed in this study. Approximately 1 g of each flowering stage sample was placed in 15 mL headspace solid-phase microextraction bottles. Subsequently, 10 µL of 2-nonanone (0.008 µL·mL^{-1}), dissolved in propylene glycol, was added to the bottom of the bottle, and the bottle cap was sealed with tin foil. We utilized the SPME extraction head from the German company Supelco, model DVB/CAR/PDMS, with a diameter of 50/30 um. After equilibrating at 45 °C for 10 min, a manual headspace sampler was inserted, and extraction was performed for 30 min. Each sample was repeated four times. The manual headspace sampler was then inserted into the inlet of the GC-MS at 250 °C for 2.5 min to completely release the aroma components in the gas chromatography-mass spectrometer [18].

Chromatographic conditions involved using an HP-5MS capillary tube (30 m × 0.25 mm × 0.25 μm, Shimadzu, Kyoto, Japan) as the chromatographic column, adopting the splitless mode. The inlet temperature was set at 250 °C, and high-purity nitrogen (99.999%) was used as the carrier gas at a flow rate of 1.0 mL·min^{-1}. The column temperature followed a heating and cooling program: the initial temperature was 40 °C, with a holding time of 3 min. The temperature was then raised to 150 °C at a rate of 5 °C·min^{-1}, followed by a further increase at a rate of 10 °C·min^{-1} to 220 °C, where it was kept for 10 min.

Mass spectrometry conditions involved the full scan mode, electron ionization (EI) as the ionization mode, an emission current of 10 Ua, electron energy of 70 eV, ion source temperature of 220 °C, transmission source temperature of 220 °C, and a mass scanning range of 35~450 amu.

For GC-MS data identification and analysis methods:

Qualitative method: The stable peak on the ion peak diagram was identified by corresponding to the first place in the preferred ranking of multiple substances under the same retention time (RT) value, starting from the first stable peak. The selected substances were compared with the four times duplicate tests to ensure they appeared at the same RT value and ranked at the top.

2.2.3. Identification and Analysis Methods

Qualitative method: Each component obtained using GC-MS analysis is searched and compared with NIST2017 mass spectrometry and a standard information database, as well as identified and analyzed together with the related literature. By using the carbon standard method, with the same column and rising and cooling procedure as GC-MS, the mixed standard of C7-C30 normal alkane is used as the standard to calculate the linear retention index (Formula (1)) of various aroma components of *Iris uniflora*, *Iris typhifolia*, and *Iris sanguinea* samples, and the results are compared with those of NIST spectrum database. Starting from the first stable peak according to the ion peak diagram, the stable peak corresponds to the first place of a variety of substances under the same RT value, and the selected substances should appear under the same RT value in the four basic biological repeats and rank in the top three.

$$LRI = 100z + 100\,(RT - RTz)/(RT\,(z + n) - RTz) \quad (1)$$

Quantitative method: Initially, 2-nonanone was chosen as the internal standard material with a density of 0.82 g/mL. The numerical equivalent of the added volume of the internal standard substance (μL) was ten times the sample mass (g). The quantification of various aroma components in the *Iris typhifolia* sample was carried out using Formula (2), and the average value was determined after four biological replicates.

$$M_i = C_0 \times V_0 \times A_i \div (A_0 \times M) \quad (2)$$

M_i is the content of each aroma component (μg/g), C_0 is the internal standard concentration (μg/μL), V_0 is the internal standard volume (μL), A_i is the peak area of the desired aroma component, A_0 is the peak area of internal standard material, and M is the sample mass (g).

OAV value: the ratio of the mass concentration of the substance (Formula (3)) to the threshold value of the substance in water, or solvent with properties similar to water (Formula (4)), is used as a standard to evaluate the contribution of the substance to the overall aroma profile of the sample, and the standard sample for aroma reconstruction is selected from it, in which the substance with OAV ≥ 1 is the characteristic aroma component.

$$C_i = C_0 \times A_i \div A_0 \quad (3)$$

C_i is the mass concentration of each aroma component (μg/mL), C_0 is the internal standard concentration (μg/μL), and A_i and A_0 are the peak area and internal standard peak area of the aroma component, respectively.

$$OAV = C_i \div OT_i \quad (4)$$

OT_i is the threshold value of the aroma component in water or solvent with properties similar to water (μg/mL). If there is no numerical value for the threshold of a compound in water, then the value of the closest medium in properties to water should be selected.

3. Results

Figure 1 illustrates the radar map depicting the characteristic response of three Iris species at each stage. The ten sensors exhibit varying responses to the aromas of the three Iris species, generally distributed between 1 and 10. Across all samples, sensors W1W (sulphur-organic), W1S (broad-methane), W2W (sulph-chlor), and W5S (broadrange) consistently display significantly higher response values compared to other sensors. The difference between the response values of these four sensors is notable, indicating that the major classes of substances corresponding to these sensors have higher relative contents in the floral fragrance of Iris plants at all stages.

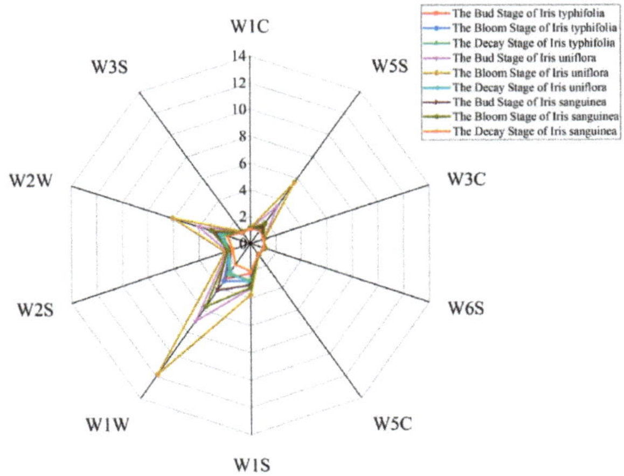

Figure 1. The radar map of the characteristic response of the volatile components in three Iris species at different stages.

During the bloom stage of *Iris uniflora*, the response values of each sensor are notably higher than those of the other eight samples. This suggests that the relative content of each type of compound in the aroma of these samples is higher compared to other stages. In the samples from various periods of *Iris typhifolia*, and the decay stage of *Iris uniflora* and *Iris sanguinea*, the response values for each sensor detecting aroma are lower, indicating a lower concentration of aromatic compounds in these samples compared to others.

In comparison with other samples, the response value of W1S replaces W5S in the top three during the decay stage of *Iris uniflora* and the decay stage of *Iris sanguinea*. This shift indicates that the relative content of methane is higher than that of nitrogen oxides in the aroma components of the samples during these three periods, while the opposite holds for other samples.

Figure 2 depicts the principal component analysis (PCA) pattern for each flowering stage of *Iris uniflora*, *Iris sanguinea*, and *Iris typhifolia*, facilitating the differentiation of floral aromas among different stages of the same Iris variety. The contribution rate of the first principal component is 97.15%, the contribution rate of the second principal component is 1.98%, and the cumulative contribution rate of the two principal components is 99.13%. This indicates that these two principal components effectively represent the main information characteristics of the samples.

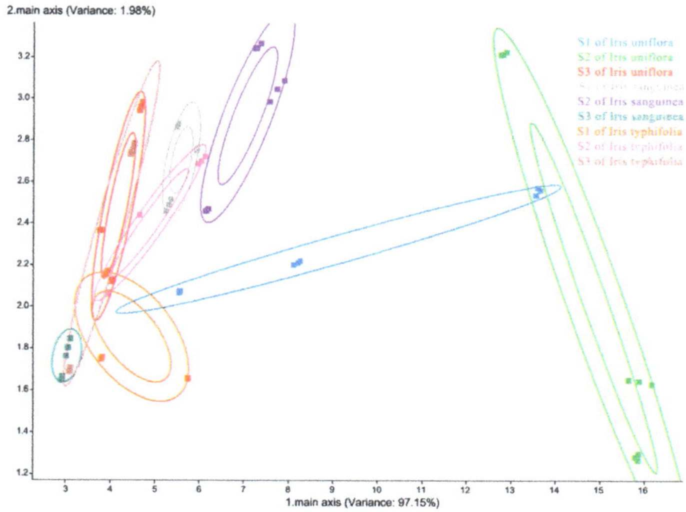

Figure 2. The PCA result of the volatile components in three Iris species at different stages. Note: S1: the bud stage, S2: the bloom stage, S3: the decay stage.

In Figure 2, it is evident that among the three types of Iris, only the bud stage, bloom stage, and decay stage of *Iris sanguinea* can be significantly distinguished. In contrast, the principal component analysis pattern diagrams for each flowering stage of *Iris uniflora* and *Iris typhifolia* exhibit overlap, indicating that they cannot be significantly differentiated in this mode. This lack of distinction is attributed to the relatively similar odors of *Iris uniflora* and *Iris typhifolia* across various periods. Principal component analysis, in this context, proves insufficient for distinguishing the internal flowering stages of *Iris uniflora* and *Iris typhifolia*.

Electronic nose loading analysis is a research method employed to distinguish volatile substances in a sample using electronic nose sensors. It primarily investigates which gas substances in the sample play a key role in this distinction and determines their contribution rates. The results of electronic nose loading analysis for each stage of *Iris uniflora*, *Iris typhifolia*, and *Iris sanguinea* are illustrated in Figure 3. The positions of sensors W1W (sulphur-organic), W2W (sulph-chlor), W1S (broad-methane), W5S (broadrange), and W2S (broad-alcohol) significantly deviate from the origin, indicating the importance of these five sensors in distinguishing the volatile components of the three Iris varieties at different stages. This observation underscores that the variations in floral composition across different periods are predominantly associated with volatile substances, such as sulphur-organic, sulph-chlor, broad-methane, broadrange, and broad-alcohol.

The results of the Linear Discriminant Analysis (LDA) for *Iris uniflora*, *Iris sanguinea*, and *Iris typhifolia* are presented in Figure 4. The contribution rate of the first discriminant factor is 76.36%, and the contribution rate of the second discriminant factor is 10.28%. The cumulative contribution rate reaches 86.64%, indicating that these two discriminant factors essentially represent the main information characteristics of the samples. Figure 4 reveals that the floral fragrances of *Iris uniflora*, *Iris sanguinea*, and *Iris typhifolia* can be significantly distinguished.

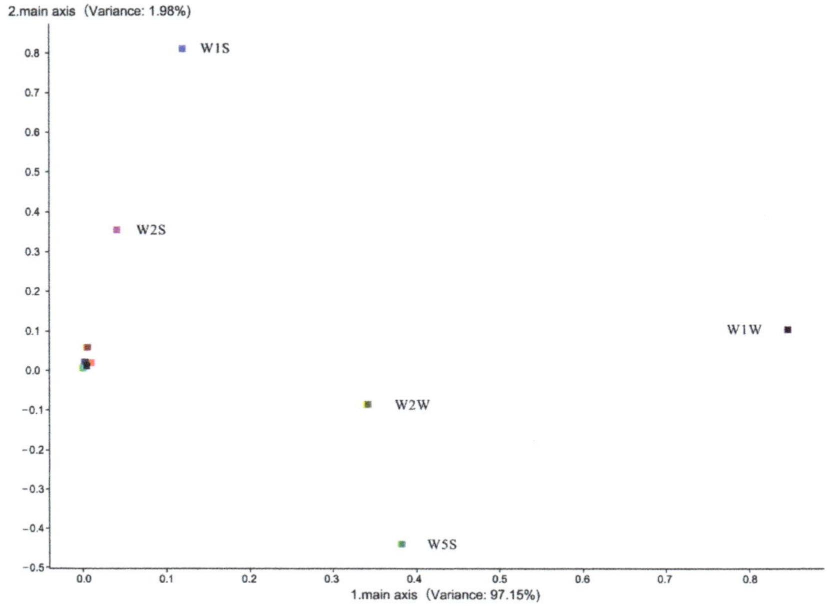

Figure 3. The loading result of the volatile components in three Iris species at different stages.

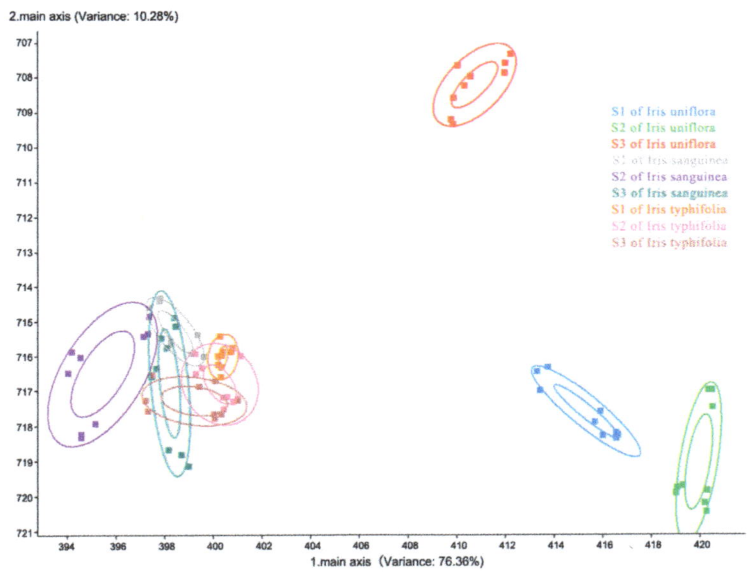

Figure 4. The LDA result of the volatile components in three Iris species at different stages. Note: S1: the bud stage, S2: the bloom stage, S3: the decay stage.

In comparison with Principal Component Analysis, Linear Discriminant Analysis demonstrates a more concentrated distribution, and its discrimination effect is notably higher than that of Principal Component Analysis.

The qualitative and quantitative determination of volatile components in *Iris typhifolia* is conducted through Gas Chromatography-Mass Spectrometry (GC-MS) and spectral library retrieval. A total of 37 species are detected in the bud stage, 43 in the bloom stage, and 39 in the decay stage. The Venn diagram in Figure 5 illustrates the floral composition across different periods, with 13 identical floral components found in all three periods, 10 unique to the bud stage, 12 in bloom, and 13 in the decay stage.

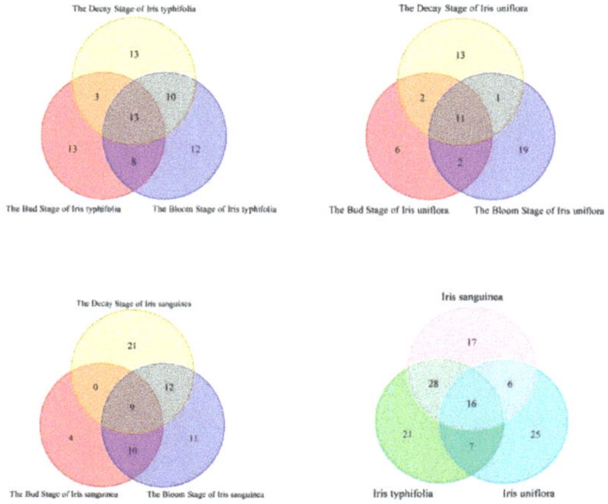

Figure 5. The Venn diagram of the numbers of volatile components in three Iris species.

Quantitative determination of volatile components in *Iris uniflora* reveals 21 species in the bud stage, 33 in the bloom stage, and 27 in the decay stage. Across the three periods, there are 11 identical floral components, 6 unique to the bud stage, 19 in bloom, and 13 in the decay stage. For *Iris sanguinea*, 23 species are detected in the bud stage, 42 in the bloom stage, and 42 in the decay stage. The three periods share 11 identical floral components, with 4 unique to the bud stage, 11 in bloom, and 21 in the decay stage.

In total, 72 species are detected in *Iris typhifolia* samples, 54 in *Iris uniflora*, and 67 in *Iris sanguinea*. The three iris species share a total of 16 identical floral components, with 21 unique to *Iris typhifolia*, 25 in *Iris uniflora*, and 17 in *Iris sanguinea*.

Figures 6 and 7 respectively depict the number and relative content of floral fragrance species in three different stages of Iris. The types of compounds present at each stage include alcohols, aromatics, aldehydes, ketones, alkanes, and esters. The relative content of aldehydes in each stage is notably high, exceeding 40% and reaching an impressive 83.23% in the flower bud stage of *Iris typhifolia*. Olefin and heterocycle samples exhibit low relative contents, less than 2%. The number of ketone species remains relatively stable across all samples over time. Terpene species are generally fewer in the bud stage but exhibit a sudden increase in the decay stage and bloom stage. Although the relative content is not high, the number of terpene species reaches 11 during the decline stage of *Iris sanguinea* and 13 during the bloom stage of *Iris typhifolia*—both having the largest number of volatile substances in each stage. However, only one species of terpene is found in *Iris uniflora* during this time, suggesting that the aroma substances in the decay stage of *Iris uniflora* do not rely on terpenes. The number of alkane species in the flower bud stage of *Iris typhifolia* is several times higher than that in the other two Iris types. In *Iris uniflora*, the number of aromaticity and alkane species gradually increases during the flower bud stage, full bloom stage, and decay stage. Alcohols are found abundantly only in *Iris uniflora*. In *Iris sanguinea*, the number of esters and ketones decreases over time, with esters having a higher quality score only in the bud stage.

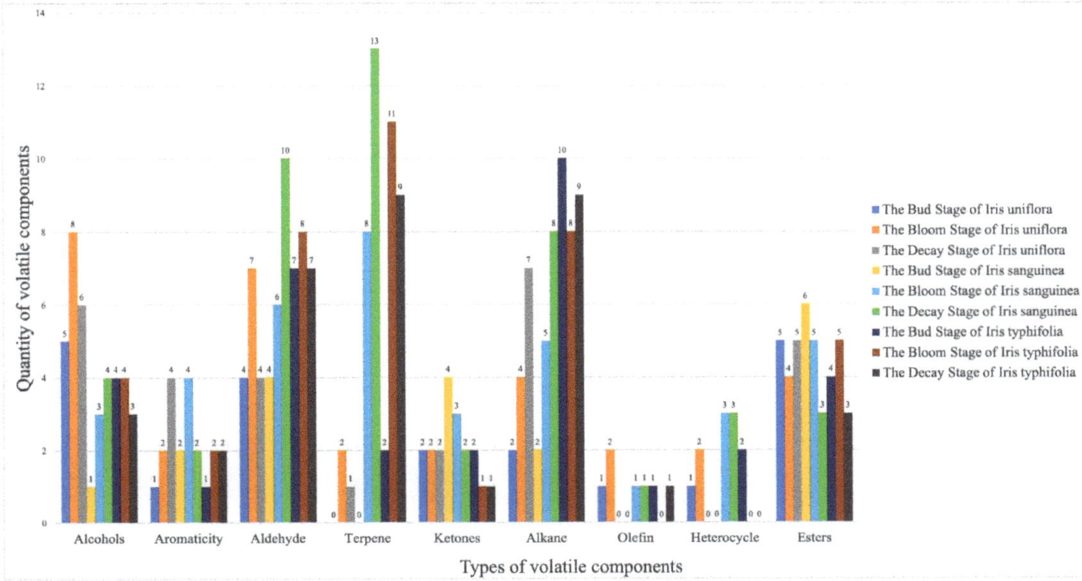

Figure 6. The number of volatile compounds in three Iris species at different stages.

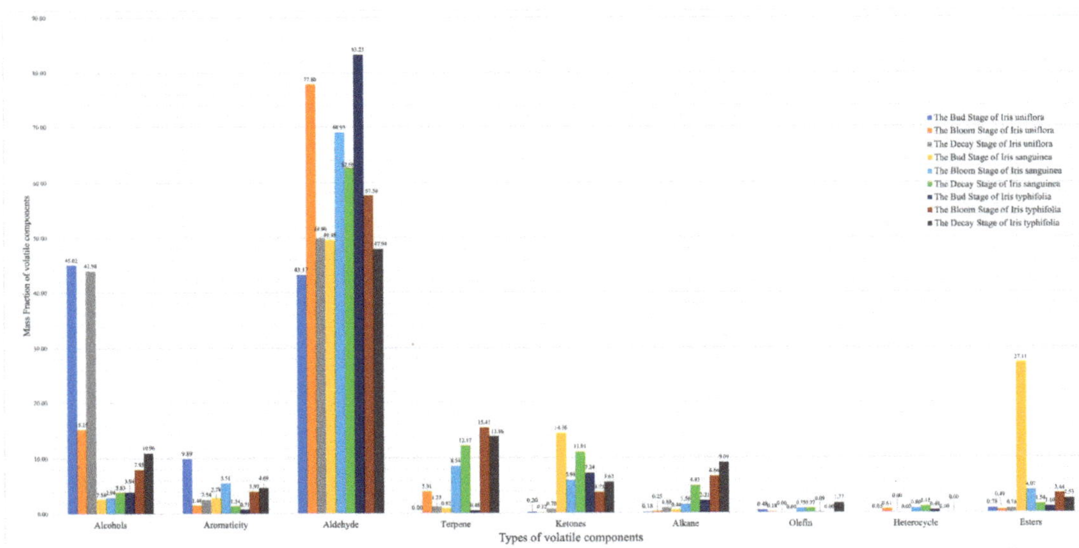

Figure 7. The relative content of volatile compounds in three Iris species at different stages.

Table 2 presents the relative content of volatile compounds from *Iris uniflora*, *Iris typhifolia*, and *Iris sanguinea* at different growth stages. Notably, nonyl aldehyde, capric aldehyde, 2,4-di-tert-butylphenol, and n-heptadecane are detected in all stages of each Iris species.

Table 2. The relative content of the compound in three Iris species at different stages (%).

Compound	Case#	The Bud Stage of *Iris uniflora*	The Bloom Stage of *Iris uniflora*	The Decay Stage of *Iris uniflora*	The Bud Stage of *Iris typhifolia*	The Bloom Stage of *Iris typhifolia*	The Decay Stage of *Iris typhifolia*	The Bud Stage of *Iris sanguinea*	The Bloom Stage of *Iris sanguinea*	The Decay Stage of *Iris sanguinea*
acetal diethyl alcohol	105-57-7	0.20		0.08						
2-hexyne-1-ol	764-60-3	8.28								
2-hexenal	505-57-7	39.02	22.38	7.30	27.06	21.02	9.11			6.09
3-hexenol	544-12-7	20.00	9.43	23.00				2.56	1.65	
N-hexyl alcohol	111-27-3	16.07	3.66	17.23					0.97	1.15
N-caprylic aldehyde	124-13-0	0.83	0.39							
Eudinol	470-82-6	0.47								
3,7-dimethyl-1,3,7-octtriene	502-99-8	0.48								
2-isopropyl-3-methoxypyrazine	25773-40-4	0.05								
nonyl aldehyde	124-19-6	2.36	1.70	1.33	1.56	2.14	8.29	1.16	0.44	3.95
capric aldehyde	112-31-2	0.95	0.42	0.16	0.10	0.65	6.83	0.75	0.24	1.31
methyl caprate	110-42-9	0.22	0.06		0.53	0.40		9.68	1.46	
ethyl caprate	110-38-3	0.22	0.32	0.58	0.14	0.96		11.60	1.54	
geranyl acetone	3796-70-1	0.14						0.34		
2,4-di-tert-butylphenol	96-76-4	9.89	1.43	1.92	0.71	3.41	1.42	2.33	1.03	0.70
2,2,4-trimethyl-1,3-pentanediol diisobutyrate	6846-50-0	0.11						0.28		
n-cetane	544-76-3	0.09	0.05	0.09	0.25	1.04	0.96			0.21
n-heptadecane	629-78-7	0.09	0.10	0.05	0.16	0.72	1.01	0.14	0.17	0.40
diisobutyl phthalate	84-69-5	0.07					0.30			0.18
7,9-di-tert-butyl-1-oxacanthin[4,5]decan-6,9-diene-2,8-dione	82304-66-3	0.05		0.03						
Dibutyl phthalate	84-74-2	0.16	0.07	0.07				0.27	0.12	
3-methyl-3-butenol	763-32-6		0.30							
Cis-2-pentenol	1576-95-0		0.42			0.61				
2-methyl-4-valeraldehyde	5187-71-3		52.68		27.06					
(1-methylamyl) cyclopropane	6976-28-9		0.06							
enanthal	111-71-7		0.21		0.29	0.26	1.56			0.98

Table 2. *Cont.*

Compound	Case#	The Bud Stage of *Iris uniflora*	The Bloom Stage of *Iris uniflora*	The Decay Stage of *Iris uniflora*	The Bud Stage of *Iris typhifolia*	The Bloom Stage of *Iris typhifolia*	The Decay Stage of *Iris typhifolia*	The Bud Stage of *Iris sanguinea*	The Bloom Stage of *Iris sanguinea*	The Decay Stage of *Iris sanguinea*
2-ethylfuran	3208-16-0		0.57		0.28				0.54	0.25
α-pinene	80-56-8		0.36		0.19		1.76		2.26	1.50
ocimene	13877-91-3		3.55	1.23		2.07				0.35
cis-α,α-5-trimethyl-5-vinyltetrahydrofuran-2-methanol	5989-33-3		0.13							
linalool	78-70-6		1.05		0.25	6.10	4.06		0.32	0.23
2,6-dimethyl-2,4,6-octtriene	3016-19-1		0.07							
2-methoxy-3-sec-butylpyrazine	24168-70-5		0.04		0.20				0.15	0.41
undecanal	112-44-7		0.02		0.12					0.30
ionone	127-41-3		0.06							
β-zirodone	79-77-6		0.07							
2,6,11-trimethyldodecane	31295-56-4		0.04		0.24		1.14			
trans-neroli tertiary alcohol	40716-66-3		0.09							
4-sec-butyl-2,6-di-tert-butylphenol	17540-75-9		0.05							
6,9-heptadecadiene	81265-03-4		0.11							
3-hexenal	4440-65-7			41.10		21.47				
trans-3-hexenol	928-97-2			0.75						2.12
trans-2-hexene-1-ol	928-95-0			0.94						
3-ethyltoluene	620-14-4			0.13						
mesitylene	108-67-8			0.30						
4-ethyltoluene	622-96-8			0.20						
decane	124-18-5			0.24						
2,6-dimethyl-nonane	17302-28-2			0.18						
n-dodecane	112-40-3			0.22	0.28	0.31	0.72		0.14	
undecane	1120-21-4			0.11						
transnon-2-enol	31502-14-4			1.99						

Table 2. *Cont.*

Compound	Case#	The Bud Stage of *Iris uniflora*	The Bloom Stage of *Iris uniflora*	The Decay Stage of *Iris uniflora*	The Bud Stage of *Iris typhifolia*	The Bloom Stage of *Iris typhifolia*	The Decay Stage of *Iris typhifolia*	The Bud Stage of *Iris sanguinea*	The Bloom Stage of *Iris sanguinea*	The Decay Stage of *Iris sanguinea*
2,2,4-trimethyl-1,3-pentanediol monoisobutyrate	25265-77-4			0.07						
2,2,4-trimethylpentanediol isobutyl ester	6846-50-0			0.04						
phytoketone	502-69-2			0.67						
pentene-3-ol	616-25-1				0.46	0.91				
1-pentene-3-ketone	1629-58-9				0.68			0.68	0.37	
cis-2-pentenol	1576-95-0				0.35					
trans-2-hexenal	6728-26-3				27.06				2.20	
leaf alcohol	928-96-1				2.88		6.32			
6-methylhept-5-en-2-one	110-93-0				6.56		5.63			7.57
basil isomer mixture	13877-91-3				0.29					
methyl caprylate	111-11-5				0.13			2.26	0.55	
tridecane	629-50-5				0.21					
1-tetradecene	1120-36-1				0.09					
tetradecane	629-59-4				0.31	0.62	1.03		0.09	0.38
n-nonadecane	629-92-5				0.11		0.58			
2,6,10-trimethyldodecane	3891-98-3				0.27					
n-pentadecane	629-62-9				0.27	0.85		0.31	0.23	
n-octadecane	593-45-3				0.12	0.59				0.35
diethylhexyl carbonate	14858-73-2				0.30					
1,2-dichloroethane	107-06-2					1.90				
chlorophyllin aldehyde	6728-26-3					10.60		6.35		
2-pinene	2437-95-8					5.05				
β-pinene	18172-67-3					0.77				
methyl heptenone	110-93-0					3.73		12.88	5.18	
D-terpenediene	5989-27-5					4.87				
(E)-B-basil	3779-61-1					0.24				

Table 2. *Cont.*

Compound	Case#	The Bud Stage of *Iris uniflora*	The Bloom Stage of *Iris uniflora*	The Decay Stage of *Iris uniflora*	The Bud Stage of *Iris typhifolia*	The Bloom Stage of *Iris typhifolia*	The Decay Stage of *Iris typhifolia*	The Bud Stage of *Iris sanguinea*	The Bloom Stage of *Iris sanguinea*	The Decay Stage of *Iris sanguinea*
phenylacetaldehyde	122-78-1					1.13	1.59		0.30	0.52
octyl formate	112-32-3					0.40				
phenylacetonitrile	140-29-4					0.32			0.13	
1-nonyl alcohol	143-08-8					0.32				
methyl salicylate	119-36-8		0.03			1.26	1.34			0.69
B-Nitrophenethane	6125-24-2					0.57			4.21	
n-tridecane	629-50-5					0.63				
alcohol ester-12	77-68-9					0.41				
alpha-pinene	3856-25-5					0.90	1.05			0.77
1-caryophyllene	87-44-5					0.36				
A-bergamonene	17699-05-7					0.50				0.46
α-caryophyllene	6753-98-6					0.25				
A-curcumene	644-30-4					0.23				
γ-juniperene	39029-41-9					0.19				
isovaleric aldehyde	590-86-3						0.60			0.33
n-caproaldehyde	66-25-1						19.97	41.22	65.62	45.94
(E,E)-2,4-hexadienal	142-83-6						1.77			
1-octen-3-ol	3391-86-4						0.57			0.33
terpinene	99-86-5						1.82			0.54
O-isopropyl benzene	527-84-4						3.27			0.64
γ-terpinene	99-85-4						1.82			
4-terpenol	562-74-3						2.84			0.67
2,7,10-trimethyldodecane	74645-98-0						1.05			
2-methyl-propionate 3-hydroxy-2,2,4-trimethyl-amyl ester	77-68-9						0.88			0.66
B-bourbon	5208-59-3						1.32			0.81
β-caryophyllene	87-44-5						1.53		0.10	1.49

Table 2. *Cont.*

Compound	Case#	The Bud Stage of *Iris uniflora*	The Bloom Stage of *Iris uniflora*	The Decay Stage of *Iris uniflora*	The Bud Stage of *Iris typhifolia*	The Bloom Stage of *Iris typhifolia*	The Decay Stage of *Iris typhifolia*	The Bud Stage of *Iris sanguinea*	The Bloom Stage of *Iris sanguinea*	The Decay Stage of *Iris sanguinea*
alpha-trachene	6753-98-6						1.00			0.90
Γ-juniperene	39029-41-9						0.74			0.54
2,6,10,15-tetramethylheptadecane	54833-48-6						1.12			0.50
N-heptacosane	593-49-7						1.48			
p-xylene	106-42-3							0.45		
acetophenone	98-86-2							0.47	0.44	3.44
ethyl caprylate	106-32-1							3.02	0.40	
N-decanoic acid	334-48-5							0.82		
3-methyl-2-butenal	107-86-8								0.14	
3-ethylthiophene	1795-01-3								0.11	
α-cresine	99-83-2								0.06	
L-beta-pinene	18172-67-3								0.43	
4-isopropyl toluene	99-87-6								0.14	
(+)-limonene	5989-27-5								5.09	1.60
3,6,6-trimethyl-bicyclic(3,1,1)hept-2-ene	4889-83-2								0.13	
(Z)-3,7-dimethyl-1,3,6-octadecatriene	3338-55-4								0.92	
allobasil	7216-56-0								0.19	
(-)-alpha-pinene	3856-25-5								0.34	
trans-squalene	111-02-4								0.06	
n-valeraldehyde	110-62-3									0.33
2-methylheptane	592-27-8									0.61
2,4-dimethylheptane	2213-23-2									2.20
narcocapsicum tomakomai	20697-20-5									0.49
dodecane	112-40-3									0.29
(-)-α-cubebeene	17699-14-8									0.77
e-β-farnesene	18794-84-8									0.93
Δ-juniperene	483-76-1									1.60

For *Iris uniflora*, 2-hexenal, 3-hexenol, and n-hexyl alcohol are consistently detected across growth stages, with relative contents exceeding 1%. Additionally, 2-methyl-4-valeraldehyde is present with a relative content of 52.68% during the bloom stage but remains undetectable during the bud and decay stages. The relative content of 3-hexenal is 41.10% during the decay stage but is not detected during the bud and bloom stages.

In *Iris typhifolia*, only nonyl aldehyde is consistently detected throughout all growth stages with relative contents exceeding 1%. 2-hexenal is abundant during the bud and bloom stages but remains undetectable during the decay stage. Both (1-methylamyl) cyclopropane and trans-2-hexenal are present in significant amounts during the bud stage but cannot be detected during the other stages.

For *Iris sanguinea*, nonyl aldehyde, capric aldehyde, n-heptadecane, 2,4-di-tert-butylphenol, and n-caproaldehyde are detected across all stages, with n-caproaldehyde consistently having relative contents above 40% throughout the stages.

The cluster analysis heatmap for the three stages of three Iris species is depicted in Figure 8, where color blocks transition from blue to red, indicating a decreasing trend in quality scores.

The nine sample groups can be classified into two major clusters. In the primary classification, the decay stage of *Iris typhifolia* and the decay stage of *Iris sanguinea* are grouped together due to their similarities in volatile terpenoids and esters, resulting in relatively high-quality scores. These two stages also exhibit distinct differences from the other samples, forming a separate cluster along with the remaining seven samples. The heatmap illustrates significant variations in both the types and concentrations of volatile substances among the nine sample groups. The clustering analysis results align with the aroma discrimination obtained from the electronic nose, indicating distinct aromatic profiles among the three Iris species at different stages.

The volatile substances in different groups show noticeable differences and can be categorized into four clusters. Cluster 1 primarily consists of terpenoid compounds, with higher concentrations observed in the bloom stage of *Iris sanguinea* and the decay stage of *Iris sanguinea* samples. Cluster 2 shows higher levels in the decay stage of *Iris typhifolia* and moderate levels in the decay stage of *Iris sanguinea*. Cluster 3 exhibits relatively higher concentrations in the bud stage of *Iris uniflora*.

The Odor Activity Value (OAV) serves as the primary criterion for determining the overall contribution of volatile compounds to a sample's aroma. Based on previously reported aroma threshold values, OAV values for three Iris species have been calculated in Table 3. This criterion is used to assess the contribution of each substance to the overall aroma profile of the sample, providing a reference for selecting aroma reconstitution standard samples in subsequent experiments.

In total, 20 compounds are found to have OAV values greater than 1 in at least one stage of Iris, including 9 aldehydes, 4 terpenes, 3 alcohols, 2 esters, 1 ketone, and 1 aromatic compound. These compounds significantly contribute to the aroma profile of the sample.

Nonyl aldehyde exhibits significantly high OAV at all stages of three Iris species, imparting distinctive oily and sweet orange aromas to the flowers. Among *Iris typhifolia*, capric aldehyde demonstrates elevated OAVs during all three stages, contributing a citrusy fragrance to the blossoms. During the bud stage, n-caprylic aldehyde displays a higher OAV, providing a fruity scent. 2-hexenal registers elevated OAV during the bloom and decay stages, giving a green leafy aroma to the flowers. The linalool attains higher OAV during the bloom and decay stages of *Iris typhifolia*. Moreover, the grassy note of n-caproaldehyde plays a contributory role in shaping the olfactory profile of Iris species during the decay stage of *Iris typhifolia*, as well as the bud and bloom stages of *Iris sanguinea*.

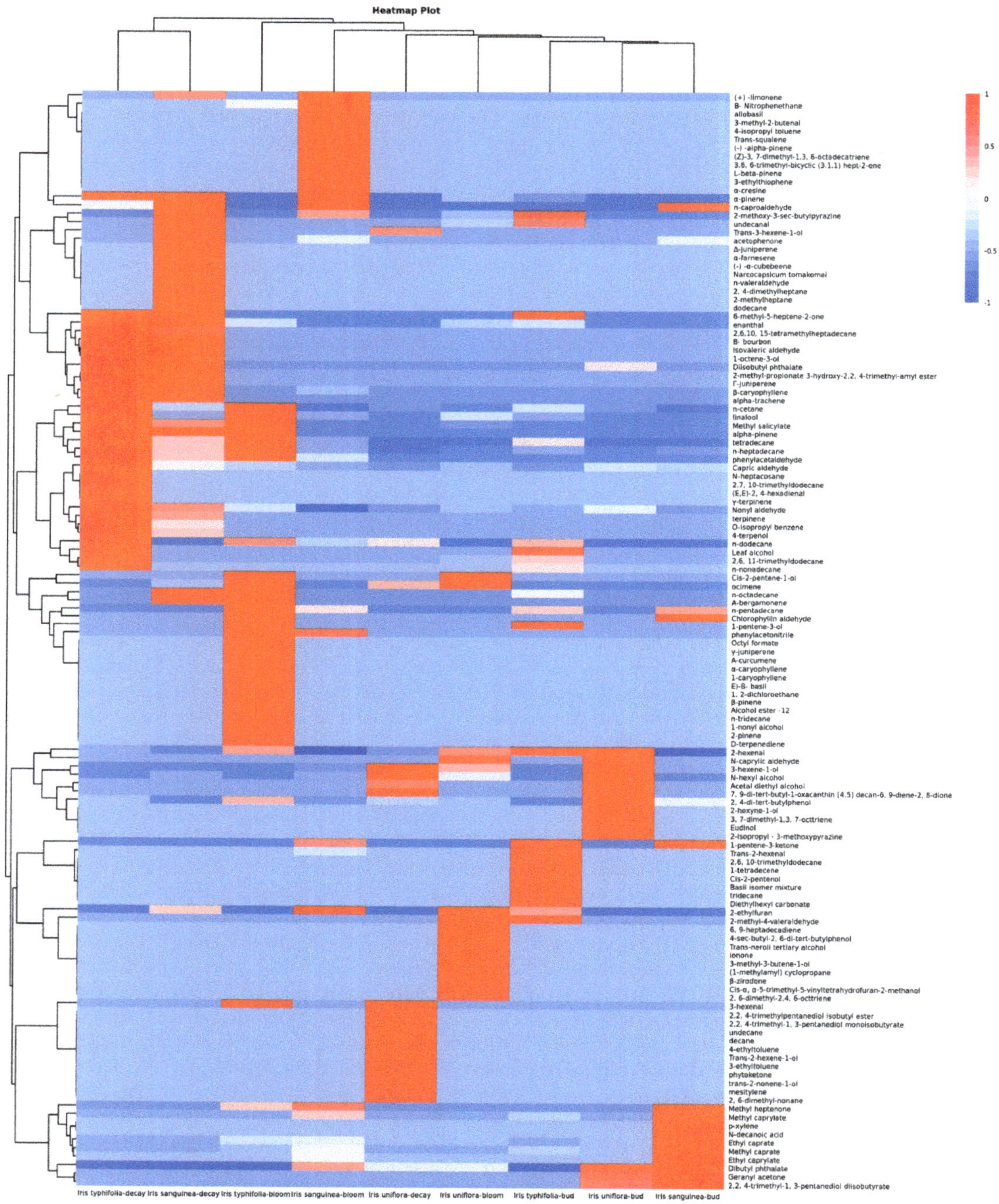

Figure 8. The cluster analysis heatmap plot of volatile compounds in three Iris species at different stages.

Table 3. The OAV of volatile compounds in three Iris species at different stages.

Compound	Threshold (µg/kg)	The Bud Stage of *Iris uniflora*	The Bloom Stage of *Iris uniflora*	The Decay Stage of *Iris uniflora*	The Bud Stage of *Iris typhifolia*	The Bloom Stage of *Iris typhifolia*	The Decay Stage of *Iris typhifolia*	The Bud Stage of *Iris sanguinea*	The Bloom Stage of *Iris sanguinea*	The Decay Stage of *Iris sanguinea*
Acetal diethyl alcohol	0.1	7.78		5.28						
2-hexenal	0.15	118.30	394.26	325.83	87.62	64.47				18.57
N-hexyl alcohol	8	7.86	1.28	14.39					0.11	0.66
N-caprylic aldehyde	0.001	3266.84	126.92							
Eudinol	0.01	185.90								
Nonyl aldehyde	0.0035	2638.26	1284.40	2547.50	217.11	281.66	475.72	298.44	117.33	516.43
Capric aldehyde	0.005	745.33	223.73	218.13	9.80	6.57	274.42	135.65	44.11	119.63
Ethyl caprate	0.2	4.29	4.28	19.32	0.35	2.21		52.34	7.94	
enanthal	0.031		17.99		4.47	3.97	1.18			14.48
linalool	0.0015		185.36		8.53	1872.43	544.47		197.87	71.36
undecanal	0.014		3.42		4.43					9.84
Chlorophyllin aldehyde	0.15					32.51		38.30		
2-pinene	0.12					19.34				
Methyl heptenone	0.1					17.15		116.18	47.82	
D-terpenediene	0.22					1.18				
phenylacetaldehyde	0.009					57.53	35.58		31.76	26.63
Methyl salicylate	0.06					9.65	4.53			5.27
alpha-pinene	0.12					3.47	1.77			2.93
n-caproaldehyde	0.0075						535.11	4958.52	878.13	28.71
alpha-trachene	0.16						1.25			2.58

The values of threshold are from Compilations of flavour threshold values in water and other media (second enlarged and revised edition) written by V. Gemert.

4. Discussion

This study investigates differences in floral scent composition among various species and developmental stages of Iris plants, specifically *Iris uniflora*, *Iris typhifolia*, and *Iris sanguinea*. PCA results indicate that the cumulative contribution rates of the first two principal components for the nine samples reach 99.13%, with only the floral scents of *Iris typhifolia* at different stages significantly differentiated. However, due to their similar fragrances, the PCA pattern plots for *Iris uniflora* and *Iris typhifolia*, overlapping in various flowering periods, do not show significant differentiation. With Linear Discriminant Analysis, the cumulative contribution rate reaches 86.64%, effectively distinguishing the floral scent samples at different stages of the same species.

Sensor contribution analysis reveals that five sensors, namely W1W, W2W, W1S, W5S, and W2S, are most sensitive to identifying floral scents from the three Iris species at various stages, playing a significant role in differentiation. The results indicate substantial differences in floral fragrances, and the variations in floral scent components among different samples are mainly related to sulfur compounds, terpenes, organic sulfur compounds, aromatic compounds, methane, nitrogen compounds, and alcohols. This conclusion provides a theoretical foundation and evidence for subsequent Gas Chromatography-Mass Spectrometry (GC-MS) analysis.

The application of headspace solid-phase microextraction-gas chromatography-mass spectrometry confirms significant differences in volatile components among the three Iris species. Throughout all stages, common volatile compounds include alcohols, aromatics, aldehydes, ketones, alkanes, and esters. Among these compounds, nonyl aldehyde, capric aldehyde, 2,4-di-tert-butylphenol, and n-heptadecane are present in all samples. Cluster analysis reveals that the decay stage samples of *Iris typhifolia* and *Iris sanguinea* are grouped together due to their high levels of terpenes and esters. This grouping stems from the abundant presence of terpenes during the decay stages of these two Iris species. Nonyl aldehyde significantly contributes to the aroma profile of the three Iris species during all stages, owing to its high odor activity value (OAV).

In the case of *Iris uniflora*, the prominent presence of 2-methyl-4-glutaraldehyde indicates its utility as an intermediate in the synthesis of the medicinal compound sleepertone and an insecticide [19]. 3-hexenol, renowned for its refreshing scent, is a crucial aromatic ingredient in perfumery production worldwide [20]. Additionally, 3-hexenal, known for its strong grassy and apple-like aroma, is approved for use as a food flavoring [21]. The inclusion of α-pinene not only offers inhibition against *Candida albicans* but also serves as a raw material for soap and detergent manufacturing [22].

Many compounds with economic and medicinal value were detected in different flower samples. In the analysis of *Iris typhifolia*, notable constituents with high relative contents include 2-hexenal, which can be applied in food additives and organic synthesis intermediates [23]. Compounds such as acetaldehyde, nonanal, trans-2-hexenal, 6-methyl-5-hepten-2-one, geraniol, decanal, and heptanal are extensively employed in edible fragrance formulations [24–30]. 2,4-di-tert-butylphenol is primarily used in the production of natural rubber and synthetic rubber antioxidants, plastic stabilizers, fuel stabilizers, UV absorbers, as well as intermediates in pesticides and dye production [31]. Linalool can be harnessed for basil alcohol production and is valued for its anti-inflammatory and antimicrobial properties [32]. The application of 2-pinenes encompasses the synthesis of pine oil alcohol, linalool, and some sandalwood-scented fragrances [33]. Methyl heptenone contributes to the synthesis of ionone and citral, being of great significance as a pharmaceutical intermediate [34]. Methyl salicylate serves as a precursor for acetylsalicylic acid synthesis [35]. Moreover, n-hexanal can be utilized as a plasticizer, an organic synthetic component in rubber and resin, an insecticide, and an edible flavoring [36]. Terpinene, recognized for its antioxidant efficacy, serves as an antioxidant in cosmetics and a food additive, which predominantly is regarded as a flavoring and preservative [37]. O-isopropylbenzene is primarily employed in creating imitation products and fixed fragrances [38]. β-Caryophyllene is mainly used for preparing fine imitation products and fragrance fixatives [39].

Regarding *Iris sanguinea*, detected volatile substances include methyl octanoate, commonly used in the production of surfactants and fragrances [40]. Ethyl octanoate gives a coconut fragrance and is predominantly utilized in the formulation of edible flavorings [41]. Methyl caproate can be applied in the preparation of pineapple, berry, and fruit-scented fragrances [42]. Ethyl octanoate serves as an enhancer for fragrances, contributing to scents like orange, strawberry, and pineapple [43]. β-Nitrophenylethane holds a pivotal role as a key chemical precursor and is utilized in numerous pharmaceuticals and dye intermediates, such as aniline, dinitrobenzene, diphenylamine, azobenzene, and m-amino benzene sulfonic acid. (+)-Limonene possesses antitussive, expectorant, and antibacterial properties [44]. n-Hexanol is instrumental in the formulation of coconut and berry-flavored fragrances [45]. 2,4-dimethyl heptane is utilized as a solvent, fragrance ingredient, organic synthesis intermediate, and herbicide [46]. Δ-Junipene offers antibacterial effects, aids in stimulating digestive enzymes, and possesses diuretic properties [47].

The high contents of certain volatile compounds in three varieties of Iris indicate that they have significant economic and medicinal potentials besides their well-known ornamental value. This provides a theoretical foundation for the development of Iris-scented perfumes, extraction of fragrance compounds, and application in aromatherapy.

5. Conclusions

This study delves into the nuanced variations in floral scent composition among three Iris plant species: *Iris uniflora*, *Iris typhifolia*, and *Iris sanguinea*, spanning various developmental stages. Utilizing advanced electronic nose technology and headspace solid-phase microextraction-gas chromatography-mass spectrometry (HS-SPME-GC-MS), the research reveals substantial transformations in the aromatic profiles of Iris flowers throughout their developmental stages.

Electronic nose measurements, complemented by principal component analysis (PCA) and linear discriminant analysis (LDA), demonstrate the ability to distinguish the floral scents of *Iris typhifolia*, *Iris sanguinea*, and *Iris uniflora* at various growth stages. Sensor contribution analysis underscores the significance of specific sensors in identifying key volatile compounds, with sulfur compounds, terpenes, organic sulfur compounds, aromatics, methane, nitrogen compounds, and alcohols emerging as pivotal contributors to these differentiating scents.

Gas chromatography-mass spectrometry (GC-MS) further elucidates significant disparities in volatile components across the three Iris species, consistently detecting common compounds, such as nonyl aldehyde, capric aldehyde, 2,4-di-tert-butylphenol, and n-heptadecane throughout the developmental stages. Cluster analysis confirms distinct aromatic profiles among the Iris species, particularly highlighting the abundance of terpenoids and esters in the decay stages of *Iris typhifolia* and *Iris sanguinea*. The study underscores the economic and medicinal potential of these Iris varieties, extending their value beyond ornamental purposes. Compounds like 2-hexenal, 3-hexenol, linalool, and n-hexyl alcohol hold promise for fragrance production, while others like nonanal, geraniol, and acetaldehyde find applications in edible fragrance formulations. Methyl salicylate, limonene, B-nitrostyrene, etc., serve as crucial raw materials for pharmaceutical synthesis.

In conclusion, this research enhances our understanding of the diverse floral scents within the Iris genus, paving the way for the development of Iris-scented products and exploring their broader economic and medicinal potential. The findings contribute to the future development of Iris's medicinal value and the application of Iris-scented perfume, essential oil, and aromatherapy.

Supplementary Materials: The following supporting information can be downloaded at: https://www.mdpi.com/article/10.3390/horticulturae10020153/s1.

Author Contributions: Conceptualization, K.C., Z.B., H.X., W.C., W.J. and H.C.; Methodology, K.C., Z.B., H.X., W.C., W.J. and H.C.; Software, W.C., W.J. and H.C.; Validation, K.C., Z.B., H.X. and W.J.; Formal analysis, K.C., Z.B. and H.X.; Resources, Y.Z.; Data curation, K.C. and W.J.; Writing—original draft, K.C., Z.B., H.X., W.C. and W.J.; Writing—review & editing, K.C., Y.Z. and H.C; Visualization, W.J.; Supervision, Y.Z. and H.C.; Project administration, H.C.; Funding acquisition, Y.Z. and H.C. All authors have read and agreed to the published version of the manuscript.

Funding: This work was financially supported by the National Natural Science Foundation of China (31701962) and the Beijing Municipal Park Management Center Science and Technology Project "Breeding and Evaluation of Drought-resistant Iris Excellent Varieties" (ZX2019010).

Data Availability Statement: Data is available in the Supplementary Materials.

Conflicts of Interest: The authors declare no conflicts of interest.

References

1. Liu, L.L.; Cao, X.D.; Xu, Z.R.; Zhang, J.F.; Wu, Y.H. Studies on Floral Syndrome and Mating System of Iris halophila. *Acta Agrestia Sinica* **2021**, *29*, 2731–2741.
2. Knudsen, J.T.; Eriksson, R.; Gershenzon, J.; Stahl, B. Diversity and distribution of floral scent. *Bot. Rev.* **2006**, *72*, 1. [CrossRef]
3. Patel, S.; Shibamoto, T. Flavor compounds in wines produced from Chardonnay grapes fermented with fruit juices. *Food Sci. Technol. Res.* **2003**, *9*, 84–86. [CrossRef]
4. Pichersky, E.; Raguso, R.A.; Lewinsohn, E.; Croteau, R. Floral scent production in Clarkia (Onagraceae) (I. Localization and developmental modulation of monoterpene emission and linalool synthase activity). *Plant Physiol.* **1994**, *106*, 1533–1540. [CrossRef] [PubMed]
5. Yuan, Y.; Sun, Y.; Zhao, Y.; Liu, C.; Chen, X.; Li, F.; Bao, J. Identification of Floral Scent Profiles in Bearded Irises. *Molecules* **2019**, *24*, 1773. [CrossRef] [PubMed]
6. Shalit, M.; Guterman, I.; Volpin, H.; Bar, E.; Tamari, T.; Menda, N.; Adam, Z.; Zamir, D.; Vainstein, A.; Weiss, D.; et al. Volatile ester formation in roses Identification of an acetyl-coenzyme A. Geraniol/Citronellol acetyltransferase in developing rose petals. *Plant Physiol.* **2003**, *131*, 1868–1876. [CrossRef] [PubMed]
7. D'Auria, J.C.; Chen, F.; Pichersky, E. Characterization of an acyltransferase capable of synthesizing benzyl-benzoate and other volatile esters in flowers and damaged leaves of *Clarkia breweri*. *Plant Physiol.* **2002**, *130*, 466–476. [CrossRef]
8. Zhang, W.; He, C.; Gong, Y. Pollinator attraction and outcrossing strategies in Iris. *Plant Sci. J.* **2019**, *37*, 672–681.
9. Li, C.; Gao, Y.; Liu, R.; Cao, Y.; Fan, Z.; Guo, L.; Zhang, Q. Analysis on the barriers of interspecific hybridization in beardless Irises. *J. Beijing For. Univ.* **2018**, *40*, 96–101.
10. Qi, J.; Zhai, Y.; Zhao, C. The Chemical Constituents of Genus Irisand Their Biological Activitie. *Nat. Prod. Res. Dev.* **2006**, *18*, 165–170.
11. Wang, W.; Wang, P.; Qiao, Q.M. Research on the Classification and Application Value of Iris Plant. *J. Anhui Agric. Sci.* **2008**, *36*, 1001–1002. (In Chinese)
12. Cai, K.; Feng, C.; Xu, S.; Sun, Y.; Lou, Q.; Sun, J.; Chen, H. The volatile components in different flowering stages of *Iris uniflora*. *J. Northeast. For. Univ.* **2023**, *51*, 53–58.
13. Cai, K.; Tian, K.; Ban, Z.; Xu, H.; Jia, W.; Zhu, Y.; Chen, H. Analysis of Floral Fragrance Components in Different Parts of *Iris typhifolia*. *Horticulturae* **2023**, *9*, 1268. [CrossRef]
14. Başer, K.H.; Demirci, B.; Orhan, I.E.; Kartal, M.; Sekeroglu, N.; Sener, B. Composition of Volatiles from Three Iris Species of Turkey. *J. Essent. Oil Res.* **2018**, *23*, 66–71. [CrossRef]
15. Zito, P.; Rosselli, S.; Bruno, M.; Maggio, A.; Sajeva, M. Floral scent in Iris planifolia (Iridaceae) suggests food reward. *Phytochemistry* **2019**, *158*, 76–90. [CrossRef]
16. Jia, X.; Deng, Q.; Yang, Y.; Xiang, X.; Zhou, X.; Tan, C.; Zhou, Q.; Huang, F. Unraveling of the Aroma-Active Compounds in Virgin Camellia Oil (*Camellia oleifera* Abel) Using Gas Chromatography–Mass Spectrometry–Olfactometry, Aroma Recombination, and Omission Studies. *J. Agric. Food Chem.* **2021**, *69*, 9043–9055. [CrossRef]
17. Zhang, T.; Bao, F.; Yang, Y.; Hu, L.; Ding, A.; Ding, A.; Ding, A.; Wang, J.; Cheng, T.; Zhang, Q. A Comparative Analysis of Floral Scent Compounds in Intraspecific Cultivars of *Prunus mume* with Different Corolla Colours. *Molecules* **2020**, *25*, 145. [CrossRef]
18. Lu, X.L.; Hai, Z.; Wang, J. Detection of Cola beverage by electronic nose. *J. Zhejiang Univ. Agric. Life Sci.* **2006**, *32*, 677–682.
19. Liu, X.; Wang, W.; Fang, S. Thermodynamic studies on NO radical with 2,3-dimethylpentantal. *Petrochem. Ind. Appl.* **2011**, *30*, 4.
20. SJaeger, R.; Pineau, B.; Bava, C.M.; Atkinson, K.R.; McRae, J.F.; Axten, L.G.; Chheang, S.L.; Beresford, M.K.; Peng, M.; Paisley, A.G.; et al. Investigation of the impact of sensitivity to cis-3-hexenol (green/grassy) on food acceptability and selection. *Food Qual. Prefer.* **2012**, *24*, 230–242. [CrossRef]
21. León, D.C.S.; Ortíz, D.K.R.; González, D.F.J. Sensory approach and chiral analysis for determination of odour active compounds from feijoa (*Acca sellowiana*). *Food Chem.* **2020**, *317*, 126383. [CrossRef]
22. Jin, K.S.; Bak, M.J.; Jun, M.; Lim, H.J.; Jo, W.K.; Jeong, W.S. α-Pinene Triggers Oxidative Stress and Related Signaling Pathways in A549 and HepG2 Cells. *Food Sci. Biotechnol.* **2010**, *19*, 1325–1332. [CrossRef]
23. Hyun, J.; Lee, J.G.; Yang, K.Y.; Lim, S.; Lee, E.J. Postharvest Fumigation of (E)-2-Hexenal on Kiwifruit (*Actinidia chinensis* cv. 'Haegeum') Enhances Resistance to Botrytis cinerea. *Postharvest Biol. Technol.* **2022**, *187*, 111854. [CrossRef]

24. Garcia, L.; Perrin, C.; Nolleau, V.; Godet, T.; Farines, V.; Garcia, F.; Caillé, S.; Saucier, C. Impact of Acetaldehyde Addition on the Sensory Perception of Syrah Red Wines. *Foods* **2022**, *11*, 1693. [CrossRef] [PubMed]
25. Han, Y.; Shen, H.; Zhao, M.; Sun, W. Flavour binding mechanism between a typical meat flavour compound (nonanal) and porcine myofibrillar proteins with consideration of conformational changes. *Int. J. Food Sci. Technol.* **2018**, *53*, 1954–1961. [CrossRef]
26. Iyer, M.M.; Sacks, G.L.; Padilla-Zakour, O.I. Impact of Harvesting and Processing Conditions on Green Leaf Volatile Development and Phenolics in Concord Grape Juice. *J. Food Sci.* **2010**, *75*, C297–C304. [CrossRef] [PubMed]
27. Kaneda, N.; Lee, I.S.; Gupta, M.P.; Soejarto, D.D.; Kinghorn, A.D. (+)-4β-Hydroxyhernandulcin, A New Sweet Sesquiterpene from the Leaves and Flowers of Lippia dulcis. *J. Nat. Prod.* **1992**, *55*, 1136–1141. [CrossRef]
28. Neelam Singh-Sangwan, N.S.S.; Sangwan, R.S.; Rajesh Luthra, R.L.; Thakur, R.S. Geraniol Dehydrogenase: A Determinant of Essential Oil Quality in Lemongrass. *Planta Medica* **1993**, *59*, 168–170. [CrossRef] [PubMed]
29. Mostafavi, A.; Shamspur, T.; Afazali, D. Chemical Composition of the Essential Oil of *Ducrosia assadii* Alava. from Kerman Province in Iran. *J. Essent. Oil Res.* **2010**, *22*, 300–302. [CrossRef]
30. Van Aardt, M.; Duncan, S.E.; Marcy, J.E.; Long, T.E.; Nielsen-Sims, S.R. Aroma Analysis of Light-Exposed Milk Stored with and Without Natural and Synthetic Antioxidants. *J. Dairy Sci.* **2005**, *88*, 881–890. [CrossRef] [PubMed]
31. Xia, H.; Gao, H.; Sun, Q.; Wu, F.; Ge, T.; Sui, K.; Wang, Z.; Song, L.; Huang, X.; Yu, Q. Puerarin, an efficient natural stabilizer for both polyethylene and polypropylene. *J. Appl. Polym. Sci.* **2020**, *137*, e49599. [CrossRef]
32. Satoh, M.; Kusumoto, N.; Matsui, N.; Makino, R.; Hashida, K.; Arai, D.; Iiduka, Y.; Ashitani, T. Antitermitic and antifungal properties of enantiopure linalool and furanoid linalool oxide confirmed in *Lindera umbellata* var. *membranacea*. *J. Wood Chem. Technol.* **2021**, *42*, 37–45. [CrossRef]
33. Castellar, A.; Oliveira, D.R.; Leitão, S.G.; Bizzo, H.R.; Soares MD, L.C.; Kinupp, V.F.; Veiga-Junior, V.F. Essential oil from Philodendron fragrantissimum, an aromatic Araceae from Amazonia, Brazil. *J. Essent. Oil Res.* **2013**, *25*, 194–197. [CrossRef]
34. Wolken, W.A.M.; Tramper, J.; van der Werf, M.J. Amino acid-catalyzed retroaldol condensation: The production of natural benzaldehyde and other flavour compounds. *Flavour Fragr. J.* **2004**, *19*, 115–120. [CrossRef]
35. Dong, Y.; Li, X.; Zhao, Y.; Ren, X.; Zheng, Y.; Song, R.; Zhong, X.; Shan, D.; Lv, F.; Deng, Q.; et al. Biotransformation and metabolism of three methyl salicylate glycosides by gut microbiota in vitro. *J. Pharm. Biomed. Anal.* **2023**, *233*, 115474. [CrossRef] [PubMed]
36. Kato, Y.; Yuasa, Y. Practical Synthesis of 3-Methylnonane-2,4-dione, an Intense Strawlike and Fruity Flavored Compound. *J. Agric. Food Chem.* **2001**, *49*, 3864–3866. [CrossRef] [PubMed]
37. Kim, H.J.; Chen, F.; Wu, C.; Wang, X.; Chung, H.Y.; Jin, Z. Evaluation of Antioxidant Activity of Australian Tea Tree (*Melaleuca alternifolia*) Oil and Its Components. *J. Agric. Food Chem.* **2004**, *52*, 2849–2854. [CrossRef] [PubMed]
38. Hou, H.; Zhang, X.; Zhao, T.; Zhou, L. Effects of Origanum vulgare essential oil and its two main components, carvacrol and thymol, on the plant pathogen Botrytis cinerea. *PeerJ* **2020**, *8*, e9626. [CrossRef] [PubMed]
39. Kim, M.G.; Lee, H.S. Growth-inhibiting Effects and Chemical Composition of Essential Oils Extracted from *Platycladus orientalis* Leaves and Stemstoward Human Intestinal Bacteria. *Food Sci. Biotechnol.* **2015**, *24*, 427–431. [CrossRef]
40. Raice, R.T.; Sjoholm, I.; Wang, H.L.; Bergenståhl, B. Characterization of volatile components extracted from *Vangueria infausta* (African medlar) by using GC–MS. *J. Essent. Oil Res.* **2015**, *27*, 76–81. [CrossRef]
41. Sun, J.; Chin, J.H.; Zhou, W.; Yu, B.; Curran, P.; Liu, S.Q. Biocatalytic Conversion of Coconut Oil to Natural Flavor Esters Optimized with Response Surface Methodology. *J. Am. Oil Chem. Soc.* **2012**, *89*, 1991–1998. [CrossRef]
42. Wang, L.; Dou, G.; Guo, H.; Zhang, Q.; Qin, X.; Yu, W.; Xiao, H. Volatile organic compounds of *Hanseniaspora uvarum* increase strawberry fruit flavor and defense during cold storage. *Food Sci. Nutr.* **2019**, *7*, 2625–2635. [CrossRef] [PubMed]
43. Kafkas, E.; Cabaroglu, T.; Selli, S.; Bozdoğan, A.; Kürkçüoğlu, M.; Paydaş, S.; Başer, K.H.C. Identification of volatile aroma compounds of strawberry wine using solid-phase microextraction techniques coupled with gas chromatography–mass spectrometry. *Flavour Fragr. J.* **2006**, *21*, 68–71. [CrossRef]
44. Plesage, L.; Candy, J.P.; Hirigoyen, C.; Humblot, F.; Basset, J.M. Selective dehydrogenation of dipentene (R-(+)-limonene) into paracymene on silica supported palladium assisted by α-olefins as hydrogen acceptor. *J. Mol. Catal. A Chem.* **1996**, *112*, 431–435. [CrossRef]
45. Procida, G.; Lagazio, C.; Cateni, F.; Zacchigna, M.; Cichelli, A. Characterization of arabica and robusta volatile coffees composition by reverse carrier gas headspace gas chromatography–mass spectrometry based on a statistical approach. *Food Sci. Biotechnol.* **2020**, *29*, 1319–1330. [CrossRef]
46. Effimia, E.; Karabagias, I.K.; Sofia, M.; Dionysios, K.; Nikolaos, K. Geographical origin discrimination of "Ntopia" olive oil cultivar from ionian islands using volatile compounds analysis and computational statistics. *Eur. Food Res. Technol. Z. Lebensm. Unters. Forschung. A* **2021**, *247*, 3083–3098.
47. Gupta, S.; Gaurav, B.; Dwivedi, D.G.; Srivastava, K. Antimycobacterial activity of fractions and isolated compounds from *Vetiveria zizanioides*. *Med. Chem. Res.* **2012**, *21*, 1283–1289. [CrossRef]

horticulturae

Article

Comparative Proteomics Analysis of *Primulina serrulata* Leaves Reveals New Insight into the Formation of White Veins

Quan-Li Dou [1,†], Da-Jun Xie [2,†], Tan Deng [1], Mo-Fang Chen [1], Zheng-Min Qian [1], Shuang-Shuang Wang [1] and Ren-Bo Zhang [1,*]

1 Department of Biology, Zunyi Normal College, Zunyi 563002, China; douquanli@163.com (Q.-L.D.); gzzydengt@163.com (T.D.); zynccmf@163.com (M.-F.C.); qianzhengminswjs@163.com (Z.-M.Q.); wangshuangshuangyo@163.com (S.-S.W.)
2 Sichuan Academy of Forestry, Chengdu 610081, China; xdj911823@126.com
* Correspondence: ddzrb@126.com
† These authors contributed equally to this work.

Abstract: *Primulina serrulata* is a valuable ornamental herb with rosette leaves and vibrant flowers. Some leaves of this species exhibit a bright and distinct white color along the upper veins, enhancing their ornamental value, while others are less white or entirely green. This variation is observed in adult leaves from natural habitats and among young leaves from seedlings grown in the laboratory. TMT-labeled proteomics technology was used to study the protein-level biogenesis of white-veined (WV) *P. serrulata* leaves. Our objective was to offer novel insight into the breeding of WV plants. Chlorophyll (Chl) content was significantly lower in the WV group than in the control group. Out of 6261 proteins identified, a mere 69 met the criteria for differentially expressed proteins (DEPs) after stringent screening for subsequent analyses. Among these DEPs, there were 44 proteins that exhibited downregulation and 25 that were upregulated in the WV plants. Some DEPs associated with chloroplasts and Chl biosynthesis were downregulated, leading to the absence of green coloration. Concurrently, Gene Ontology enrichment analysis further emphasized an insufficiency of magnesium, the key element in Chl biosynthesis. Many DEPs associated with abiotic or biotic stressors were downregulated, suggesting an overall weakening of stress resistance with certain compensatory mechanisms. Similarly, many DEPs related to modifying biomacromolecules were downregulated, possibly affected by the decrease in proteins involved in photosynthesis and stress resistance. Some DEPs containing iron were upregulated, indicating that iron is mainly used to synthesize heme and ferritin rather than Chl. Additionally, several DEPs related to sulfur or sulfate were upregulated, suggesting strengthened respiration. Expansin-A4 and pectinesterase were upregulated, coinciding with the emergence of a rough and bright surface in the white area of leaves, indicative of the elongation and gelation processes in the cell walls. These findings provide new insight for future studies to explore the mechanism of color formation in WV leaves.

Keywords: gesneriad; coloration mechanism; chlorophyll metabolism; differentially expressed proteins; functional enrichment analysis; protein interaction network

Citation: Dou, Q.-L.; Xie, D.-J.; Deng, T.; Chen, M.-F.; Qian, Z.-M.; Wang, S.-S.; Zhang, R.-B. Comparative Proteomics Analysis of *Primulina serrulata* Leaves Reveals New Insight into the Formation of White Veins. *Horticulturae* **2024**, *10*, 19. https://doi.org/10.3390/horticulturae10010019

Academic Editor: Jinzhi Zhang

Received: 21 November 2023
Revised: 15 December 2023
Accepted: 20 December 2023
Published: 23 December 2023

1. Introduction

Variations in leaf color are widely observed in the natural world, encompassing a spectrum of patterns and hues. Among the variations, the presence of white sections in leaves is particularly notable. Plants exhibiting such white variegation, including those specifically bred to possess albino leaves, are highly prized in ornamental horticulture. While the presence of albino sections can affect photosynthetic efficiency, this unique coloration contributes to ornamental appeal, thereby increasing the economic value of these plants.

Many causes have been linked with white-leaf veins. This phenomenon is not limited to a single factor, as it can result from various conditions, including genetic mutations, environmental stressors, nutritional deficiencies, or diseases such as viral or fungal infections and insect nibbling. Each of these factors affects the pigmentation or the structural integrity of the leaf veins, leading to their white appearance.

The absence of chloroplast activity in white leaves leads to an increase in mitochondrial gene copy number and elevated mitochondrial transcription levels [1], suggesting a compensatory response within the cells, where the reduced functionality of chloroplasts, typically responsible for photosynthesis, prompts upregulation of mitochondrial activity. This adaptation may be a way to manage energy production and cellular respiration under the changing conditions caused by the lack of chlorophyll (Chl) in white leaves [2].

miRNAs play a role in the formation of white leaves and the development of white cells. miRNAs are involved in the regulation of gene expression and affect the development and differentiation of plant tissues. These miRNAs have been implicated in the modulation of genes responsible for pigment production, chloroplast development, and other related processes, contributing to the altered pigmentation observed in these leaves [3].

Some proteins affect the white coloration in leaves. The IM (IMMUTANS) protein is involved in the formation and functioning of chloroplasts. Mutations that occur in the IM gene result in areas of leaves lacking Chl and appearing white [4]. When a fully functional plastid terminal oxidase (PTOX) is non-functional or absent, the disruption results in the formation of white leaves, as key processes during chloroplast development and function are impaired, leading to the lack of normal leaf coloration [5].

Some metabolites also affect coloration. An increased concentration of 12-hydroxyjasmonic acid (OPDA) in white leaves suggests that the white cells are utilizing OPDA as a signaling molecule to upregulate genes associated with the stress response and scavenging activities [6]. This mechanism helps to reduce the levels of reactive oxygen species, thereby conferring a survival advantage [6].

Several studies have shown that yellowing or albinism of plant leaves is related to poor development of the chloroplast structure and obstruction of Chl synthesis [7] stemming from direct or indirect effects of genetic mutations [8,9]. Leaves with less Chl become light green or lose their green color [10]. The expression of Chl biosynthetic genes is downregulated in yellow or albino leaves, whereas Chl degradation-related enzymes and carotenoid-related genes are highly expressed [11,12].

The ability to accurately identify and locate leaf-color mutant genes is increasing, given the continuous improvements in genomics, transcriptomics, and molecular marker technology [13]. Proteomics analysis contributes to our understanding of how plants respond to various environmental changes and is an effective analytical method for determing the relationships between changes in the abundances of proteins and their biological functions [9].

Primulina serrulata R.B. Zhang & F. Wen is a gesneriad with high ornamental value because of its rosette leaves and colorful flowers [14]. Some white-veined (WV) leaves in the natural habitat have bright WVs, whereas others have fewer WVs or are almost completely green (Figure 1A,B). The leaves of seedlings grown in the laboratory also vary in the ratio of WVs (Figure 1C). Here, we determined the Chl content of leaves and carried out a proteomics experiment to identify the differentially expressed proteins (DEPs) between WV plants and the control and characterized the role of DEPs in WV coloration. Our data provide new insight to aid in the breeding of WV plants.

Figure 1. Variation in the ratio of white veins in *Primulina serrulata* leaves. (**A**) Bright and distinct white veins in the natural habitat; (**B**) few, indistinct white veins in the natural habitat; (**C**) compared seedling leaves grown in the laboratory.

2. Materials and Methods

2.1. Plant Materials and Chl Determination

The plant materials were obtained from *P. serrulata* seedlings cultivated in the laboratory at Zunyi Normal College (Zunyi, China). The parent plants, consisting of only a few fruiting individuals (from which we collected seeds) and not posing any harm to wild populations, were collected from the *P. serrulata* type locality in a karst gorge located in Qiandongnan Prefecture, Guizhou Province.

Leaf blades (2 × 2 cm) were categorized into two distinct types. One type had bright and distinct WVs, and the other had few WVs (Figure 1C). Chl *a* and Chl *b* were determined using spectrophotometry, following Zhang, Qu, and Li [15]:

$$m_T(mg/g) = m_a + m_b$$

where m_T represents total Chl content and m_a and m_b denote the contents of Chl *a* and *b*, respectively.

$$m_a(mg/g) = \frac{Ca(mg/L) \times V}{1000 \times W}$$

$$m_b(mg/g) = \frac{Cb(mg/L) \times V}{1000 \times W}$$

V refers to the total volume of the extract (L) and W indicates the fresh leaf weight (0.5 g).

$$Ca(mg/L) = 12.7OD_{663} - 2.69OD_{645}$$

$$Cb(mg/L) = 22.9OD_{645} - 4.68OD_{663}$$

OD_{663} and OD_{645} denote the absorbance values at wavelengths of 663 nm and 645 nm, respectively.

Leaves from WV and control plants were promptly harvested (weight 0.18–0.30 g) and immediately submerged in liquid nitrogen. Three biological replicates were used for each group. These samples were sent for proteomics analysis at BioNovoGene Co. (Suzhou, China).

2.2. Protein Extraction, Digestion, and TMT Labeling

First, 600 μL of lysis solution (lysis buffer composition: 7 M urea, 4% SDS, 30 mM HEPES, 1 mM PMSF, 2 mM EDTA, 10 mM DTT, 1× protease inhibitor) was added to each sample and the tissues were dissociated by ultrasound on ice for 10 min. The mixture was centrifuged at 10,000× *g* for 10 min, and the supernatant was collected. Proteins were quantified using the BCA method. Next, 100 μg of protein solution was added to DTT at a final concentration of 10 mM; the solution was placed in a 55 °C water bath for 1 h, indole acetic acid was quickly added to a final concentration of 55 mM, and the solution was left in a dark room for 30 min. Acetone at four times the volume of the sample solution was added and precipitated at −20 °C for at least 3 h. After centrifugation at 4 °C and 20,000× *g* for 30 min, the precipitate was removed; this step was repeated twice. Finally, 100 µL of 100 mM TEAB solution was added, mixed, and centrifuged.

A total of 1.5 µg of trypsin was added to each sample. Next, 1.0 µg of enzyme was added for each 100 µg of protein substrate, and the samples were placed in a water bath at 37 °C for 4 h. After centrifugation at 5000 × g, one sample was selected from each group, and 1 µg of enzyme-dissolved sample was analyzed by liquid chromatography–mass spectrometry (LC-MS) for 1 h.

The peptide solution was cold-dried and then dissolved in 100 µL of 0.1 M TEAB to a concentration of 1 µg/µL. The TMT-labeling reagent was stabilized at room temperature, and 41 µL of isopropanol was added, followed by vortexing for 1 min and centrifugation. Labeling reagents were added to the peptide solution. The peptides were fully mixed with the labeling reagents, centrifuged, and left at room temperature for 1 h. Next, 5% hydroxylamine was added to terminate the reactions for 15 min. Five µg of material from each sample was mixed to test labeling efficiency. The labeled samples were mixed with one sample and transferred to a new EP tube. The samples were cold-dried.

2.3. High pH Phase Separation

The mixed labeled peptides were desalted using 0.1% TFA in a 0.5% and 60% acetonitrile solution. The labeled peptides were separated using high pH and reverse-phase LC. The pH was 10.0; 10 mM ammonium formate was used for mobile phase A, and 10 mM ammonium formate and 90% acetonitrile were used for mobile phase B. The labeled peptides were separated using an H-Class LC system (Waters Corp., Miford, MA, USA): sample volume was 50 µL, flow velocity was 250 µL/min, and the samples were tested at 215 nm. One tube of the distillation fraction was collected each minute starting at the second minute. The distillation fractions were pooled into six shares according to the chromatographic peak pattern.

2.4. MS analysis

MS analysis was performed using an Easy-nLC 1200 liquid chromatography system and an Orbitrap Fusion Lumos mass spectrometer (Thermo Fisher Scientific, Waltham, MA, USA). The ion source spray voltage was 2.2 kV. The ion transfer tube temperature of the Orbitrap Fusion Lumos mass spectrometer was maintained at 320 °C. The system automatically switched between MS and MS/MS acquisition in data-dependent mode. Full-scan MS was performed using the Orbitrap for primary scanning, with a scan range of m/z 350–1550 and a resolution of 120,000 (at m/z 200). The maximum ion injection time was 50 ms, and the automatic gain control (AGC) was set to 50,000 ions. Subsequently, parent ions meeting the criteria for MS/MS fragmentation were fragmented using 38% higher energy C-trap dissociation and scanned with the Orbitrap at a resolution of 15,000. The scan range was automatically controlled based on the parent ion m/z ratio, with the minimum scan range fixed at $m/z = 100$. The minimum ion intensity for MS/MS was set to 50,000. The maximum ion injection time was 22 ms, AGC was set to 50,000 ions, and the parent ion selection window was set to 1.6 Da. Ions with two to seven charge states were collected for MS/MS, with dynamic exclusion set to prevent MS/MS acquisition of the same parent ion for 30 s after one MS/MS acquisition.

2.5. Protein Identification and Quantification

The total ion flow chromatogram of the MS signal was derived from MS scanning; the abscissa was the elution time, and the ordinate was peak intensity. The MS data were inputted into Proteome Discoverer software (version 2.4, Thermo Fisher Scientific) for sieving. The embedded Sequest program was used to analyze the results of the sieved spectrogram. The protein quantification value was determined based on the signal value from the extracted TMT-reported ions; the peptide quantification value was indicated by the median, and the protein quantification value was indicated by the accumulation of the peptide quantification value.

2.6. Bioinformatics Analysis

The original RNA-seq transcripts (no. SRR1184436) of closely related species (*P. fimbrisepala*) were downloaded from the SRA database. The sequences were assembled to construct the predicted transcriptional sequences and translated to form a reference protein pool for the species in this experiment. Sequencing quality was assessed with FastQC software (version 0.11.9), and Trimmomatic software (version 0.36) was used to filter low-quality sequences and putative chimeras. The high-quality sequences were assembled using Trinity (version 2.10.0) to obtain the transcripts. The coding proteins were predetermined for the obtained transcript sequences using TransDecoder (version 5.5.0).

The search results were filtered using Proteome Discoverer 2.4 to reduce the false positive rate. Peptide spectra that matched with reliability > 99% and proteins that contained at least one unique peptide were considered credible. Only credible proteins and peptide spectrum matches were retained and verified by the false discovery rate (FDR). Peptides and proteins with an FDR > 1% were removed.

DEP screening: The ratio of the average values of each protein in the two sample groups was considered as the fold change (FC). A T-test was conducted on the quantitative values of each protein in the two sample groups, and the corresponding *p*-value was calculated to detect differences. The BH FDR method was used to adjust the *p*-values. Proteins with a fold-change > 1.5 and a *p*-value < 0.05 were selected as DEPs. The fold change for each protein was logarithmically transformed to base 2, and the *p*-value was converted to the negative logarithm to base 10. These data were used to create a volcano plot.

3. Results

3.1. Chl Contents in WV Plants and the Control

Chl content ($a + b$; mg g^{-1}) was significantly lower ($p < 0.001$) in the WV group (0.64 ± 0.16) than in the control group (1.40 ± 0.12) (Figure 2), suggesting that Chl synthesis was lower in WV leaves than in the control group.

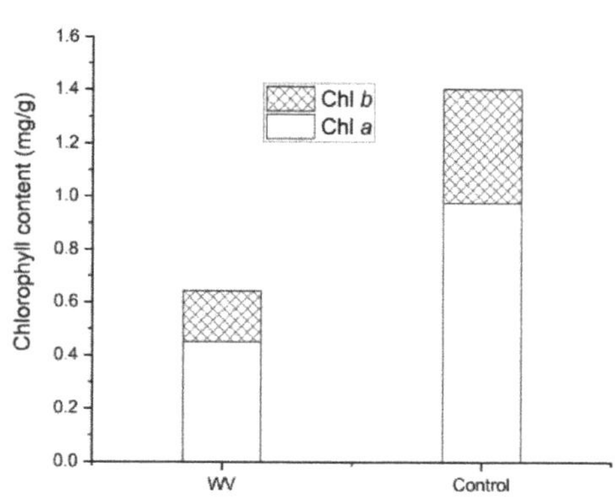

Figure 2. Chl *a* and Chl *b* contents in the WV and control groups.

3.2. Functional Analysis of Total Proteins

Our proteomics analysis successfully identified 6261 proteins (Table S1). Figure 3 illustrates the results through Gene Ontology (GO) database annotation and shows that the *P. serrulata* leaf proteins are primarily associated with diverse biological processes (BPs). These processes encompass metabolic activities, responses to stimuli, biological regulation, organization of cellular components, and other processes. The proteins also exhibited molecular function (MF) activities such as ion binding, hydrolase activity, and protein binding. The proteins were enriched in cellular components (CCs) such as the membrane, the nucleus, and the endomembrane system.

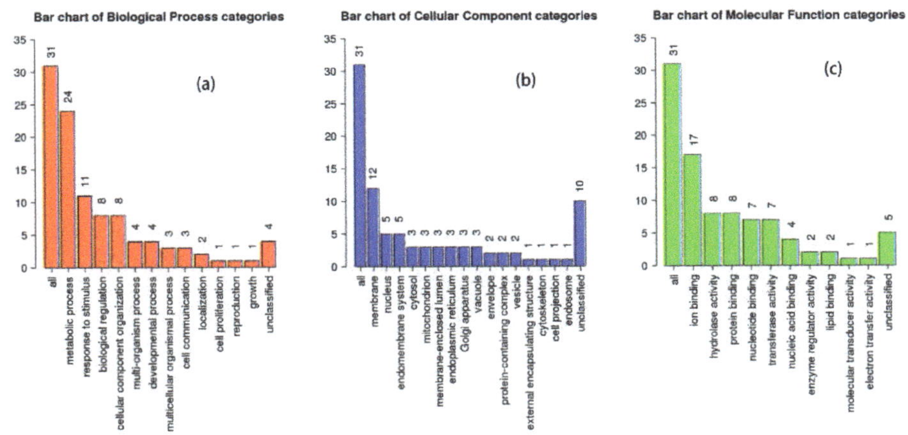

Figure 3. Gene Ontology annotation of the total proteins. (**a**) Bar chart of Biological Process categories; (**b**) Bar chart of Cellular Component categories; (**c**) Bar chart of Molecular Function categories.

3.3. *Quantity of DEPs*

In this study, a total of 6261 DEPs was identified (Table S1). Of these, 399 DEPs were selected based on log2 fold change (log2FC) criteria of less than −0.58 or greater than 0.58 (Table S2). For a more accurate and realistic analysis, the BH FDR method was applied to adjust the *p*-values of these 399 DEPs. Ultimately, only 69 DEPs (Table S3) were chosen for further analysis.

The volcano plot (Figure 4) visualizes the differential expression of proteins between the WV plants and the control group, as assessed by our proteomics analysis. The plot highlights 44 proteins that were downregulated in WV plants, as shown by the blue points, and 25 proteins that were upregulated, indicated by the red points. These DEPs were significantly distant from zero on the Log2FC axis, which denotes the magnitude of the expression change. The vertical axis represents the negative logarithm to the base 10 of the *p*-value (−log10 [*p*-value]), illustrating the difference in the expression of each protein. Proteins that surpassed the threshold for statistical significance, as denoted by the horizontal dotted line, demonstrated substantial changes in expression and high statistical significance. The clustering of red points to the right suggests a group of proteins that were more abundant in the WV plants than in the control, while the cluster of blue points to the left indicates a set of less-abundant proteins.

3.4. *Functional Enrichment Analysis of the DEPs*

3.4.1. Functional GO Enrichment Analysis

The DEPs with the highest enrichment ratios were predominantly associated with magnesium metabolism, despite a medium level of statistical significance (Figure 5). Specifically, these results encompassed cellular responses to magnesium starvation (1, with an enriched quantity of DEPs following suit), response to magnesium ions (1), and cellular response to magnesium ions (1). As magnesium is an essential element for Chl synthesis, these results align with the decrease in Chl content and the downregulation of DEPs related to Chl synthesis observed in the WV group.

Other BP DEPs exhibiting notable enrichment ratios were linked to cellular responses to a mannitol stimulus (1) or sorbitol (1), as well as the biosynthetic processes of quinolinate (1) or glycerol-3-phosphate (1). Despite a relatively low enrichment ratio, these DEPs were associated with metabolic processes (24), cellular metabolic processes (22), and cellular processes (24).

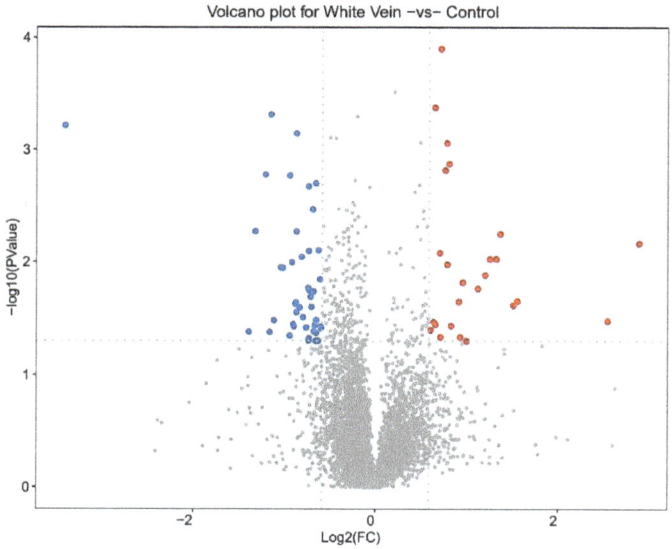

Figure 4. Volcano plot of the DEPs between the WV plants and the control. Red: upregulated DEPs; blue: downregulated DEPs; grey: no significant difference.

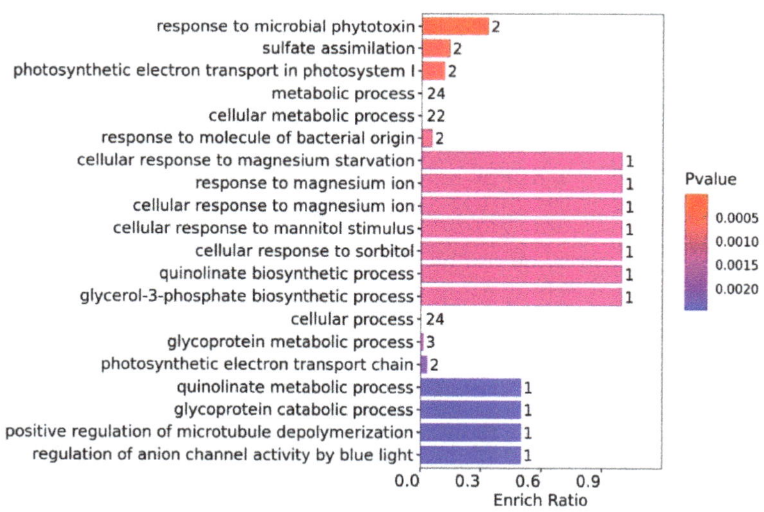

Figure 5. Bar chart of the GO-BP enrichment results.

From the GO-CC enrichment analysis (Figure 6), the DEPs with the highest enrichment ratios were associated with chromoplasts [chromoplast envelope (1) and chromoplast membrane (1)], and statistical significance was high. Although the enrichment ratio was low, the DEPs were primarily related to the cytoplasm, organelles, and chloroplasts, including the cytoplasm (23), cytoplasmic parts (21), chloroplasts (10), plastids (10), membrane-bounded organelles (24), intracellular organelles (24), organelles (24), and intracellular membrane-bounded organelles (23). These results agree with the downregulation of DEPs related to chloroplast synthesis observed in the WV group.

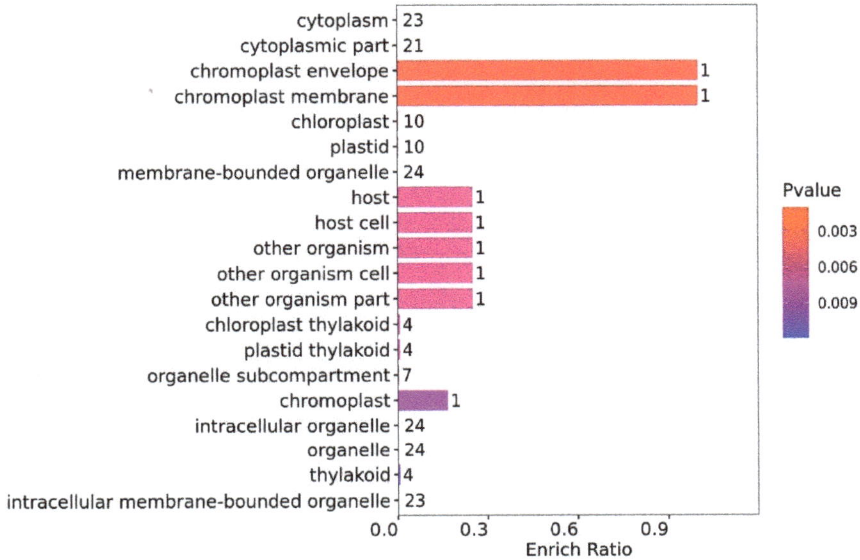

Figure 6. Bar chart of GO-CC enrichment analysis.

The DEPs with the highest enrichment ratios from the GO-MF enrichment analysis (Figure 7) were involved in material binding (arginine (1) and phosphatidylinositol-3,4,5-triphosphate (1)) and enzyme activity (seven enzymes and one DEP for each). Although the enrichment ratios were low, the DEPs were related to ion binding (17) and cation binding (12), with three DEPs involved in calcium ion binding.

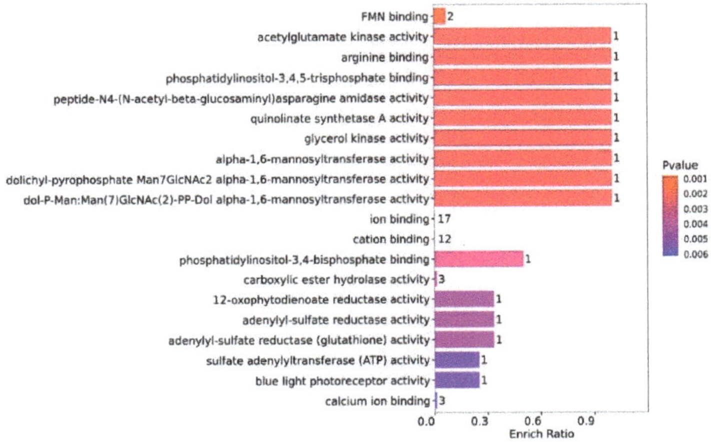

Figure 7. Bar chart of the GO-MF enrichment results.

3.4.2. Functional KEGG Enrichment Analysis

The DEPs were significantly enriched in "metabolic pathways" (9) in the KEGG enrichment analysis (Figure 8). Other counts (one or two DEPs) were enriched in nicotinate and nicotinamide metabolism, monobactam biosynthesis, selenocompound metabolism, sulfur metabolism, *N*-glycan biosynthesis, protein processing in the endoplasmic reticulum (ER), and biosynthesis of amino acids.

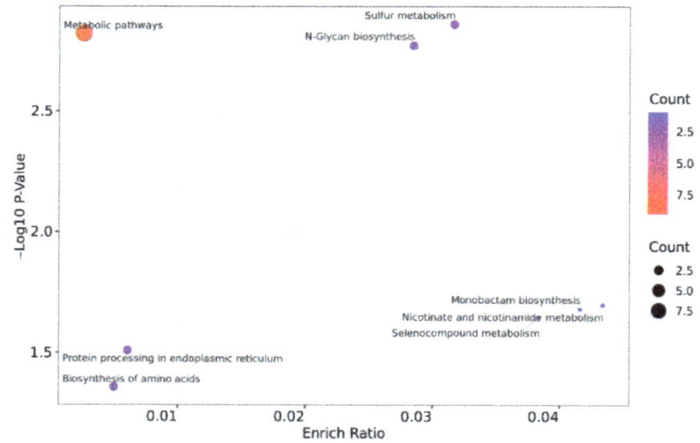

Figure 8. Bubble chart of the KEGG enrichment results.

3.5. Protein Interaction Network Analysis

Strong correlations were detected (Figure 9) between phosphoadenosine phosphosulfate reductase (APR3) and adenylyl phosphosulfate (APS1) and between the peptide: *N*-glycanase (PNG1) and glycerol kinase (NHO1). Interactions occurred between the pentatricopeptide repeat-containing (PPR) protein (AT4G36680) and PPR336 (At1g61870), mannosyl-oligosaccharide 1,2-alpha-mannosidase IA (MNS2) and the asparagine-linked glycosylation (ALG) protein 12 homolog (ALG12), quinolinate synthase (QS) A and probable phosphoglycerate mutase (PGAM) GpmB (AT3G05350), and phototropin (PHOT2) and photosynthetic NAD(P)H dehydrogenase (NDH) subunit of subcomplex B 4 (PnsB4).

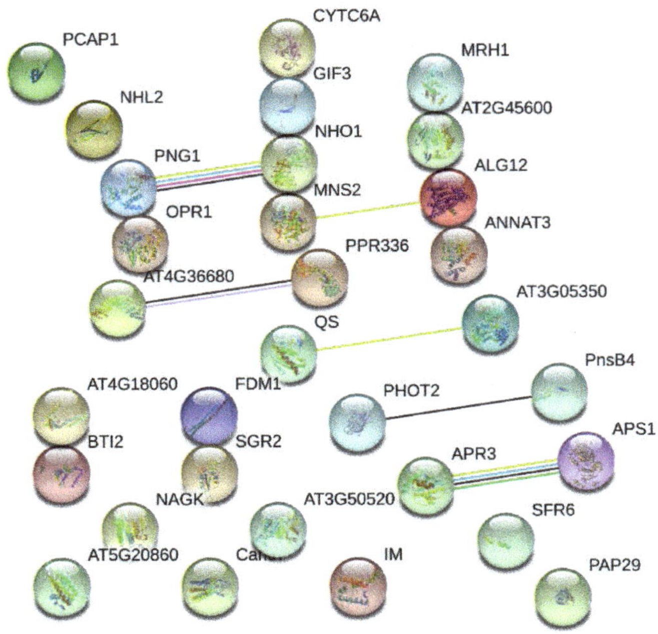

Figure 9. Protein interaction network in *Primulina serrulata* leaves. Different colored lines between the proteins indicate ofdifferent interactions.

4. Discussion

4.1. Proteins Affecting the Formation of Chloroplast and Chl

Some downregulated DEPs in the WV group were associated with the formation of chloroplasts, including reticulon-3, allene oxide synthase (AOS), and cytochrome c6. Reticulons are ubiquitous ER-localized proteins in eukaryotic organisms [16]. Inactivation of the plant reticulon gene leads to severe defects in chloroplast functioning and oxidative stress, resulting in pale-green leaves [17]. AOS is targeted to the chloroplast envelope through a pathway that involves ATP and proteins [18]. Cytochrome c6 is a soluble metalloprotein involved in the reduction of photosystem I [19]. The downregulation of these DEPs suggests a weakening of chloroplast biosynthesis in WV leaves.

Some downregulated DEPs in the WV group were related to Chl biosynthesis, including PHOT2 and guanine nucleotide-binding protein (G-protein) subunit beta. Chl biosynthesis may be controlled by the joint action of phytochromes and PHOTs, which regulate the formation of different chlorophyllide forms [20]. In addition to GTPase activity, G-proteins exhibit various specific molecular functions [21] and are involved in activating the phytochrome-mediated Chl *a*/*b*-binding proteins (cab) gene [22]. Downregulation of these DEPs suggests a weakening of chlorophyll biosynthesis in WV leaves.

4.2. Proteins Related to Stress Resistance

Some downregulated DEPs enhance resistance to environmental stressors, such as drought (late embryogenesis abundant protein [23] and maspardin [24]), salt (salt stress root protein RS1 (conjectured from the literal meaning), maspardin [24] and MNS2 [25]), heat shock (general stress protein CTC [26] and retrovirus-related Pol polyprotein from transposon [27]), oxidation (putative oxidoreductase GLYR1 [28], SIA1 [29] and reticulon [17]), toxins (GPR107 [30] and cell number regulator [31]), misfolded glycoprotein (peptide: *N*-glycanase 1 [32]), herbivore (AOS [33]), phytopathogenic fungi (polygalacturonase inhibitor [34]), and other abiotic and biotic stressors (alcohol dehydrogenase 1 [35] and beta-glucosidase [36]).

Fewer DEPs related to stress resistance were upregulated in the WV group than in the control group, such as acetylglutamate kinase, carboxylesteraseisozyme, major pollen allergen Bet v 1, purple acid phosphatase, and ferritin-1.

In general, stress resistance was weak in the WV leaves but there were some compensatory mechanisms against particular stressors.

4.3. Proteins Related to Biomacromolecule Modification

Some downregulated DEPs in the WV group were related to biomacromolecule modification, such as factor of DNA methylation 1, retrovirus-related Pol polyprotein from transposon 297, GPI transamidase component GAA1, polyadenylate-binding protein 2, and pentatricopeptide repeat-containing protein At1g61870.

Many DEPs related to modifying biomacromolecules were downregulated in the WV group. This was likely attributable to the downregulation of many DEPs involved in photosynthesis and stress resistance in the WV group.

4.4. Proteins Related to Iron and/or Sulfur Increase in WV Leaves

Some upregulated DEPs in the WV group contain iron and/or sulfur, such as QS, APR3, ferritin-1, acid phosphatase, sulfate adenylyl transferase, and prolyl endopeptidase-like protein. QS and APR are iron-sulfur enzymes. Most of the cellular iron is stored in ferritin, which is both the acceptor and donor of iron for metabolic processes [37].

Two upregulated proteins contain heme, whose porphyrin ring surrounds an iron ion [38]. The nitrate reductase monomer is composed of a $\sim$100-kD polypeptide and FAD, heme-iron, and molybdenum-molybdopterin molecules [39]. A heme-heme binuclear center is in the cytochrome *bd* ubiquinol oxidase [40].

One upregulated protein catalyzes the reaction related to iron. Acid phosphatase catalyzes the hydrolysis of phosphate monoesters under acidic condition and most of the isoforms exhibit a characteristic purple color due to charge transfer from a tyrosine residue to Fe (III) [41].

Although iron is an essential element for Chl synthesis [42], it was likely utilized for synthesizing heme and ferritin in WV leaves. This is further supported by the downregulation of several DEPs involved in Chl synthesis and the upregulation of several DEPs containing heme or ferritin in this study. Iron-rich ferritins are thought to hinder Chl biosynthesis because of the mutual inhibition between heme and Chl [13]. Heme and Chl contain a porphyrin ring structure; in heme, the porphyrin ring surrounds an iron ion, while in Chl, it surrounds a magnesium ion [38].

Some upregulated DEPs in the WV group are related to sulfate and respiration. Heme catalyzes respiration to release the energy stored in chemical bonds, while Chl catalyzes to convert the energy of sunlight into the stored chemical energy of bonds [43]. Sulfate is taken up by the cell and is activated by sulfate adenylyl transferase to adenylyl phosphosulfate and inorganic pyrophosphate [44]. Prolyl endopeptidase-like is a (thio)esterase involved in mitochondrial respiratory chain function [45].

Two pairs of proteins related to sulfate or respiration interacted in this study, such as APR3 and APS1 and QS and AT3G05350 (Figure 9). APR controls the sulfate assimilation pathway, which provides reduced sulfur for synthesizing the amino acids cysteine and methionine [46]. APS is reduced to sulfite and adenosine monophosphate [44], so it is related to sulfate assimilation pathway and energy metabolism.

QS is an iron–sulfur enzyme involved in the biosynthesis of nicotinamide adenine dinucleotide (NAD) and plays a crucial role as a cofactor in many essential redox biological reactions [47], such as the tricarboxylic acid cycle (TCA). 2,3-Biphophoglycerate-independent PGAM AT3G05350 is a key glycolytic enzyme that catalyzes the reversible interconversion of 3-phosphoglycerate (3-PGA) to 2-phosphoglycerate (2-PGA) [48]. Because TCA and glycolysis are crucial processes in energy metabolism, the interaction between QS and AT3G05350 in this study indicated that the formation of WV was related to energy metabolism.

The upregulated DEPs containing iron or sulfur in the WV group, along with the interacting proteins related to the sulfate assimilation pathway and energy metabolism, indicate that photosynthesis was weakened, and respiration was enhanced in the WV leaves. Our findings are consistent with the result that the absence of chloroplast activity leads to an increase in mitochondrial gene copy number and elevated mitochondrial transcription levels in white leaves [10].

4.5. DEPs Affecting Cell-Wall Formation

Expansin-A4, PnsB4, and pectinesterase were upregulated in the WV group. Expansin-A4 is involved in cell elongation and cell-wall modifications by loosening and extending plant cell walls [49]. PnsB4 is also related to cell elongation [50]. Pectinesterase acts on pectins in the cell wall in vivo, causing gel formation [51]. The upregulation of these three DEPs in *P. serrulata* suggests ongoing cell-wall elongation and gelation within the WVs, which agrees with the visual traits of the WVs, where cells appear more protrusive and irregular (Figure 10A) compared to the smoother cells in the green areas (Figure 10B). These differences are potentially ascribed to the extension of the cell walls in the white region. Moreover, the enhanced refractivity noted in the white area suggests a greater accumulation of gels within these cellular structures.

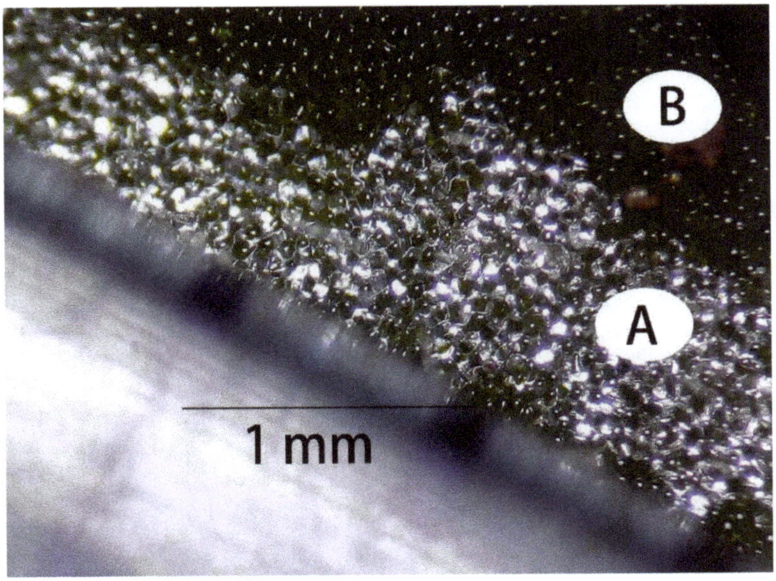

Figure 10. Top view of the white vein of *Primulina serrulata*. (**A**) White area; (**B**) green area.

5. Conclusions

In this study, we performed an in-depth proteomic analysis using a TMT-based approach, specifically focusing on contrasting WV and green-leaved *P. serrulata*. This proteomics landscape offered several key insights into the physiological changes occurring in WV leaves, including: (1) a notable decrease in chloroplast and Chl biosynthesis, pointing to fundamental changes in photosynthetic machinery; (2) a shift in iron usage, favoring heme and ferritin synthesis over Chl, suggesting altered metabolic priorities; (3) enhanced respiratory activity, which could be a compensatory response to a low photosynthetic rate; and (4) changes in cell-wall properties, as evidenced by elongation and gelling, indicating structural adaptations. These proteomic variations underscore the complex interplay of molecular processes in WV leaves. Importantly, our study highlights the potential for targeted manipulation of these proteomic pathways to optimize the cultivation of WV *P. serrulata* plants, opening up avenues for further research in plant variegation and horticulture.

Supplementary Materials: The following supporting information can be downloaded at: https://www.mdpi.com/article/10.3390/horticulturae10010019/s1, Table S1: All differentially expressed proteins in this study; Table S2: DEPs with log2FC less than −0.58 or greater than 0.58; Table S3: DEPs with fold change greater than 1.5.

Author Contributions: Conceptualization and methodology, Q.-L.D. and R.-B.Z.; formal analysis, M.-F.C.; investigation, D.-J.X. and T.D.; software, Z.-M.Q. and S.-S.W.; writing—original draft preparation, Q.-L.D. and D.-J.X.; writing—review and editing, R.-B.Z.; funding acquisition, D.-J.X. All authors have read and agreed to the published version of the manuscript.

Funding: This research was funded by the National Natural Science Foundation of China (Grant No. 31860162), the Science and Technology Bureau Scientific Research Project of Zunyi City (Grant No. 2305011203), and the Science & Technology Fundamental Resources Investigation Program (Grant No. 2022FY100201).

Data Availability Statement: Data are contained within the article and supplementary materials.

Acknowledgments: We thank Suzhou Bionovogene Company, China, for helping with the experiments.

Conflicts of Interest: The authors declare no conflict of interest.

References

1. Hedtke, B.; Wagner, I.; Börner, T.; Hess, W. Inter-organellar crosstalk in higher plants: Impaired chloroplast development affects mitochondrial gene and transcript levels. *Plant J. Cell Mol. Biol.* **1999**, *19*, 635–643. [CrossRef] [PubMed]
2. Toshoji, H.; Katsumata, T.; Takusagawa, M.; Yusa, Y.; Sakai, A. Effects of chloroplast dysfunction on mitochondria: White sectors in variegated leaves have higher mitochondrial DNA levels and lower dark respiration rates than green sectors. *Protoplasma* **2011**, *249*, 805–817. [CrossRef] [PubMed]
3. Xiong, Y.; Ma, J.; He, Y.; Lin, Z.; Li, X.; Yu, S.; Li, R.; Jiang, F.; Li, X.; Huang, Z.; et al. High-throughput sequencing analysis revealed the regulation patterns of small RNAs on the development of *A. comosus* var. *bracteatus leaves. Sci. Rep.* **2018**, *8*, 1947. [CrossRef] [PubMed]
4. Aluru, M.; Bae, H.; Wu, D.; Rodermel, S. The *Arabidopsis* immutans mutation affects plastid differentiation and the morphogenesis of white and green sectors in variegated plants. *Plant Physiol.* **2001**, *127*, 67–77. [CrossRef] [PubMed]
5. Foudree, A.; Putarjunan, A.; Kambakam, S.; Nolan, T.; Fussell, J.; Pogorelko, G.; Rodermel, S. The Mechanism of Variegation in immutans Provides Insight into Chloroplast Biogenesis. *Front. Plant Sci.* **2012**, *3*, 260. [CrossRef] [PubMed]
6. Sun, Y.; Hung, C.; Qiu, J.; Chen, J.; Kittur, F.; Oldham, C.; Henny, R.; Burkey, K.; Fan, L.; Xie, J. Accumulation of high OPDA level correlates with reduced ROS and elevated GSH benefiting white cell survival in variegated leaves. *Sci. Rep.* **2017**, *7*, 44158. [CrossRef] [PubMed]
7. Chen, L.Y.; He, L.T.; Lai, J.L.; He, S.T.; Wu, Y.X.; Zheng, Y.S. The variation of chlorophyll biosynthesis and the structure in different color leaves of *Bambusa* multiplex 'Silverstripe'. *J. For. Environ.* **2017**, 37, 385–391.
8. Wei, Y.L.; Shi, Y.F.; Wu, J.L. Rice (*Oryza sativa* L.) leaf color mutants. *J. Nucl. Agric. Sci.* **2011**, *25*, 1169–1178.
9. Li, S.M.; Wang, S.; Wang, P.; Gao, L.L.; Yang, R.T.; Li, Y. Label-free comparative proteomic and physiological analysis provides insight into leaf color variation of the golden-yellow leaf mutant of *Lagerstroemia indica*. *J. Proteom.* **2020**, *228*, 103942. [CrossRef]
10. Xu, M.Y.; He, P.; Lai, W.; Chen, L.H.; Ge, L.L.; Liu, S.Q.; Yang, Y.J.; Advances in Molecular Mechanism of Plant Leaf Color Variation. Molecular Plant Breeding 2021. Available online: https://kns.cnki.net/kcms/detail/46.1068.S.20210112.1619.016.html (accessed on 22 October 2023).
11. Dong, X.Y.; Huang, L.B.; Chen, Q.S.; Lyu, Y.Z.; Sun, H.N.; Liang, Z.H. Physiological and anatomical differences and differentially expressed genes reveal yellow leaf coloration in Shumard Oak. *Plants* **2020**, *9*, 169. [CrossRef]
12. Li, C.F.; Xu, Y.; Ma, J.Q.; Jin, J.Q.; Huang, D.; Yao, M.; Ma, C.L.; Chen, L. Biochemical and transcriptomic analyses reveal different metabolite biosynthesis profiles among three color and developmental stages in 'Anji Baicha' (*Camellia sinensis*). *BMC Plant Biol.* **2016**, *16*, 195. [CrossRef] [PubMed]
13. Du, W.K.; Yuan, S.X.; Hu, F.R. Research Progress on Molecular Mechanisms of the Leaf Color Mutation. *Mol. Plant Breed.* **2019**, *17*, 1888–1897.
14. Jiang, H.; Deng, T.; Lv, X.-Y.; Zhang, R.-B.; Wen, F. *Primulina serrulata* (Gesneriaceae), a new species from southeastern Guizhou, China. *PhytoKeys* **2019**, *132*, 11–18. [CrossRef] [PubMed]
15. Zhang, Z.L.; Qu, W.J.; Li, X.F. *Experimental Guidance of Plant Biology*, 4th ed.; Higher Education Press: Beijing, China, 2009.
16. Tolley, N.J. *Functional Analysis of a Reticulon Protein from Arabidopsis thaliana*; Department of Biological Sciences, University of Warwick: Coventry, UK, 2010.

17. Tarasenko, V.I.; Garnik, E.Y.; Katyshec, A.I.; Subota, I.Y.; Konstantinov, Y.M. Disruption of *Arabidopsis* reticulon gene RTNLB16 results in chloroplast dysfunction and oxidative stress. *J. Stress Physiol. Biochem.* **2012**, *8*, S21.
18. Knopf, R.R.; Feder, A.; Mayer, K.; Lin, A.; Rozenberg, M.; Schaller, A.; Adam, Z. Rhomboid proteins in the chloroplast envelope affect the level of allene oxide synthase in *Arabidopsis thaliana*. *Plant J.* **2012**, *72*, 559–571. [CrossRef] [PubMed]
19. De la Rosa, M.A.; Navarro, J.A.; Dıaz-Quintana, A.; De la Cerda, B.; Molina-Heredia, F.P.; Balme, A.; del S Murdoch, P.; Dıaz-Moreno, I.; Durán, R.V.; Hervás, M. An evolutionary analysis of the reaction mechanisms of photosystem I reduction by cytochrome c6 and plastocyanin. *Bioelectrochemistry* **2002**, *55*, 41–45. [CrossRef] [PubMed]
20. Kong, Y.; Zheng, Y. Phototropin is partly involved in blue-light-mediated stem elongation, flower initiation, and leaf expansion: A comparison of phenotypic responses between wild *Arabidopsis* and its phototropin mutants. *Environ. Exp. Bot.* **2020**, *171*, 103967. [CrossRef]
21. Millner, P.A. Are guanine nucleotide-binding proteins involved in regulation of thylakoid protein kinase activity? *FEBS Lett.* **1987**, *226*, 155–160. [CrossRef]
22. Romerro, L.C.; Lam, E. Guanine nucleotide binding protein involvement in early steps of phytochrome-regulated gene expression. *Proc. Natl. Acad. Sci. USA* **1993**, *90*, 1465–1469. [CrossRef]
23. Shih, M.-D.; Hoekstra, F.A.; Hsing, Y.-I.C. Late embryogenesis abundant proteins. In *Advances in Botanical Research*; Elsevier: Amsterdam, The Netherlands, 2008; pp. 211–255.
24. Wang, F.-B.; Wan, C.-Z.; Niu, H.-F.; Qi, M.-Y.; Gang, L.; Zhang, F.; Hu, L.-B.; Ye, Y.-X.; Wang, Z.-X.; Pei, B.-L. OsMas1, a novel maspardin protein gene, confers tolerance to salt and drought stresses by regulating ABA signaling in rice. *J. Integr. Agric.* **2023**, *22*, 341–359. [CrossRef]
25. Liu, C.; Niu, G.; Zhang, H.; Sun, Y.; Sun, S.; Yu, F.; Lu, S.; Yang, Y.; Li, J.; Hong, Z. Trimming of N-glycans by the Golgi-localized α-1, 2-mannosidases, MNS1 and MNS2, is crucial for maintaining RSW2 protein abundance during salt stress in *Arabidopsis*. *Mol. Plant* **2018**, *11*, 678–690. [CrossRef] [PubMed]
26. Völker, U.; Engelmann, S.; Maul, B.; Riethdorf, S.; Völker, A.; Schmid, R.; Mach, H.; Hecker, M. Analysis of the induction of general stress proteins of *Bacillus subtilis*. *J. Microbiol.* **1994**, *140*, 741–752. [CrossRef] [PubMed]
27. Traylor-Knowles, N.; Rose, N.H.; Sheets, E.A.; Palumbi, S.R. Early transcriptional responses during heat stress in the coral Acropora hyacinthus. *Biol. Bull.* **2017**, *232*, 91–100. [CrossRef] [PubMed]
28. Margaria, P.; Abbà, S.; Palmano, S. Novel aspects of grapevine response to phytoplasma infection investigated by a proteomic and phospho-proteomic approach with data integration into functional networks. *BMC Genom.* **2013**, *14*, 1–15. [CrossRef] [PubMed]
29. Manara, A.; DalCorso, G.; Leister, D.; Jahns, P.; Baldan, B.; Furini, A. At SIA 1 AND At OSA 1: Two Abc1 proteins involved in oxidative stress responses and iron distribution within chloroplasts. *New Phytol.* **2014**, *201*, 452–465. [CrossRef] [PubMed]
30. Tafesse, F.G.; Guimaraes, C.P.; Maruyama, T.; Carette, J.E.; Lory, S.; Brummelkamp, T.R.; Ploegh, H.L. GPR107, a G-protein-coupled Receptor Essential for Intoxication by Pseudomonas aeruginosa Exotoxin A, Localizes to the Golgi and Is Cleaved by Furin. *J. Biol. Chem.* **2014**, *289*, 24005–24018. [CrossRef]
31. Qiao, K.; Tian, Y.; Hu, Z.; Chai, T. Wheat cell number regulator CNR10 enhances the tolerance, translocation, and accumulation of heavy metals in plants. *Environ. Sci. Technol.* **2018**, *53*, 860–867. [CrossRef]
32. Suzuki, T.; Park, H.; Hollingsworth, N.M.; Sternglanz, R.; Lennarz, W.J. PNG1, a yeast gene encoding a highly conserved peptide: N-glycanase. *J. Cell Biol.* **2000**, *149*, 1039–1052. [CrossRef]
33. Farmer, E.E.; Goossens, A. Jasmonates: What ALLENE OXIDE SYNTHASE does for plants. *J. Exp. Bot.* **2019**, *70*, 3373–3378. [CrossRef]
34. De Lorenzo, G.; Ferrari, S. Polygalacturonase-inhibiting proteins in defense against phytopathogenic fungi. *Curr. Opin. Plant Biol.* **2002**, *5*, 295–299. [CrossRef]
35. Shi, H.; Liu, W.; Yao, Y.; Wei, Y.; Chan, Z. Alcohol dehydrogenase 1 (ADH1) confers both abiotic and biotic stress resistance in Arabidopsis. *Plant Sci.* **2017**, *262*, 24–31. [CrossRef] [PubMed]
36. Gómez-Anduro, G.; Ceniceros-Ojeda, E.A.; Casados-Vázquez, L.E.; Bencivenni, C.; Sierra-Beltrán, A.; Murillo-Amador, B.; Tiessen, A. Genome-wide analysis of the beta-glucosidase gene family in maize (*Zea mays* L. var B73). *Plant Mol. Biol.* **2011**, *77*, 159–183. [CrossRef] [PubMed]
37. Zielińska-Dawidziak, M. Plant ferritin—A source of iron to prevent its deficiency. *Nutrients* **2015**, *7*, 1184–1201. [CrossRef] [PubMed]
38. Alberts, B. *Molecular Biology of the Cell*; Garland Science: New York, NY, USA, 2017.
39. Campbell, W.H. Nitrate reductase structure, function and regulation: Bridging the gap between biochemistry and physiology. *Annu. Rev. Plant Biol.* **1999**, *50*, 277–303. [CrossRef] [PubMed]
40. Hill, J.J.; Alben, J.O.; Gennis, R.B. Spectroscopic evidence for a heme-heme binuclear center in the cytochrome bd ubiquinol oxidase from *Escherichia coli*. *Proc. Natl. Acad. Sci. USA* **1993**, *90*, 5863–5867. [CrossRef]
41. Anand, A.; Srivastava, P.K. A molecular description of acid phosphatase. *Appl. Biochem.* **2012**, *167*, 2174–2197. [CrossRef]
42. Marsh, H., Jr.; Evans, H.; Matrone, G. Investigations of the role of iron in chlorophyll metabolism. II. Effect of iron deficiency on chlorophyll synthesis. *Plant Physiol.* **1963**, *38*, 638. [CrossRef]
43. Granick, S. *Evolution of Heme and Chlorophyll, Evolving Genes and Proteins*; Elsevier: Amsterdam, The Netherlands, 1965; pp. 67–88.

44. Christensen, G.; Zane, G.; Kazakov, A.; Li, X.; Rodionov, D.; Novichkov, P.; Dubchak, I.; Arkin, A.; Wall, J. Rex (encoded by DVU_0916) in Desulfovibrio vulgaris Hildenborough is a repressor of sulfate adenylyl transferase and is regulated by NADH. *J. Bacteriol.* **2015**, *197*, 29–39. [CrossRef]
45. Rosier, K.; McDevitt, M.T.; Smet, J.; Floyd, B.J.; Verschoore, M.; Marcaida, M.J.; Bingman, C.A.; Lemmens, I.; Peraro, M.D.; Tavernier, J. Prolyl endopeptidase-like is a (thio) esterase involved in mitochondrial respiratory chain function. *Iscience* **2021**, *24*, 103460. [CrossRef]
46. Kopriva, S.; Koprivova, A. Plant adenosine 5′-phosphosulphate reductase: The past, the present, and the future. *J. Exp. Bot.* **2004**, *55*, 1775–1783. [CrossRef]
47. Choudens, S.O.-D.; Loiseau, L.; Sanakis, Y.; Barras, F.; Fontecave, M. Quinolinate synthetase, an iron–sulfur enzyme in NAD biosynthesis. *FEBS Lett.* **2005**, *579*, 3737–3743. [CrossRef] [PubMed]
48. Zhao, Z.; Assmann, S.M. The glycolytic enzyme, phosphoglycerate mutase, has critical roles in stomatal movement, vegetative growth, and pollen production in *Arabidopsis thaliana*. *J. Exp. Bot.* **2011**, *62*, 5179–5189. [CrossRef] [PubMed]
49. Lin, Z.; Li, Q.; Yin, Q.; Wang, J.; Zhang, B.; Gan, S.; Wu, A.-M. Identification of novel miRNAs and their target genes in Eucalyptus grandis. *Tree Genet. Genomes* **2018**, *14*, 1–9. [CrossRef]
50. Costacurta, A.; Vanderleyden, J. Synthesis of phytohormones by plant-associated bacteria. *Crit. Rev. Microbiol.* **1995**, *21*, 1–18. [CrossRef]
51. Zhou, H.-W.; Ben-Arie, R.; Lurie, S. Pectin esterase, polygalacturonase and gel formation in peach pectin fractions. *Phytochemistry* **2000**, *55*, 191–195. [CrossRef]

horticulturae

MDPI

Article

UPLC–MS/MS and Gene Expression Research to Distinguish the Colour Differences of *Rhododendron liliiflorum* H. Lév

Jin Dai [1,†], Xinglin Wang [1,†], Xingpan Meng [1], Xu Zhang [2], Qihang Zhou [3], Zhengdong Zhang [4], Ximin Zhang [1], Yin Yi [1], Lunxian Liu [1,*] and Tie Shen [1,3,*]

1 Key Laboratory of National Forestry and Grassland Administration on Biodiversity Conservation in Karst Mountainous Areas of Southwestern China, Engineering Research Center of Carbon Neutrality in Karst Areas, Ministry of Education, Key Laboratory of Environment Friendly Management on High Altitude Rhododendron Diseases and Pests, Institutions of Higher Learning in Guizhou Province, School of Life Science, Guizhou Normal University, Guiyang 550025, China; 21010100373@gznu.edu.cn (J.D.); 232100100407@gznu.edu.cn (X.W.); 222100100438@gznu.edu.cn (X.M.); zhxm409@gznu.edu.cn (X.Z.); gzklppdr@gznu.edu.cn (Y.Y.)

2 Guizhou Caohai Wetland Ecosystem National Observation and Research Station, Guizhou Academy of Forestry Sciences, Guiyang 550001, China; lkyzx163@163.com

3 Key Laboratory of Information and Computing Science Guizhou Province, School of Mathematical Sciences, Guizhou Normal University, Guizhou 550025, China; 19010210498@gznu.edu.cn

4 College of Mathematics and Information Science, Guiyang University, Guiyang 550001, China; zzd@gznu.edu.cn

* Correspondence: llx@gznu.edu.cn (L.L.); shentie@gznu.edu.cn (T.S.)

† These authors contributed equally to this work.

Abstract: Among ornamental plants, the colour of the petals is an important feature. However, the reason for the colour differences of *Rhododendron liliiflorum* remains unclear. To reveal the differences in the colour of *R. liliiflorum*, high-efficiency liquid chromatographic collar (UPLC–MS/MS) technology was used to study the yellow and white parts of *R. liliiflorum*. A total of 1187 metabolites were identified in *R. liliiflorum* petals, including 339 flavonoid metabolites. Seventy-eight types of flavonoids in these metabolites were found in the yellow and white parts of *R. liliiflorum* petals, along with 11 other significantly enriched substances. Combining gene expression-related data with differential metabolite data demonstrated effects of enrichment in the flavanonols (fustin), flavonols (epiafzelechin and afzelechin), and flavanones (pinocembrin) of flavonoid biosynthesis; glyccitin, 6″-O-malonylgenistin, and 6-hydroxydaidzein of isoflavonoid biosynthesis; and anthocyanin biosynthesis of malvidin-3-O-galactoside (primulin), delphinidin-3-O-rutinoside, cyanidin-3-O-glucoside (kuromanin), and cyanidin-3-O-rutinoside (keracyanin), which are potentially the contributing factors responsible for the differences in petal colour in *R. liliiflorum*. This study establishes a connection between the differential metabolites underlying the color differences in the petals of *R. liliiflorum* and the gene expression in *R. liliiflorum*. This will provide a foundation for subsequent research on the regulation of flower color in *R. liliiflorum* and have profound implications for horticultural applications of *R. liliiflorum*.

Keywords: colouration; petal; metabolites; flavonoids

Citation: Dai, J.; Wang, X.; Meng, X.; Zhang, X.; Zhou, Q.; Zhang, Z.; Zhang, X.; Yi, Y.; Liu, L.; Shen, T. UPLC–MS/MS and Gene Expression Research to Distinguish the Colour Differences of *Rhododendron liliiflorum* H. Lév. *Horticulturae* **2023**, *9*, 1351. https://doi.org/10.3390/horticulturae9121351

Academic Editor: Rosario Paolo Mauro

Received: 20 November 2023
Revised: 14 December 2023
Accepted: 15 December 2023
Published: 18 December 2023

1. Introduction

Rhododendron is the largest genus of the Rhododendron family, with wild rhododendrons found all over the world [1]. There are traces of azalea discovered in low latitudes from tropical areas to high latitudes. Wild *Rhododendron* genus communities have contributed to the economic development of local communities, such as the Baili Rhododendron Nature Reserve in northwestern Guizhou Province, China [2]. Rhododendron flower colour is highly diverse, from purple to red, pink, and even white [3]; thus, its extensive usage in research on plant flower colours [4–6]. *Rhododendron Liliiflorum* belongs to the Rhododendron family and exists in southeast China (southeast of Yunnan to Hunan). The base of the petals in the late flowering period is yellow and the upper part is white (Figure 1a,b).

Figure 1. *R. liliiflorum* phenotype and overall characteristics. (**a**). Wild *R. liliiflorum.* (**b**). The morphology of *R. liliiflorum*, *R. liliiflorum* petals overall diagram, *R. liliiflorum* petal inner diagram. The part marked by the white square border is the white part of the *R. liliiflorum* (W). The yellow elliptical border bezel. The marked part is the yellow part (Y) of *R. liliiflorum.* (**c**). MWs vs. MYs overall clustering heatmap; horizontal is sample name, longitudinal information is metabolites, groups are packets, and class classification is material. (**d**). MW vs. MY correlation analysis diagram.

Among ornamental plants, the colour of the petals is an important feature. Studies have shown that colour characteristics are an important factor in the interaction between plants and their powder [7,8]. The potential factors affecting the colour of plants include hydrogen, metal ions, pigment ingredients, and petal structure [9]. The most critical factor is pigment composition. The pigment ingredients that affect the formation of colours include flavonoids, carotene, betaine, etc. Each of these combinations adds to the diversity of colour [10]. Flavonoids are a class of polyphenol secondary metabolites, which are natural pigments widely present in plants. Based on the phenolic molecular structure with the heterocyclic ring and conformation, these metabolites can be divided into flavones, flavonols, isoflavonoids, flavanones, flavanols, anthocyanidins, and so on [11–13]. Flavonoids are widely present in plants, and their contents in flowers [14], leaves [15], fruits [16], and other parts of plants are high. The colour changes of petals may be caused by different flavonoids, while the proportional changes of various types of flavonoids may be a pertinent factor in the change in colour. This includes anthocyanidins, flavanones, chalcones, flavonols, isoflavonoids, and flavanones [17,18]. In a study of lotus petals, it was found that the content of anthocyanins was positively correlated with red petals, and flavonols contributed to the yellow part of the auxiliary pigment [19]. After comparing seven types of plants with red flowers and white flowers, it was found that the content of anthocyanins in safflower was significantly higher than that in white flowers. In white flowers, flavonol is the main pigment type [20].

As discovered through many studies, anthocyanins, a downstream material for flavonoid biosynthesis, is considered to be the main factor stimulating colour changes in plant flowers [21]. For example, a study by Mizuta et al. [22] on the pattern of flower colour and the anthocyanin composition of evergreen rhododendrons found that purple rhododendrons had more anthocyanins in the red, purple, and white groups of samples, while no anthocyanin content was detected in white rhododendrons [22]. Sun et al. [23] used 'Yanzhi Mi' (pink azalea) and the wild-type (WT) cultivar 'Dayuanyangjin' (white azalea with pink stripes) as their study subjects and found that pink-petalled azalea showed higher anthocyanin contents, while white-petalled azalea contained mainly flavonoids and a small amount of anthocyanin, which may be attributable to the presence of stripes on white-petalled azalea [24]. In a study of the wild type of primrose (yellow petals) and its variants (yellow transformed to red), it was found that the orange and red primrose types contain anthocyanins such as cyanidin-3-O-glucoside [23]. Most of the current research on rhododendrons focuses on white rhododendrons and red rhododendrons, and many studies on white rhododendrons have shown that white rhododendrons have lower anthocyanin contents [22,24,25]. Additionally, the enzyme genes CHS, F3′5′H, FLS, I2′H, HID, DFR, and LAR may have a role in the alteration of petal colour in *R. liliiflorum* [21,26–28]. However, few studies have revealed colour differences within the same petal colour.

Metabolomics is widely used in research on plant colour formation [25,29,30]. Metabolic group studies have studied four noncolours of wheat, explaining that the reason for the colour difference of wheat is potentially the accumulation of flavonoid compounds [31]. Guan et al. [30] investigated the differentiation mechanism of the colouration of sophora flowers and identified the key metabolite responsible for the red colour as anthocyanins of the delphinidin type using metabolomics approaches. Combining the related genes of plant traits can reflect the reasons for changes in the organism. At present, related research on the colour metabolic pathway has been combined with metabolic gene research. For example, Wang et al. [31] combined metabolic group and gene expression analyses and explored phenylpropanoid biosynthesis and flavonoid biosynthesis in wheat of different colours. Jiang [32] comprehensively analysed anthocyanins in salvin.

Previously, we used the transcriptome to study *R. liliiflorum*, and the differentially expressed genes of the two parts of *R. liliiflorum* petals were selected. The expression of related enzymes may affect the difference in the colour of *R. liliiflorum* [21]. This study used high-efficiency liquid chromatography–coupled mass spectrometry (UPLC–MS/MS) to study the two parts of *R. liliiflorum*, yellow and white, and screened differentiated metabolites during biological synthesis, such as phenylpropanoids, flavonoids, and anthocyanins. Gene expression was verified to better explain the formation mechanism of *R. liliiflorum* colour.

2. Materials and Methods

2.1. Plant Materials

Wild *R. liliiflorum* (Figure 1a) was picked in the Baili Rhododendron Nature Reserve (above sea level: 1060~2200 m; precipitation: 1.6 °C; mean temperature of air: 1180.8 mm; N 27°12′54″, E 105°55′5″) in northwestern Guizhou Province and was subsequently planted in the nursery of Guizhou Normal University (above sea level: 1100 m; precipitation: 24 °C; mean temperature of air: 1129.5 mm; N 26°35′18″, E 106°43′18″). In April 2021, fresh *R. liliiflorum* was collected from the nursery of Guizhou Normal University. After collection, each *R. liliiflorum* petal was divided into two parts: yellow (Y) and white (W). Subsequently, the samples were quickly placed in liquid nitrogen and frozen at −80 °C to provide experimental materials for subsequent experiments.

2.2. Experimental Method

2.2.1. Sample Preparation and Extraction

This experiment divided *R. liliiflorum* into white (W) and yellow (Y) parts, divided into two groups (MW, MY), and every group of samples had 3 biological repeats. Biological samples were freeze-dried with a vacuum freeze-dryer (Scientz-100F). Freeze-dried samples were crushed using a mixer mill (MM 400, Retsch, Shanghai, China) with a zirconia bead for 1.5 min at 30 Hz. Then, 100 mg of lyophilized powder was dissolved in 1.2 mL of a 70% methanol solution, vortexed for 30 s every 30 min 6 times in total, and placed in a refrigerator at 4 °C overnight. Following centrifugation at 12,000× *g* rpm for 10 min, the extracts were filtered (SCAA-104, 0.22 μm pore size; ANPEL, Shanghai, China) before UPLC–MS/MS analysis.

2.2.2. UPLC Conditions

The sample extracts were analysed using a UPLC–ESI–MS/MS system (UPLC, SHIMADZU Nexera X2; MS, Applied Biosystems 4500 Q TRAP). The analytical conditions were as follows: UPLC—Agilent SB-C18 column (1.8 μm, 2.1 mm × 100 mm). The mobile phase consisted of solvent A, pure water with 0.1% formic acid, and solvent B, acetonitrile with 0.1% formic acid. Sample measurements were performed with a gradient programme that employed the starting conditions of 95% A, 5% B. Within 9 min, a linear gradient to 5% A, 95% B was programmed, and a composition of 5% A, 95% B was kept for 1 min. Subsequently, a composition of 95% A and 5.0% B was adjusted within 1.1 min and kept for 2.9 min. The flow velocity was set as 0.35 mL per minute, the column oven was set to 40 °C, and the injection volume was 4 μL.

2.2.3. ESI-Q TRAP-MS/MS

LIT and triple quadrupole (QQQ) scans were acquired on a triple quadrupole-linear ion trap mass spectrometer (Q TRAP) of the AB4500 Q TRAP UPLC/MS/MS system, equipped with an ESI turbo ion spray interface operating in positive and negative ion mode, and controlled by Analyst 1.6.3 software (AB Sciex). The ESI source operation parameters were as follows: ion source, turbo spray; source temperature, 550 °C; ion spray voltage (IS), 5500 V (positive ion mode)/−4500 V (negative ion mode); ion source gas I (GSI), gas II (GSII), and curtain gas (CUR) set at 50, 60, and 25.0 psi, respectively; and collision-activated dissociation (CAD), high. Instrument tuning and mass calibration were performed with 10 and 100 μmol/L polypropylene glycol solutions in QQQ and LIT modes, respectively. QQQ scans were acquired as MRM experiments with collision gas (nitrogen) set to medium. DP and CE for individual MRM transitions were performed with further DP and CE optimization. A specific set of MRM transitions was monitored for each period according to the metabolites eluted within this period.

2.2.4. Data Quality Control and Statistics Analysis

To test the reproducibility of the samples during the extraction and detection process, we performed and obtained quality control (QC) on the samples; we then analysed the occurrence frequency of the CV (coefficient of variation) for substances smaller than the reference value using the empirical cumulative distribution function (ECDF). After showing the stability of the experimental data, we first used R software's (www.r-project.org/; accessed on 15 July 2021) built-in statistics for the two sets of samples to perform the main component analysis method (PCA). The structure compared the differences between the yellow and white parts of the *R. liliiflorum* petal and the three parallel experimental groups. Using the software R Metaboanalystr (V1.0.1) to perform orthogonal signal correction and partial two-multiplication analysis, the OPLS-DA model was established by the *R. liliiflorum* yellow and white parts (orthogonal bias minimum multiplication judgement analysis). The metabolic content data used normalized processing (UV Scaling), and the clustering heatmaps were drawn using R software's Complexheatmap (V 2.8.0).

To further find the cause of the colour differences between the white part (W) and the yellow (Y) section of *R. liliiflorum*, taking $|\log2\text{fold change}| \geq 1$ and VIP ≥ 1 as the screening conditions, the different metabolites were screened using the KEGG Compound database (http://www.kegg.jp/kegg/compound/; accessed on 17 July 2021) and then the annotated metabolites were mapped to the KEGG Pathway database (http://www.kegg.jp/kegg/pathway.html; accessed on 19 July 2021). Whereafter, using a Pearson's correlation coefficient ($|r|$) greater than 0.8 and $p < 0.05$ as the screening conditions, the difference between the yellow and white parts of the R. lili-iflorum showed the correlation between the screened material. Correlation analysis can help determine the metabolic closeness between metabolic proximities, which is conducive to further understanding the mutually regulating relationship between metabolites. Common analysis of genes and metabolic groups represents the correlation between metabolites and genes through network diagrams. Regarding Pearson's correlation coefficient ($|r|$), the closer it was to 1, the stronger the correlation. We used $|r| > 0.8$. Unigene and different metabolites to draw in the flavonoid, anthocyanin, and isoflavonoid biosyn-thesis. Finally, the pathway of flavonoid biosynthesis of differential metabolites was obtained. According to the front and back relationships of flavonoids, we used enzymes that are closely related to flavonoids as intermediaries and used in plants to synthesize the path diagram.

3. Results

3.1. Morphological Characteristics and Overall Characteristics of Metabolites

The wild *R. liliiflorum* is shown in Figure 1a. The upper base part of the *R. liliiflorum* petals was yellow and white. This experiment divided *R. liliiflorum* into the white part (MW) and yellow part (MY) (Figure 1b). Using principal component analysis (PCA), the sample was divided into three parts: *R. liliiflorum* petals, yellow and white parts, and quality control samples. The quality control samples were prepared by mixing them with sample extracts. PCA scoring graphs are shown in Figure S1A. PC1 and PC2 explained 42.00% and 21.00% of the variance in the total samples, which effectively separated the yellow and white parts. The cover inspection of the quality control (QC) samples had a good overlap, which proved that the experimental conditions were stable. In this study, we built an OPLS-DA model for the white part (MW) and yellow part (MY) of *R. liliiflorum* (Figure S1B). OPLS-DA is a diverse statistical analysis method with supervision mode recognition. It can effectively eliminate the impact, which is irrelevant to research screening the differences in differential metabolites. In this model, R^2X and R^2Y represented the interpretation rate of the X and Y matrix of the models built and Q^2 represented the predictive ability of the model. The closer the matrixes to the indicators of R^2X and R^2Y, the more reliable the model, and Q^2 was greater than 0.9, so the model was very good. From the verification diagram (Figure S1B), it is demonstrated that $Q^2 = 0.966$, $R^2X = 0.668$, and $R^2Y = 1$ for the yellow part (MY), and the white part (MW) indicated that the OPLS-DA model established by the *R. liliiflorum* was good. The prediction ability was reliable. After performing the overall process of the metabolite data analysis, the difference in the accumulation of metabolites in the yellow and white parts of *R. liliiflorum* petals was displayed through heatmaps. The results showed (Figure 1c) that the material classification of the two parts of *R. liliiflorum*, yellow and white, was concentrated in more flavonoids, followed by phenolic acids, lipids, others, amino acids, and derivatives. Figure 1d shows that the three groups of samples of the white and yellow parts of the *R. liliiflorum* petals and the yellow parts $|r|$ were greater than 0.9.

3.2. Identification and Comparison of Metabolites and Differentiated Metabolites

To better understand the changes in metabolites in the two parts of *R. liliiflorum* petals, we identified the primary metabolites and secondary metabolites in the sample through the database established by the UPLC–MS/MS. After identification, a total of 1187 types of metabolites were detected, including 339 kinds of flavonoids, 177 kinds of phenolic acids, 124 varieties of lipids, 100 amino acids and derivatives, 75 organic acids, 67 terpenoids, 58 nucleotides and derivatives, 57 kinds of alkaloids, 38 kinds of lignans and coumarins, 26 tannins, 7 kinds of quinones, and 119 types of other types (Table 1). We found that the detected flavonoid metabolites were the most abundant. Among the flavonoid metabolites were chalcone, flavanones, flavanonols, flavones, flavonols, flavonoid carbonoside, flavanols, anthocyanidins, isoflavones, and dihydroisoflavones. There were 20, 44, 13, 54, 122, 16, 23, 27, 18, and 2 species, respectively.

Table 1. Flavonoid metabolite secondary classification.

Secondary Classification of Flavonoids	Quantity	Percentage (%)
Chalcone	20	5.8%
Flavanones	44	12.9%
Flavanonols	13	3.8%
Flavones	54	15.9%
Flavonols	122	35.9%
Flavonoid carbonoside	16	4.7%
Flavanols	23	6.7%
Anthocyanidins	27	7.9%
Isoflavones	18	5.3%
Dihydroisoflavones	2	0.5%

Note: A total of 339 flavonoid metabolites were screened.

3.3. Flavonoid Differential Metabolites

With $|\text{Log2Fold Change}| \geq 1$, VIP ≥ 1 was used as a condition to screen the differential metabolites (Figure 2a). The results demonstrated that the comparison of the white part (MW) and yellow part (My) consisted of a total of 197 different metabolites, 78 types of flavonoids, and metabolites (Table S1). Of these, 82 differential metabolites were upregulated and 31 flavonoid-differentiated metabolite expressions were upregulated. There were 115 differential metabolites that were downregulated and 31 flavonoid-differentiated metabolite expressions were downregulated.

The 78 metabolites were flavanones, chalcones, flavanonols, anthocyanidins, flavones, isoflavones, flavonols, flavanols, and flavonoid carbonoside, of which there were 5, 4, 1, 12, 9, 7, 26, 8, and 6 of each type. Among the substances that were screened, some of the differences were detected in the yellow flower part of *R. liliiflorum* (Figure 2b), such as chalcone isosalipurposide-6″-O-p-coumaric acid, anthocyanidin delphinidin-3-O-rutinoside, flavone baicalein and chrysoeriol-7-O-(6″-malonyl)glucoside, isoflavone calycosin-7-O-glucoside, 6″-O-malonylgenistin, and 6-hydroxydaidzein. Flavonoid carbonoside chrysoeriol-6-C-glucoside-4′-O-glucoside, flavonol 6-C-methylquercetin-3-O-rutinoside, isorhamnetin-3-O-neohesperidoside, isorhamnetin-3-O-galactoide-7-O-rhamnoside, and isorhamnetin-3-O-glucoside-7-O-rhamnoside were also detected.

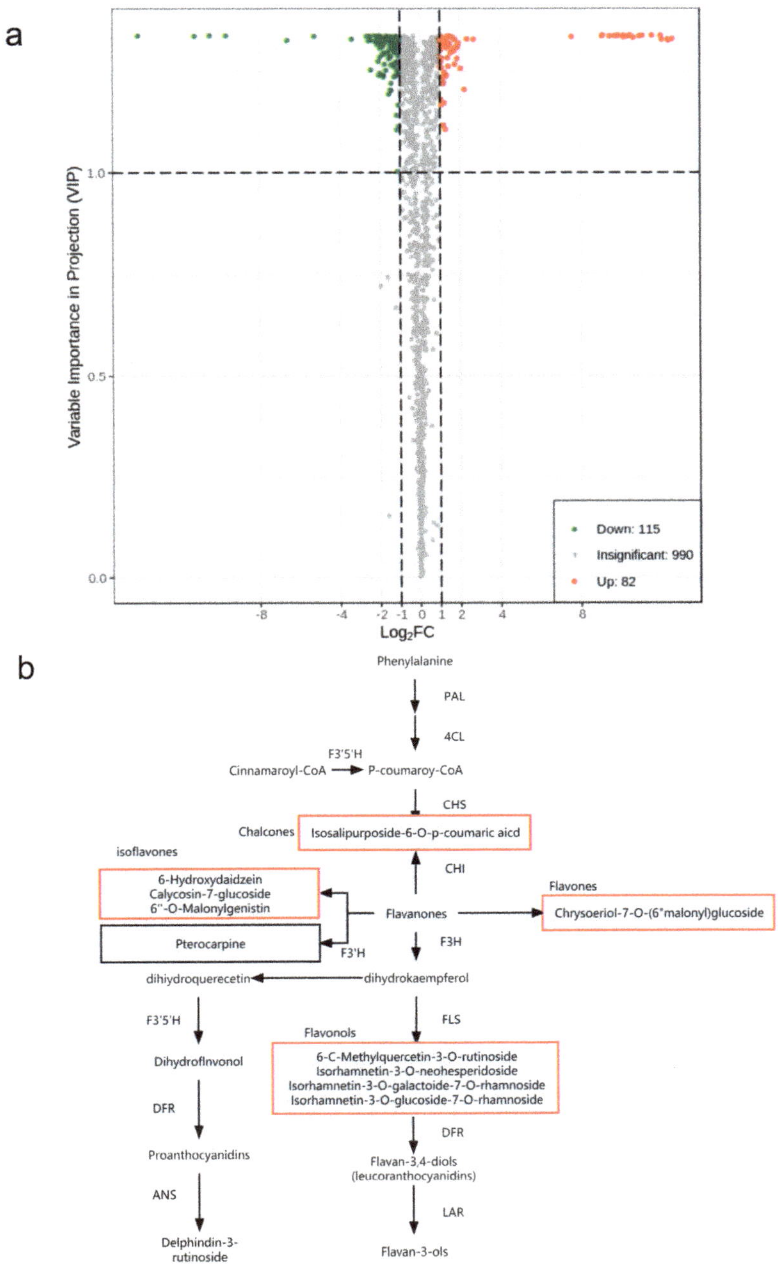

Figure 2. (**a**). Volcanic map of the differences in metabolites in *R. liliiflorum* flowers in the yellow (Y) and white (W) parts. Red dots represent the differential metabolites, green dots represent the differential metabolites, and grey represents the differences. (**b**). The flavonoid synthesis chart shows the specific differences in the white or yellow parts of *R. liliiflorum*. The substance in the black square was significantly expressed in the white part. The substance in the red square was significantly expressed in the yellow part.

3.4. KEGG Enrichment Analysis of Metabolites

First, we used |Log2Fold Change| ≥ 1 and VIP ≥ 1 to screen metabolites and match the selected differential metabolites with the KEGG database to obtain the annotation path of the differential metabolites. Subsequently, we chose the top 20 prominent pathways and then made bubble drawings (Figure 3). We found that the differences between the yellow parts of *R. liliiflorum* and the white part were mainly enriched in isoflavonoid biosynthesis, flavonoid biosynthesis, and anthocyanin biosynthesis (Table 2). Here, we found that six different metabolites were enriched in flavonoid biosynthesis, namely, with pinocembrin (dihydrochrysin) in flavanones, epiafzelechin and afzelechin (3,5,7,4′-tetrahydroxyflavan) in flavanols, fustin in flavanonols, and 5-o-caffeoyl shikimic acid and chlorogenic acid (3-O-caffeoylquinic acid) in phenolic acids. Four different metabolites were enriched in anthocyanin biosynthesis, namely, cyanidin-3-O-glucoside (kuromanin), malvidin-3-O-glucoside (oenin), cyanidin-3-O-rutinoside (keracyanin), and delphinidin-3-O-rutinoside. 6′-O-malonylgenistin, glycitin (glycitein 7-O-glucoside), and 6-hydroxydaidzein were enriched in isoflavonoid biosynthesis.

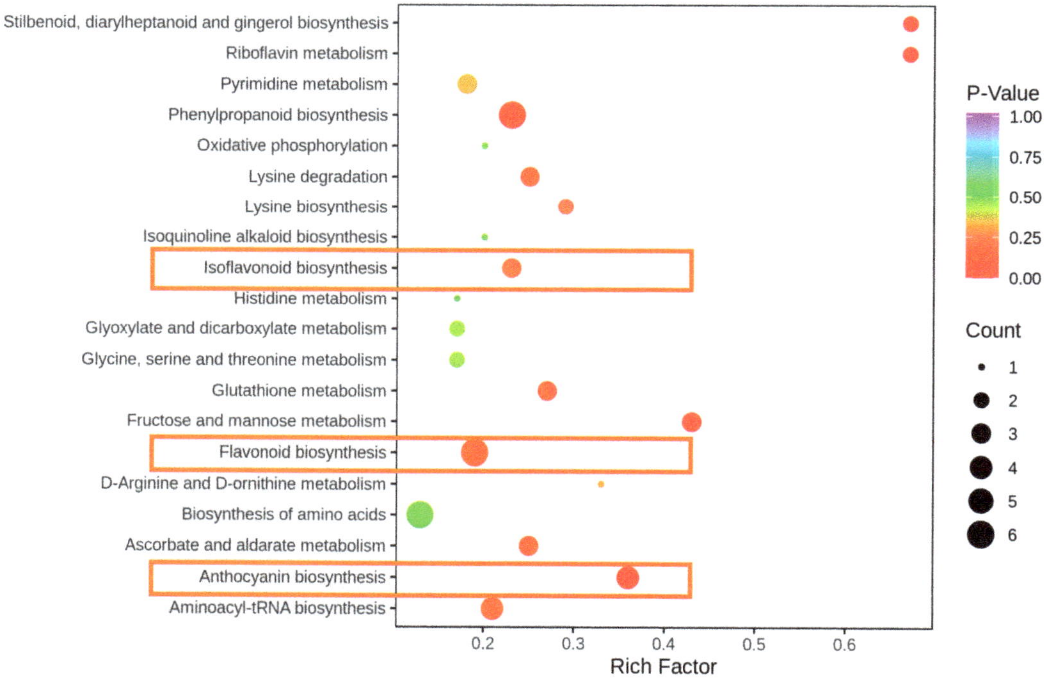

Figure 3. Kyoto Encyclopedia of Genes and Genomes (KEGG) pathway enrichment analysis of differentially expressed metabolites. The horizontal coordinate indicates the rich factor corresponding to each path. The vertical coordinate is the KEGG name. The greater the rich factor is, the greater the degree of enrichment. The larger the point, the greater the number of metabolites that are enriched in this way. The redder the colour of the dot, the more obvious the degree of enrichment. The yellow box represents the significantly enriched metabolic pathway that we are particularly interested in.

Table 2. The metabolites that are enriched in the KEGG pathway.

Formula	Compounds	Class II	Type	CAS
$C_{15}H_{12}O_4$	Pinocembrin (Dihydrochrysin)	Flavanones	down	480-39-7
$C_{15}H_{10}O_5$	6-Hydroxydaidzein	Isoflavones	up	17817-31-1
$C_{15}H_{14}O_5$	Epiafzelechin	Flavanols	up	24808-04-6
$C_{15}H_{14}O_5$	Afzelechin (3,5,7,4′-Tetrahydroxyflavan)	Flavanols	up	2545-00-8
$C_{15}H_{12}O_6$	Fustin	Flavanonols	up	20725-03-5
$C_{22}H_{22}O_{10}$	Glycitin (Glycitein 7-O-Glucoside)	Isoflavones	down	40246-10-4
$C_{21}H_{21}O_{11}+$	Cyanidin-3-O-glucoside (Kuromanin)	Anthocyanidins	down	47705-70-4
$C_{23}H_{25}O_{12}+$	Malvidin-3-O-glucoside (Oenin)	Anthocyanidins	down	18470-06-9
$C_{24}H_{22}O_{13}$	6″-O-Malonylgenistin	Isoflavones	up	51011-05-3
$C_{27}H_{31}O_{15}+$	Cyanidin-3-O-rutinoside (Keracyanin)	Anthocyanidins	up	28338-59-2
$C_{27}H_{31}O_{16}+$	Delphinidin-3-O-rutinoside	Anthocyanidins	up	15674-58-5
$C_{16}H_{16}O_8$	5-O-Caffeoylshikimic acid	Phenolic acids	down	180981-12-8
$C_{16}H_{18}O_9$	Chlorogenic acid (3-O-Caffeoylquinic acid)	Phenolic acids	down	327-97-9

Note: + represents the substance as a positive optical substance. Down represents the content of the substance in the white part (W) of *R. liliiflorum* compared to the yellow part (Y). UP indicates that the content of the species in the white part of *R. liliiflorum* (W) was lower than that in the yellow part (Y).

3.5. Flavonoid, Isoflavone, Anthocyanin Biosynthesis-Related, and Metabolite-Related Networks and Metabolite and Gene Correlation Networks

Eleven types of metabolites were filtered out, including alkaloids, amino acids and derivatives, flavonoids, lignans and coumarins, lipids, nucleotides and derivatives, organic acids, phenolic acids, quinones, tannins, and terpenoids (Figure 4). These 11 types of substances included 196 different metabolites, of which 78 types of flavonoids had different metabolites. The 78 types of flavonoids had a strong correlation. Four different metabolites were rich in flavonoid biosynthesis. Pinocembrin (dihydrochrysin) and epiafzelechin, afzelechin (3,5,7,4′-testrahydrovone), and fustin were negatively related. Pinocembrin (dihydrochrysin) decreased in the yellow part of the *R. liliiflorum* petals and epiafzelechin, afzelechin (3,5,7,4′-testrahydrovan), and fustin increased in the yellow part. The changes in correlations and contents were consistent. In anthocyanin biosynthesis, malvidin-3-O-glucoside (oenin), cyanidin-3-O-glucoside (kuromanin), delphinidin-3-O-rutinoside, and cyanidin-3-O-rutinoside (keracyanin) presented negative correlations, malvidin-3-O-glucoside (oenin) and cyanidin-3-O-glucoside (kuromanin) decreased in the yellow part of *R. liliiflorum* petals, and delphinidin-3-O-rutinoside and cyanidin-3-O-rutinoside (keracyanin) increased in the yellow part. The changes in correlations and contents were consistent. In the isoflavone synthesis pathway, the 6′-o-malonyl genistin and 6-hydroxydaidzein contents in the yellow part of *R. liliiflorum* were higher than those in the white part, and the Glycitin (glycitein 7-O-glucoside) content in the yellow part was lower than that in the white part. 6″-O-Malonylgenistin, 6-hydroxydaidzein, and glycitin (glycitein 7-O-glucoside) were negatively correlated. The changes in correlations and contents were consistent.

The correlation between metabolites and unigene is represented as a network diagram (Figure 5). The above figure shows the correlation between metabolites and genes in flavonoid biosynthesis, anthocyanin biosynthesis, and isoflavonoid biosynthesis. The metabolites pinocembrin (dihydrochrysin), chlorogenic acid (3-O-caffeoylquinic acid), epiafzelechin, and afzelechin (3,5,7,4′-tetrahydroxyflavan), as well as the genes associated with them, played a significant role in flavonoid biosynthesis. The production of differential metabolites is likely related to these genes.

Figure 4. Differential metabolites and chord diagrams. The outermost number in the figure is the difference in metabolites. The size of the middle point represents the log_2FC value. The larger the point, the larger the corresponding log_2FC value. Blue lines represent a negative correlation and red lines represent a positive correlation.

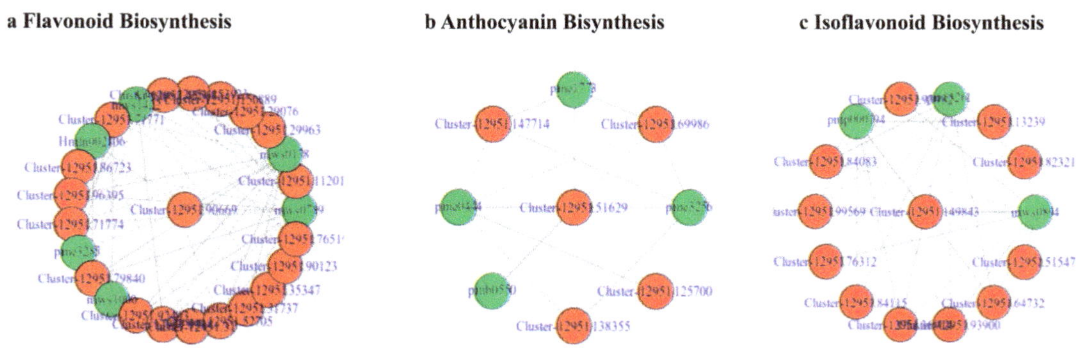

Figure 5. The correlation diagram of metabolites and genes is labelled green and the gene is marked red. The solid line represents a positive correlation and the dotted line represents a negative correlation. (**a**). Genetics and metabolites related to flavonoid biosynthesis. (**b**). Anthocyanin biosynthesis genes and metabolites. (**c**). Genes and metabolites related to isoflavonoid biosynthesis.

In the synthetic pathway of flavonoids (Figure 6), we found that pinocembrin (dihydrochrysin) in the flavanones, epiafzelechin and afzelechin (3,5,7,4′-tetrahydroxyflavan) in the flavanols, fustin in the flavanonols, malvidin-3-O-glucoside (Oenin), cyanidin-3-O-rutinoside (keracyanin) and delphinidin-3-O-rutinoside, 6″-O-malonylgenistin, and 6-hydroxydaidzein in MY content were higher than in MW content. Glycitin (GlyCitein 7-O-glucoside) and cyanidin-3-O-glucoside (kuromanin) had lower MWs than MYs.

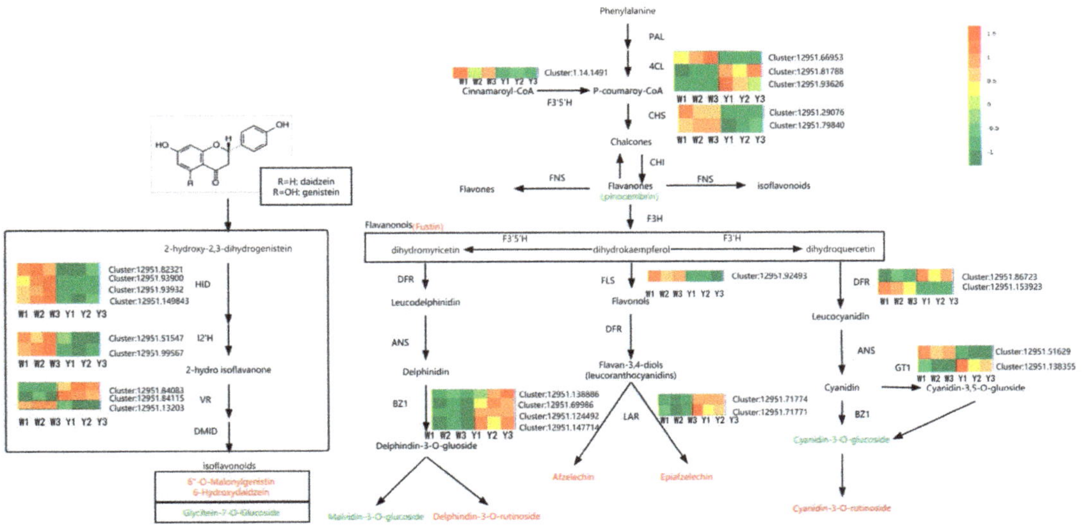

Figure 6. Significant diagram of flavonoid biological synthesis of *R. liliiflorum*. The shrinking words and expression patterns of enzymes are displayed next to each metabolic step and the synthetic direction is represented by a black arrow. The RNA-Seq expression mode of the gene is displayed in the heat map. The colour was obtained by the FPKM value converted from log2. Some data used to generate the gene heat maps are reported by [21].

In Figure 6, we found that CHS, FLS, DFR, LAR, BZ1, GT1, 4CL, HID, I2′H, VR, etc., played a role in connection. The correspondence between unigene and enzymes can be seen in Table S2. In the correlation between genes and metabolites (Figure 6), we found that the expression of these enzymes and pinocembrin (dihydrochrysin), epiafzelechin, afzelechin (3,5,7,4′-tetrahydroxyflavan), fustin, malvidin-3-O-glucoside (oenin), cyanidin-3-O-rutinoside (keracyanin), delphinidin-3-O-rutinoside, 6″-O-malonylgenistin, 6-hydroxydaidzein, cyanidin-3-O-glucoside (kuromanin), and glycitin (glycitein 7-O-glucoside) showed different positive and negative correlations. The different positive correlations presented by the enzyme unigene further verified that more than 10 metabolites were accumulated in MW and MY. The difference between the colour of the *R. liliiflorum* petal may be caused by the cumulative accumulation of more than 10 metabolites in MW and MY.

4. Discussion

Most rhododendrons are bright red, purple, etc. In the late stage of *R. liliiflorum*, the petals are composed of yellow and upper white colours at the base and contain ornamental value. By analysing the data of the metabolic group, we obtained the metabolites of *R. liliiflorum* petals and used the Kyoto genes and the genome of the encyclopedia (KEGG) database to comment on the two parts of *R. liliiflorum* yellow and white petals. Some of the differences in metabolites were significantly enriched in flavonoid biosynthesis (KO00941), anthocyanin biosynthesis (KO00942), and isoflavone biosynthesis (KO00943). In the above three biological biosynthesis pathways, only 11 different metabolites were enriched. The pinocembrin (dihydrochrysin) in flavanones, epiafzelechin and afzelechin (3,5,7,4′-tetrahydroxyflavan) in flavanols, fustin in flavanonols, cyanidin-3-O-glucoside (kuromanin), malvidin-3-O-glucoside (oenin), cyanidin-3-O-rutinoside (keracyanin) and delphinidin-3-O-rutinoside in anthocyanin biosynthesis, 6′-O-malonylgenistin, glycitein 7-O-glucoside, and 6-hydroxydaidzein were enriched in isoflavonoid biosynthesis. The above 11 substances are conditionally attributable to the colour differences caused in *R. liliiflorum*. In the previous transcription group article, we found that the genes related to CHS, F3′5′h, FLS, I2′H, HID, DFR, and LAR were key to the colour differences in *R. liliiflorum* petals [19].

In this article, we used widely targeted metabolomics to further reveal the reasons for the colour differences in *R. liliiflorum* petals.

4.1. Analysis of Metabolites and Related Enzymes Enriched in Flavonoid Biosynthesis

In synthetic biosynthesis, flavanones are very important intermediate substrates [33]. Flavanones can participate in the biosynthesis of chalcones, flavonoids, and isoflavones under the action of various enzymes [34]. Substances such as chalcones, isoflavones, and flavonoids have been reported in related documents to play a key role in the formation of yellow flowers [19,35]. Pinocembrin was detected in the skin of red and orange species in related studies of Sorghum bicolor. It was conditionally found that pinocembrin is the reason for the colour differences affecting sweet sorghum seeds [36]. Common forefronts of a flavanone compound are also a flavonol compound, anthocyanin, and procyanidins [37,38]. In this process, pinocembrin (dihydrochrysin), an important metabolite in dihydrogen flavonoids, will also convert to other metabolites [39]. The genes that show a correlation with pinocembrin are CHS, FLS, DFR, and LAR. CHS is the first rate-limiting enzyme in flavonoid biosynthesis, and FLS is a bridge between the synthesis path of flavonoids and the synthesis of catechins. The expression of its related genes will affect the biological synthesis of flavanones [40–44]. In this study, we found that CHS- and FLS-related genes were downregulated in the yellow part of *R. liliiflorum* (Figure 6). The pinocembrin (dihydrochrysin) content in the yellow petal part of *R. liliiflorum* also decreased. The most important thing was that the synthesis of the expression of CHS and FLS related to pinocembrin (dihydrochrysin) showed a positive relationship (Figure 5a).

Flavanonols are an important intermediate metabolic product and a key branch point in flavonoid biosynthesis [45]. The cumulative accumulation of flavanol compounds in different parts of the petals also causes colour differences [46]. Epiafzelechin and afzelechin (3,5,7,4′-Tetrahydroxyflavan) are flavanol compounds, and fustin is a flavananol. In Figure 6, we noticed that by participating in epiafzelechin, afzelechin mainly influenced the biological synthesis of flavonoids through CHS-, LAR-, and CHS-related enzymes and then regulated the synthesis of flavanonols and catechin synthesis downstream [40–42]. LAR-related genes form a flavanol compound by regulating leucocyanidin [47]. As the enzyme downstream of fustin, LAR participates in the regulation of fustin. The expression of LAR-related genes did not cause a decline in the content of fustin in the upstream flavanonols. This may have been due to competition between DFR and FLS for the common substrate flavanonols, as DFR and FLS suppress each other's transcription [48]. If LAR wants to convert flavanonols into a flavanol compound, first, under the action of DFR, the flavanonols are converted into leucocyanidin, and then flavanols are generated under the action of LAR [43,44]. The regulation of epiafzelechin, afzelechin, and fustin showed a positive relationship (Figure 4). This is consistent with the increase in the contents of epiafzelechin and afzelechin (3,5,7,4′-tetrahydroxyflavan) in flavanols in the yellow portion of the petals of *R. liliiflorum.*

4.2. Analysis of Metabolites and Related Enzymes Enriched in Isoflavonoid Biosynthesis

In isoflavonoid biosynthesis, the expression levels of isoflavones in yellow flowers were higher than those in white flowers [21,49]. In this study, the two isoflavone metabolites, 6″-O-malonylgenistin and 6-hydroxydaidzein, were higher in the yellow part than in the white part of *R. liliiflorum*, and only the yellow part was detected. GlyCitin (glyCitein 7-O-glucoside) in the yellow part of *R. liliiflorum* was lower than that in the white part (Figure 6). Relevant studies have shown that glyCitin (glyCitein 7-O-glucoside) will form isoflavonoid compounds such as malonylglycitin [50]. This shows that in the process of regulating glycitin (glycitein 7-O-glucoside) synthesis, I2′H and VR may transform glycitin into other isoflavones, which may cause the glycitin (glycitein 7-O-glucoside) content of the yellow part of *R. liliiflorum* to be lower than that of the white part.

4.3. Analysis of Metabolites and Related Enzymes Enriched in Anthocyanin Biosynthesis

We believe that malvidin-3-O-glucoside (oenin), delphinidin-3-O-rutinoside, cyanidin-3-O-glucoside (kuromanin), and cyanidin-3-O-rutinoside (keracyanin) accumulation in different parts of *R. liliiflorum* may be a factor that leads to its colour differences. In related studies, anthocyanins have been found to be the most important types of colourful agents in flavonoid biosynthesis [51]. Their accumulation in different parts of the petals differentiate the colour of the petals [21,24]. Cyanidin-3-O-rutinoside (keracyanin) is found in yellow cherry and apricot. A higher accumulation of related anthocyanins increases the chances that the colour will change from yellow to red [52,53]. Figure 6 shows that GT1 and BZ1 are enzymes that regulate anthocyanins, and the expression of related genes may affect the accumulation of anthocyanin compounds [54,55].

5. Conclusions

This article used widely targeted metabolomics to study the differences in unigene expression between the yellow part of *R. liliiflorum* and the white part of the *R. liliiflorum* flower cuckoo. We compared the different metabolites in the KEGG database and found numerous differences in anthocyanin biosynthesis, flavonoid biosynthesis, and isoflavone biosynthesis. Different metabolites in the enrichment of flavonoid synthesis and isoflavone synthesis pathways are flavanonols (fustin), flavanols (epiafzelechin, afzelechin), flavanones (pinocembrin), and isoflavones (6″-O-malonylgenistin, glycitin, 6-hydroxydaidzein). The differences between each part of these seven substances in the yellow part of the *R. liliiflorum* petal and the white part may be a factor that causes the colour differences between the colours of *R. liliiflorum* petals. Additionally, we also found that malvidin-3-O-galactoside (primulin), delphinidin-3-O-rutinoside, cyanidin-3-O-glucoside (kuromanin), and cyanidin-3-O-rutinoside (keracyanin) may cause colour differences between the colours of *R. liliiflorum* petals. We also found that CHS, FLS, LAR, DFR, HID, and I2′H play a key role in *R. Liliiflorum* colour formation while using differential metabolite and unigene coanalysis. This study reveals the differential metabolites and genes involved in the biosynthesis pathways of flavonoids that contribute to the colour differences between the yellow and white parts of *R. liliiflorum*. These findings provide fundamental data for subsequent research on the regulation of flower colour in *R. liliiflorum* and have implications for the horticultural applications of *R. liliiflorum*.

Supplementary Materials: The online version contains supplementary material available at https://www.mdpi.com/article/10.3390/horticulturae9121351/s1: Figure S1: The main component analysis (PCA), orthogonal signal correction, and OPLS-DA model plots; Table S1: Differential flavonoid metabolites in the white and yellow fractions of *R. liliiflorum*; Table S2: The corresponding relationship between unigene and enzyme.

Author Contributions: Conceptualization, L.L. and T.S.; data collection, Q.Z. and X.M.; data anlysis, J.D. and X.W.; article writing, J.D. and X.W.; funding acquisition, X.Z. (Xu Zhang); supervision, X.Z. (Ximin Zhang), Z.Z. and Y.Y. All authors have read and agreed to the published version of the manuscript.

Funding: This research was funded by the Guizhou Provincial Science and Technology Projects [QIANKEHEJICHU-ZK [2021] Key 038, QIANKEHEZHICHENG [2022] Key 017, QIANKEHEPINGTAIRENCAI [2017] 5726-15]; Guizhou Provincial Basic Research Program (Natural Science) [QIANKEHEJICHU-ZK [2023] Key 268]; National Science Foundation of China NSFC [32260225]; Natural Science Foundation of China and the Karst Science Research Center of Guizhou Province, China (U1812401); Higher Education Science and Research Youth Project of Guizhou Education Department (Qianjiaoji [2022]130); and Key Laboratory of Environment Friendly Management on Alpine Rhododendron Diseases and Pests of Institutions of Higher Learning in Guizhou Province, Guizhou Normal University (Qianjiaoji [2022]044). The Research Foundation for Science & Technology Innovation Team of Guizhou Province (Grant No. QJJ[2023]063).

Data Availability Statement: The data that support the findings of this study are available from the corresponding author upon reasonable request. The data are not publicly available due to no public repository for raw data is provided.

Conflicts of Interest: The authors declare no conflict of interest.

References

1. Xia, X.-M.; Yang, M.-Q.; Li, C.-L.; Huang, S.-X.; Jin, W.-T.; Shen, T.-T.; Wang, F.; Li, X.-H.; Yoichi, W.; Zhang, L.-H.; et al. Spatiotemporal Evolution of the Global Species Diversity of Rhododendron. *Mol. Biol. Evol.* **2021**, *39*, msab314. [CrossRef] [PubMed]
2. Duan, S.-G.; Hong, K.; Tang, M.; Tang, J.; Liu, L.-X.; Gao, G.-F.; Shen, Z.-J.; Zhang, X.-M.; Yi, Y. Untargeted metabolite profiling of petal blight in field-grown *Rhododendron agastum* using GC-TOF-MS and UHPLC-QTOF-MS/MS. *Phytochemistry* **2021**, *184*, 112655. [CrossRef] [PubMed]
3. De Keyser, E.; Lootens, P.; Van Bockstaele, E.; De Riek, J. Image analysis for QTL mapping of flower colour and leaf characteristics in pot azalea (*Rhododendron simsii* hybrids). *Euphytica* **2013**, *189*, 445–460. [CrossRef]
4. Park, C.; Yeo, H.; Kim, N.; Park, Y.; Park, S.-Y.; Kim, J.; Park, S. Metabolomic Profiling of the White, Violet, and Red Flowers of *Rhododendron schlippenbachii* Maxim. *Molecules* **2018**, *23*, 827. [CrossRef] [PubMed]
5. Zhiliang, L.; Qiaofeng, Y.; Xue, D.; Yu, Z.; Shuang, Z.; Wenying, Z.; Shuzhen, W. Transcriptome analysis of flower color variation in five *Rhododendron species* (*Ericaceae*). *Rev. Bras. Bot.* **2021**, *44*, 685–695. [CrossRef]
6. Du, H.; Lai, L.; Wang, F.; Sun, W.; Zhang, L.; Li, X.; Zheng, Y. Characterisation of flower colouration in 30 Rhododendron species via anthocyanin and flavonol identification and quantitative traits. *Plant Biol.* **2018**, *20*, 121–129. [CrossRef]
7. Ruxton, G.D.; Schaefer, H.M. Floral colour change as a potential signal to pollinators. *Curr. Opin. Plant Biol.* **2016**, *32*, 96–100. [CrossRef]
8. Morote, L.; Rubio-Moraga, Á.; López-Jiménez, A.J.; Argandoña, J.; Niza, E.; Ahrazem, O.; Gómez-Gómez, L. A carotenoid cleavage dioxygenase 4 from *Paulownia tomentosa* determines visual and aroma signals in flowers. *Plant Sci.* **2023**, *329*, 111609. [CrossRef]
9. Kumi, Y.; Yuki, T.-K.; Kiyoshi, K.; Tadao, K. Sepal color variation of Hydrangea macrophylla and vacuolar pH measured with a proton-selective microelectrode. *Plant Cell Physiol.* **2003**, *44*, 262–268. [CrossRef]
10. Yoshikazu, T.; Nobuhiro, S.; Akemi, O. Biosynthesis of plant pigments: Anthocyanins, betalains and carotenoids. *Plant J.* **2008**, *54*, 733–749. [CrossRef]
11. Wang, T.-y.; Li, Q.; Bi, K.-s. Bioactive flavonoids in medicinal plants: Structure, activity and biological fate. *Asian J. Pharm. Sci.* **2018**, *13*, 12–23. [CrossRef] [PubMed]
12. Zhuang, W.; Li, Y.; Shu, X.; Pu, Y.; Wang, X.; Wang, T.; Wang, Z. The Classification, Molecular Structure and Biological Biosynthesis of Flavonoids, and Their Roles in Biotic and Abiotic Stresses. *Molecules* **2023**, *28*, 3599. [CrossRef] [PubMed]
13. Wu, J.; Lv, S.; Zhao, L.; Gao, T.; Yu, C.; Hu, J.; Ma, F. Advances in the study of the function and mechanism of the action of flavonoids in plants under environmental stresses. *Planta* **2023**, *257*, 108. [CrossRef] [PubMed]
14. Wan, H.; Yu, C.; Han, Y.; Guo, X.; Luo, L.; Pan, H.; Zheng, T.; Wang, J.; Cheng, T.; Zhang, Q. Determination of flavonoids and carotenoids and their contributions to various colors of rose cultivars (*Rosa* spp.). *Front. Plant Sci.* **2019**, *10*, 123. [CrossRef] [PubMed]
15. Chen, M.; Chang, C.; Li, H.; Huang, L.; Zhou, Z.; Zhu, J.; Liu, D. Metabolome Analysis Reveals Flavonoid Changes During Leaf Colour Transition in *Populus × euramericana* 'Zhonghuahongye'. *Front. Plant Sci.* **2023**, *14*, 1595. [CrossRef]
16. Wang, Z.; Cui, Y.; Vainstein, A.; Chen, S.; Ma, H. Regulation of fig (*Ficus carica* L.) fruit color: Metabolomic and transcriptomic analyses of the flavonoid biosynthetic pathway. *Front. Plant Sci.* **2017**, *8*, 1990. [CrossRef] [PubMed]
17. Lou, Q.; Liu, Y.; Qi, Y.; Jiao, S.; Tian, F.; Jiang, L.; Wang, Y. Transcriptome sequencing and metabolite analysis reveals the role of delphinidin metabolism in flower colour in grape hyacinth. *J. Exp. Bot.* **2014**, *65*, 3157–3164. [CrossRef]
18. Sendri, N.; Singh, S.; Sharma, B.; Purohit, R.; Bhandari, P. Effect of co-pigments on anthocyanins of *Rhododendron arboreum* and insights into interaction mechanism. *Food Chem.* **2023**, *426*, 136571. [CrossRef]
19. Deng, J.; Chen, S.; Yin, X.; Wang, K.; Liu, Y.; Li, S.; Yang, P. Systematic qualitative and quantitative assessment of anthocyanins, flavones and flavonols in the petals of 108 lotus (*Nelumbo nucifera*) cultivars. *Food Chem.* **2013**, *139*, 307–312. [CrossRef]
20. Luo, P.; Ning, G.; Wang, Z.; Shen, Y.; Jin, H.; Li, P.; Huang, S.; Zhao, J.; Bao, M. Disequilibrium of Flavonol Synthase and Dihydroflavonol-4-Reductase Expression Associated Tightly to White vs. Red Color Flower Formation in Plants. *Front. Plant Sci.* **2015**, *6*, 1257. [CrossRef]
21. Zhang, H.; Chen, M.; Wang, X.; Dai, J.; Zhang, X.; Zhang, Z.; Zhang, X.; Tang, M.; Tang, J.; Gong, J.; et al. Transcriptome Analysis of *Rhododendron liliiflorum* H. Lév. Flower Colour Differences. *Horticulturae* **2023**, *9*, 82. [CrossRef]
22. Mizuta, D.; Ban, T.; Miyajima, I.; Nakatsuka, A.; Kobayashi, N. Comparison of flower color with anthocyanin composition patterns in evergreen azalea. *Sci. Hortic.* **2009**, *122*, 594–602. [CrossRef]
23. Apel, L.; Kammerer, D.R.; Stintzing, F.C.; Spring, O.; Roessner, U. Comparative Metabolite Profiling of Triterpenoid Saponins and Flavonoids in Flower Color Mutations of *Primula veris* L. *Int. J. Mol. Sci.* **2017**, *18*, 153. [CrossRef] [PubMed]

24. Sun, X.; He, L.; Guo, Z.; Xiao, Z.; Su, J.; Liu, X.; Zhou, H.; Li, C.; Gao, H. Comparative transcriptome analyses reveal genes related to pigmentation in the petals of a flower color variation cultivar of *Rhododendron obtusum*. *Mol. Biol. Rep.* **2022**, *49*, 2641–2653. [CrossRef]
25. Mu, H.; Jin, X.; Zhang, Y.; Lin, L. Anthocyanin Metabolite and Transcriptome Sequencing Analysis Reveals White Flowers in *Rhododendron dauricum* var. *albiflorum*. *Horticulturae* **2023**, *9*, 1071. [CrossRef]
26. Chen, X.; Tao, Y.; Ali, A.; Zhuang, Z.; Guo, D.; Guo, Q.; Riaz, A.; Zhang, H.; Xu, P.; Liao, Y.; et al. Transcriptome and Proteome Profiling of Different Colored Rice Reveals Physiological Dynamics Involved in the Flavonoid Pathway. *Int. J. Mol. Sci.* **2019**, *20*, 2463. [CrossRef]
27. Li, Y.; Li, H.; Wang, S.; Li, J.; Bacha, S.A.S.; Xu, G.; Li, J. Metabolomic and transcriptomic analyses of the flavonoid biosynthetic pathway in blueberry (*Vaccinium* spp.). *Front. Plant Sci.* **2023**, *14*, 1082245. [CrossRef]
28. Pu, Y.; Wang, C.; Jiang, Y.; Wang, X.; Ai, Y.; Zhuang, W. Metabolic profiling and transcriptome analysis provide insights into the accumulation of flavonoids in chayote fruit during storage. *Front. Nutr.* **2023**, *10*, 1029745. [CrossRef]
29. Ai, Y.; Zheng, Q.-D.; Wang, M.-J.; Xiong, L.-W.; Li, P.; Guo, L.-T.; Wang, M.-Y.; Peng, D.-H.; Lan, S.-R.; Liu, Z.-J. Molecular mechanism of different flower color formation of *Cymbidium ensifolium*. *Plant Mol. Biol.* **2023**, *113*, 193–204. [CrossRef]
30. Guan, L.; Liu, J.; Wang, R.; Mu, Y.; Sun, T.; Wang, L.; Zhao, Y.; Zhu, N.; Ji, X.; Lu, Y.; et al. Metabolome and Transcriptome Analyses Reveal Flower Color Differentiation Mechanisms in Various *Sophora japonica* L. Petal Types. *Biology* **2023**, *12*, 1466. [CrossRef]
31. Wang, X.; Zhang, X.; Hou, H.; Ma, X.; Sun, S.; Wang, H.; Kong, L. Metabolomics and gene expression analysis reveal the accumulation patterns of phenylpropanoids and flavonoids in different colored-grain wheats (*Triticum aestivum* L.). *Food Res. Int.* **2020**, *138*, 109711. [CrossRef] [PubMed]
32. Jiang, T.; Zhang, M.; Wen, C.; Xie, X.; Tian, W.; Wen, S.; Lu, R.; Liu, L. Integrated metabolomic and transcriptomic analysis of the anthocyanin regulatory networks in *Salvia miltiorrhiza* Bge. flowers. *BMC Plant Biol.* **2020**, *20*, 349. [CrossRef] [PubMed]
33. Casas, M.I.; Falcone-Ferreyra, M.L.; Jiang, N.; Mejía-Guerra, M.K.; Rodríguez, E.; Wilson, T.; Engelmeier, J.; Casati, P.; Grotewold, E. Identification and Characterization of Maize salmon silks Genes Involved in Insecticidal Maysin Biosynthesis. *Plant Cell* **2016**, *28*, 1297–1309. [CrossRef] [PubMed]
34. Gebhardt, Y.; Witte, S.; Forkmann, G.; Lukačin, R.; Matern, U.; Martens, S. Molecular evolution of flavonoid dioxygenases in the family Apiaceae. *Phytochemistry* **2005**, *66*, 1273–1284. [CrossRef] [PubMed]
35. Lin, Z.; Yan, W.; Zhenhua, P. Advances in study on formation mechanism and genetic engineering of yellow flowers. *Sci. Silvae Sin.* **2009**, *45*, 111–119.
36. Zhou, Y.; Lv, J.; Yu, Z.; Wang, Z.; Li, Y.; Li, M.; Deng, Z.; Xu, Q.; Cui, F.; Zhou, W. Integrated metabolomics and transcriptomic analysis of the flavonoid regulatory networks in Sorghum bicolor seeds. *BMC Genom.* **2022**, *23*, 619. [CrossRef] [PubMed]
37. Hondo, K.; Sukhumpinij, P.; Kakihara, F. Flower color and pigments in yellow-flowered hybrid progeny raised from the interspecific cross *Pelargonium quinquelobatum* × white-flowered geraniums. *Sci. Hortic.* **2015**, *195*, 145–153. [CrossRef]
38. Tu, Y.; Liu, F.; Guo, D.; Fan, L.; Zhu, Z.; Xue, Y.; Gao, Y.; Guo, M. Molecular characterization of flavanone 3-hydroxylase gene and flavonoid accumulation in two chemotyped safflower lines in response to methyl jasmonate stimulation. *BMC Plant Biol.* **2016**, *16*, 132. [CrossRef]
39. Zhao, Q.; Zhang, Y.; Wang, G.; Hill, L.; Weng, J.-K.; Chen, X.-Y.; Xue, H.; Martin, C. A specialized flavone biosynthetic pathway has evolved in the medicinal plant, *Scutellaria baicalensis*. *Sci. Adv.* **2016**, *2*, e1501780. [CrossRef]
40. Deng, X.; Bashandy, H.; Ainasoja, M.; Kontturi, J.; Pietiäinen, M.; Laitinen, R.A.E.; Albert, V.A.; Valkonen, J.P.T.; Elomaa, P.; Teeri, T.H. Functional diversification of duplicated chalcone synthase genes in anthocyanin biosynthesis of *Gerbera hybrida*. *New Phytol.* **2013**, *201*, 1469–1483. [CrossRef]
41. Zhang, X.; Abrahan, C.; Colquhoun, T.A.; Liu, C.-J. A Proteolytic Regulator Controlling Chalcone Synthase Stability and Flavonoid Biosynthesis in Arabidopsis. *Plant Cell* **2017**, *29*, 1157–1174. [CrossRef] [PubMed]
42. Schijlen, E.G.W.M.; de Vos, C.H.R.; Martens, S.; Jonker, H.H.; Rosin, F.M.; Molthoff, J.W.; Tikunov, Y.M.; Angenent, G.C.; van Tunen, A.J.; Bovy, A.G. RNA interference silencing of chalcone synthase, the first step in the flavonoid biosynthesis pathway, leads to parthenocarpic tomato fruits. *Plant Physiol.* **2007**, *144*, 1520–1530. [CrossRef] [PubMed]
43. Xu, F.; Li, L.; Zhang, W.; Cheng, H.; Sun, N.; Cheng, S.; Wang, Y. Isolation, characterization, and function analysis of a flavonol synthase gene from Ginkgo biloba. *Mol. Biol. Rep.* **2011**, *39*, 2285–2296. [CrossRef] [PubMed]
44. Forkmann, G.; Martens, S. Metabolic engineering and applications of flavonoids. *Curr. Opin. Biotechnol.* **2001**, *12*, 155–160. [CrossRef] [PubMed]
45. Liu, W.; Feng, Y.; Yu, S.; Fan, Z.; Li, X.; Li, J.; Yin, H. The Flavonoid Biosynthesis Network in Plants. *Int. J. Mol. Sci.* **2021**, *22*, 12824. [CrossRef] [PubMed]
46. Walliser, B.; Lucaciu, C.R.; Molitor, C.; Marinovic, S.; Nitarska, D.A.; Aktaş, D.; Rattei, T.; Kampatsikas, I.; Stich, K.; Haselmair-Gosch, C.; et al. *Dahlia variabilis* cultivar 'Seattle' as a model plant for anthochlor biosynthesis. *Plant Physiol. Biochem.* **2021**, *159*, 193–201. [CrossRef] [PubMed]
47. Wang, L.; Jiang, Y.; Yuan, L.; Lu, W.; Yang, L.; Karim, A.; Luo, K. Isolation and characterization of cDNAs encoding leucoanthocyanidin reductase and anthocyanidin reductase from *Populus trichocarpa*. *PLoS ONE* **2013**, *8*, e64664. [CrossRef]
48. Solfanelli, C.; Poggi, A.; Loreti, E.; Alpi, A.; Perata, P. Sucrose-specific induction of the anthocyanin biosynthetic pathway in Arabidopsis. *Plant Physiol.* **2005**, *140*, 637–646. [CrossRef]

49. Wang, R.; Ren, C.; Dong, S.; Chen, C.; Xian, B.; Wu, Q.; Wang, J.; Pei, J.; Chen, J. Integrated metabolomics and transcriptome analysis of flavonoid biosynthesis in safflower (*Carthamus tinctorius* L.) with different colors. *Front. Plant Sci.* **2021**, *12*, 712038. [CrossRef]
50. Arun Kumaran Anguraj, V.; Tim, M.; Justin, B.R.; Sangeeta, D. A combinatorial action of GmMYB176 and GmbZIP5 controls isoflavonoid biosynthesis in soybean (*Glycine max*). *Commun. Biol.* **2021**, *19*, 356. [CrossRef]
51. Stintzing, F.C.; Carle, R. Functional properties of anthocyanins and betalains in plants, food, and in human nutrition. *Trends Food Sci. Technol.* **2004**, *15*, 19–38. [CrossRef]
52. Wang, Y.; Wang, Z.Y.; Zhang, J.; Liu, Z.S.; Wang, H.; Tu, H.X.; Zhou, J.T.; Luo, X.R.; Chen, Q.; He, W.; et al. Integrated Transcriptome and Metabolome Analyses Provide Insights into the Coloring Mechanism of Dark-red and Yellow Fruits in Chinese Cherry [*Cerasus pseudocerasus* (Lindl.) G. Don]. *Int. J. Mol. Sci.* **2023**, *24*, 3471. [CrossRef] [PubMed]
53. Xi, W.P.; Feng, J.; Liu, Y.; Zhang, S.K.; Zhao, G.H. The R2R3-MYB transcription factor PaMYB10 is involved in anthocyanin biosynthesis in apricots and determines red blushed skin. *BMC Plant Biol.* **2019**, *19*, 287. [CrossRef] [PubMed]
54. Ogata, J.; Kanno, Y.; Itoh, Y.; Tsugawa, H.; Suzuki, M. Plant biochemistry: Anthocyanin biosynthesis in roses. *Nature* **2005**, *435*, 757–758. [CrossRef]
55. Offen, W.; Martinez-Fleites, C.; Yang, M.; Kiat-Lim, E.; Davis, B.G.; Tarling, C.A.; Ford, C.M.; Bowles, D.J.; Davies, G.J. Structure of a flavonoid glucosyltransferase reveals the basis for plant natural product modification. *EMBO J.* **2006**, *25*, 1396–1405. [CrossRef]

horticulturae

MDPI

Article

The Variations of C/N/P Stoichiometry, Endogenous Hormones, and Non-Structural Carbohydrate Contents in *Micheliamaudiae* 'Rubicunda' Flower at Five Development Stages

Ting Yu [1], Yao Yang [2], Hongrui Wang [2], Wenzhang Qian [2], Yunyi Hu [2], Shun Gao [2,*] and Hai Liao [1,*]

1 School of Life Science and Engineering, Southwest Jiaotong University, Chengdu 611756, China; yuting212916@my.swjtu.edu.cn

2 Department of Forestry, Faculty of Forestry, Sichuan Agricultural University, Chengdu 611130, China; yangyao@stu.sicau.edu.cn (Y.Y.); 201905757@stu.sicau.edu.cn (H.W.); 202001247@stu.sicau.edu.cn (W.Q.); huyunyi@stu.sicau.edu.cn (Y.H.)

* Correspondence: shungao@sicau.edu.cn (S.G.); ddliaohai@home.swjtu.edu.cn (H.L.)

Abstract: *Michelia maudiae* 'Rubicunda' (*M. maudiae* 'Rubicunda') is one of the most popular ornamental plants. However, relatively little is known regarding its floral development. Here, the variations of the mineral, endogenous hormone, and non-structural carbohydrate (NSC) contents in the petals and gynoecium and androecium (GA) at five developmental stages during *M. maudiae* 'Rubicunda' flower development were analyzed. The results suggested that the carbon (C), nitrogen (N), and phosphorus (P) endogenous hormones, NSC contents, and C/N/P stoichiometric ratios exhibit large variations during flower development. There were significant differences in N and P contents in the GA and petals among the five growth stages, while C contents did not change significantly. In the five flower development stages, the average N and P contents in the GA were higher than those in the petals. The maximum C/N and N/P ratios in the GA and petals were foundat the senescent flower stage (stage5) and green bud stage (stage1), respectively. The C/P ratio in petals reached its maximum value at the mature bud stage (stage 2), and the C/P ratio in the GA reached its maximum value at the senescent flower stage (stage 5). The C/N/P stoichiometric ratios in the petals were more sensitive to development stages than those in the GA. There were highly significant correlations between the NSC and C/N/P stoichiometric ratios in the GAand petals. Moreover, acetic acid (IAA), abscisic acid (ABA), gibberellic acid 3 (GA3), and cytokinin (CTK) contents in the petals exhibited significant changes in response to development stages. Principal component analysis (PCA) revealed significant correlations and clear differences in the test indexes between the development stage and organs, and the variation was explained by PC-1 (55.6%), PC-2 (23.3%), and the cumulative contribution percentage of the total biplot (78.9%). These studies can lay the foundation for elucidating the requirements and dynamic balance among C, N, P, NSC, and hormone contents during the flower development of *M. maudiae*' Rubicunda'.

Keywords: *M. maudiae* 'Rubicunda'; flower development; non-structural carbohydrates; stoichiometry; phenological growth stages

Citation: Yu, T.; Yang, Y.; Wang, H.; Qian, W.; Hu, Y.; Gao, S.; Liao, H. The Variations of C/N/P Stoichiometry, Endogenous Hormones, and Non-Structural Carbohydrate Contents in *Micheliamaudiae* 'Rubicunda' Flower at Five Development Stages. *Horticulturae* **2023**, *9*, 1198. https://doi.org/10.3390/horticulturae9111198

Academic Editor: Jinzhi Zhang

Received: 13 September 2023
Revised: 18 October 2023
Accepted: 23 October 2023
Published: 3 November 2023

1. Introduction

Flowers are reproductiveand temporal organs in plants and are also highly prized objects of beauty with commercial and industrial value [1]. Flower development is a highly complex process characterized by two distinct physiological stages: bud initiation and floral bud development [2]. These two stages include growth, development, senescence, and abscission, which are events that can be affected by internal and external factors. The internal factors includeminerals, endogenous hormones, carbohydrates, and transcription factors, while the external factors includetemperature, water status, light, etc. [3,4]. Thus, understanding flower development is very important for ensuring successful production and improving orchard management, which would help enhance the visual quality and vase life of flowers, thus increasing their commercial value.

Carbon (C), nitrogen (N), and phosphorus (P) are essential components for plant growth and plant metabolism, including flower development [5,6]. C, N, and P contents, as well as C/N/P stoichiometry, are the most investigated nutrient relationships in the study of plants because these elements often limit organism growth andinevitably affect the growth rate and primary production of plants [7]. To date, several studies have been conducted on a wide-scale basis to explain many phenomena at all levels of biology, from genes and molecules to whole organisms and even to ecosystems, which largely reflect the vegetation composition, dynamics, and nutrient limitation of plants. The C/N/P stoichiometry of terrestrial plants can reflect how plants adjust themselvesto adapt to growth conditions [5,6]. Recently, many researchers have reported the distribution of mineral nutrients and some metabolites in many plant species [8], and significantly changed mineral nutrients and metaboliteswere also observedduring the flower development of *Hibiscus rosa-sinensis* and *Camellia sinensis* [9,10]. Moreover, some measurements have already been recorded on the spatiotemporal variations, biological regulation mechanisms, and ecological implications of C/N/P stoichiometric ratios during flower development [11,12].These studies help to offer beneficial information on the distribution and changes in nutrient contents and stoichiometry, as well as their relationships during flower development.

Non-structural carbohydrates (NSCs) are important energy sources and signaling and regulatory molecules for plant growth, reproduction, and metabolism. NSC contents and composition are measures of the relationship between plant carbon uptake (photosynthesis) and carbon consumption (growth and respiration) in plants [13,14]. Carbohydrates perform various functions during flower development, such as acting as energy sources, osmotic regulators, and precursors of metabolic processes. Thus, the lack of carbohydrates will lead to undersized petal development or stop flower development [10,15,16]. During bud break, high carbon demand may exceed the carbon supplied by current photosynthesis in evergreen species that also partly depend on the NSC pool [17,18]. Earlier studies showed that the changes in glucose and fructose content are associated with inflorescence development, differentiation, blooming, and color formation in different plant species, including *Asiatic lilies* [19], *Campanula rapunculoides* [20], *Borago officinalis*, and *Centaurea cyanus* [21]. Thus, there are complex changes in carbohydrates during flower development, which have an important rolein the nutritional composition, taste, texture, and other indicators of flowers. Moreover, flower development is affected by a complex network of genes that integrate multiple endogenous signals, thus ensuring that flowering occurs at the right time. Hormones, their content, and ratios are involved in the formation of flowers [2,9]. Accumulated studies have shown that flower development showed a positive and/or negative correlation with the contents of endogenous hormones—for example, acetic acid (IAA), abscisic acid (ABA), gibberellic acid 3 (GA3), and cytokinin (CTK) [22,23]. It has been shown that ABAand GAs are crucial regulators for plant growth, and these play an important role in flower development, affecting floral transition [9,23]. From the results obtained, it was clear that changes in carbohydrate and endogenous hormone content play an important role during flower development.

Michelia maudiae 'Rubicunda' (*M. maudiae* 'Rubicunda'), a member of the Magnoliaceae family, is mainly cultivated in subtropical China. It is a very ornamental and aromatic plant due to its long flowering period anddistinctive flower colors (Figures 1 and 2). The tree can reach a height of 3–8 m with extended branches and can blossom 4–5 years after planting. The flowering period is fromEarlyFebruary to LateApril, and the fruiting period is from September to October [24,25]. Its flower color is purplish red, which differs from the original subspecies of the Magnoliaceae family. It is also an excellent plant resource for studying flower color gene regulation and for breeding new varieties of Magnoliaceae. Apart from these remarkable features, the flowercontains proteins, minerals, carbohydrates, vitamins, amino acids, essential oils, etc. It has alsobeen used to prepare various traditional foods such as cakes, casseroles, herbal teas, and drinks [26]. However, studies on flower development in *M. maudiae* 'Rubicunda' are rather limited. Thus, the main objective of our study was toclarifythe changes in C, N, and P, endogenous hormones, non-structural carbohydrates (NSCs) content, and C/N/P stoichiometry in the gynoecium and androecium (GA) and petals of *M. maudiae* 'Rubicunda' flower at five developmental stages.

Figure 1. Morphological changes in flower development in *M. maudiae*' Rubicunda' at five development stages. (**A**) Whole flower, (**B**) Petals, (**C**) Gynoecium, and androecium (GA). S1: green bud stage/flower bud differentiation (stage 1). S2: mature flower budstage/floral bud elongation (stage 2). S3: early bloomingstage(stage 3). S4: fully opened flower stage (stage 4). S5: senescent flowerstage (stage 5). Scale bar = 1 cm.

Figure 2. The general look of *M. maudiae* 'Rubicunda' tree (**A**) and its flower development (**B**). Scale bar = 1 cm. a (Stage 50), the first tender green flower buds are visible on the main stem. b (Stage 52), the flower bud breaks through the first layer of bracts; the outer bracts are light brown, and the inner flower buds are green. c (Stage 54), flower buds and peduncles are elongated, and the first bract shedding marks are clearly visible. d (Stage 57), pedicels continue to elongate; second bract shedding mark is clearly visible. e (Stage 58), flower buds break through the third bract, purple petals visible. f (Stage 59), the third layer of bracts falls off, the whole petal closes, and flower bud begins to open. The first flower will blossom the next day. g (Stage 62), the petals are purple-red, and the stamens are close to the pistils. h (Stage 65), the petals are erect, the upper part is whitish, the lower part is purple-red, and the stamens are slightly open. i (Stage 67), the petals wither, the stamens fully unravel, and the flower begins to decay. j (Stage 69), end of flowering, and fruit set visible.

2. Materials and Methods

2.1. Study Site

The experimental site is located in Xie-yuan Town, Da-yi County, Chengdu, China (30°37′ N, 103°20′ E, altitude: ≈771 m). The regionhas a classic subtropical monsoon climate with an average temperature of 15.1 °C, an annual rainfall of 1095.5 mm, an average relative air humidity of 83%, a frost-free period of 284 days, and total annual sunshine of 1076.5 h. The weather data were obtained from the China Meteorological Administration.

2.2. Flowers Collection and Parameters Measurement

Ten six-year-old *M. maudiae* 'Rubicunda' trees with similar height were marked for flower harvesting. The dates of the beginning and end of flowering were recorded for each tree from February to April 2023. The fresh flowers were harvested in the early morning (08:00−11:00am). Referring to the Biologische Bundesantalt, Bundes-sortenamt, and Chemische Industrie(BBCH) scales [27,28], *M. maudiae* 'Rubicunda' flower development is divided into five stages, including green bud stage (S1), mature flower budstage (S2), early bloomingstage (S3), fully opened flower stage (S4), and senescent flowerstage(S5) (Figures 1A and 2B). At each stage, buds and flowersfrom ten trees were randomly selected, numbered, photographed, and measured. Each flower at five developmental stages was separated as petal and gynoecium and androecium (GA) (Figure 1B,C) and immediately weighed. Samples weredivided into two parts, one of which was oven-dried at 65 °C for 24 h, weighed, ground, and sieved for chemical analysis. The other was immediately frozen in liquid nitrogen and then stored at −80 °C for analysis of endogenous hormones.

2.3. Determination of C, N, and P Contents

A total of 500mg of sample was added into polytetrafluoroethylene (PTFE, Teflon®) vessels, and then 10 mL of concentrated HNO_3 and 2 mL of H_2O_2 were added. The samples were mixed and digestedcompletely, and 1 mL of digested solution was adjusted up to 25 mL with Milli-Q deionized water for further analysis. The total C content (g/Kg) was determined by the $H_2SO_4/K_2Cr_2O_7$ oxidization-FeSO4 titration method. Total N concentration (g/Kg) was analyzed by the Kjeldahl determination method. Total P content (g/Kg) was determined and analyzed using the Mo-Sb colorimetric method [29]. The concentrationwas expressed as g/Kg dry weight (DW). The C/N, N/P, and C/P ratios were calculated from the content ratio.

2.4. Determination of NSC Content

Soluble sugar and starch contents were determined according to theanthrone colorimetric method [30]. A 0.1 g sample wasmixed with 10 mL of 80% ethanol and incubated in a boiling water bath for 10 min. The extracts were centrifuged at 2380× *g* for 10 min, and the supernatant was collected. Theaforementioned process was performed intriplicate to ensure complete extraction. Then, the precipitate was further resuspended in 30% perchloric acid (10 mL), and the samples sat undisturbed for 12 h. Then, the samples wereplaced in an 80 °C water bath for 10 min and were centrifuged at 5000 rpm for 10 min.The supernatant was diluted to achieve a 50 mL constant volume for measuring starch contents. For glucose and starch measurements, 0.1 mL of soluble sugar extraction and 5 mL of anthrone reagent were mixed, and the solution was carried out in a 90 °C water bath for 15 min. For sucrose analysis, 0.1 mL of sugar extraction and 0.1 mL of 7.6 mol/L KOH solution were mixed and then incubated in a boiling water bath for 15 min. After cooling to room temperature, 5 mL of anthrone solution was added, and mixed, placed in a 90 °Cwater bath for 15 minuntil analysis. For fructose analysis, 0.1 mL of sugar extract and 5 mL of anthrone solution were mixed and kept for 90 min at room temperature. The absorption values at 620 nmof the above solutions were measured and were calculated according to the standard curve ofglucose, sucrose, fructose, and starch, respectively. The NSC contents were calculated as the sum ofglucose, sucrose, fructose, and starch contents, and the results were representedas mg per g dry weight (DW).

2.5. Endogenous Hormone Measurement

IAA, IBA, ABA, and GA3 were detected using enzyme-linked immunosorbent assay (ELISA) method [23]. In brief, 500 mg offresh sampleswere crushed into powder in liquid nitrogen and dissolved in 4 mL of 95% methanol. The mixture was sonicated for 30 min and kept at 4 °C overnight, and then the supernatant was collected by centrifuging at 4000 rpm for 10 min at 4 °C for the next use. This step was performed three times to ensure complete extraction. These supernatants were retained, combined, andevaporated at 40 °C until the organic phase was removed.The concentrated samples were filtered with a 0.22 µm filter membrane, and the hormone contents were measured byELISA according to the manufacturer's protocol. Every sample was measured in triplicate.

2.6. Statistical Analysis

Data were expressed as the mean ± standard deviation (SD) with at least three replications. Data were analyzed with SPSS 20.0 (SPSS, Inc., Chicago, IL, USA) and Microsoft Excel 2007 (IBM Corp., Armonk, NY, USA). One-way analyses of variance (ANOVAs) and two-way ANOVA were used to assess the effects of development stages and organs as well as their interactionson the C, N, and P contents, C/N/P stoichiometric ratios, endogenous hormones, and NSC contents. The mean values of measured parameters at five development stages were used for principal component analysis (PCA). Differences were considered significant for all statistical tests at $p \leq 0.05$. Figures were drawn using Origin 18.0 (Origin Lab., Northampton, MA, USA).

3. Results

3.1. Flower Developmentand Morphological Features

As shown in Figure 2, the flowers of *M. Maudiae* 'Rubicunda' are showy red solitary flowers with eight petals. The flowers are mostly foundon axillary leaves or rarely on the tops of branches (Figure 2A). Inflorescence emergence occurs from early to midAugust, depending on temperature and growing conditions.In the early stages of development (Figure 2B, a, stage 50), it was difficult to distinguish the floral bud or leaf bud until the air temperature gradually increased in early February of the following year. Flower differentiation resumed, and rapid development took place; the floral buds began to swell and open, the scales loosened and separated, and then the buds broke (Figure 2B, b, stage 52). As the temperature increasedtowards the middle/end of February, the flower buds developed rapidlyvia three dehiscence steps, and three rings of bract shedding marks were observed on the pedicel (Figure 2B, c–f, stages 54, 57, 58, and 59). When the bractsare completely abscised, the flowers with 7–9 petals are about to bloom. Moreover, the outer ring of petals is obovate, and the inner two rings become narrower. When the flowerfirst opens, the petals are purplish red, and the stamens are close to the pistil (Figure 2B, g, stage 62). The flowers are fully unfolded, the petals are erect, the upper part is whitish, the lower part is purplish red, and the stamens are slightly open (Figure 2B, h, stage 65). Asthe petals begin to wither, the stamens arefully spread out, and the pistils are clearly visible (Figure 2B, i, stage 67), followed by fruit development (Figure 2B, j, stage 69). During flower development, longer periods and partial/total overlap are observedbetween inflorescence development and flowering within individual trees (Figure 2A). Moreover, morphological parameters were also measured, and the results showed that flower length ranged from 38.62 mm to 71.78 mm and width ranged from 11.75 mm to 45.22 mm at five developmental stages, respectively. The length of petals ranged from 39.24 mm to 78.96 mm at five developmental stages, and the width ranged from 23.16 mm to 33.85 mm, respectively. Moreover, the relative water content (RWC) in the petals varied from 73% to 85% (Table S1).

3.2. C, N, and Pcontents

As shown in Figure 3, developmental stages and organs have significant effects on C, N, and P contents, but no significant differences were found in the interaction effects between developmental stages and organs ($p > 0.05$). The C contents in the petal ranged from 452.7 to 490.9 g/kg and 454.1 to 482.4 g/kg in the GA, respectively. The maximum C concentration in the petal and GA was found in the S4 and S5, respectively. There were no significant differences among the five development stages (Figure 3A). The N contents in the petal ranged from 10.8 to 12.7 g/kg, butthe values showed slightly decreasing trends from S1 to S5. Similar to the patterns in the petal, the N contents in the GA also displayed a decrease from S1 toS5, and the values ranged from 16.4 to 23.0 g/kg over the flower development stages. In addition, the values in the GA were significantly higher than those in the petal at five developmental stages (Figure 3B). The P contents in the petal ranged from 0.533 to 0.652 g/kg, and the values showed increasing trends from S1 to S5. However, the P contents in the GA tended to decrease from S1 to S5, and the values ranged from 0.773 to 0.848 g/kg duringflower development. Moreover, the values in the GA were significantly higher than those in the petals at five development stages (Figure 3C). These results showed that the responses of N and P contents in flowers were more sensitive to developmental stages and organs than those of C contents.

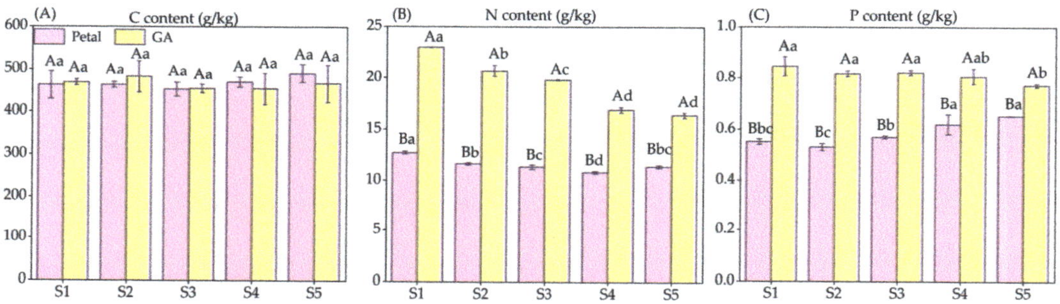

Figure 3. Changes in C (**A**), N (**B**), and P (**C**) contents atfive developmental stages of *M. maudiae*'Rubicunda' flowers. Values represent the mean ± standard deviation ($n = 3$). Different lowercase letters above bars indicatesignificant differences among stages at $p \leq 0.05$. Different uppercase letters above the bars indicate significant differences among organs at $p \leq 0.05$.

3.3. C/N/P Stoichiometry

As shown in Figure 4, developmental stages and organs significantly affected the patterns of C/N, C/P, and N/P ratios, but no significant differences in the interaction effects of developmental stage and organ were found ($p > 0.05$). The C/N ratios in the petal ranged from 36.3 to 43.3,andthose in the GA ranged from 20.4 to 28.5 at five developmental stages, respectively. The maximum C/N ratios in the petal and GA were found in S5 (Figure 4A). The C/P ratios were significantly different at five developmental stages, and the values ranged from 752.3 to 869.0in the petal and from 551.7 to 603.7in the GA, respectively (Figure 4B). The N/P ratios in the petal and GA ranged from 17.5 to 23.1 and 20.9 to 27.2 over the flower development stages, respectively. The N/P ratios in the petal tended to decrease from S1 to S5, and the N/P ratios in the GA also showed a slight decreasing trend from S1 to S5. The values of N/P ratios in the petal were significantly lower than those in the GA atfive developmental stages (Figure 4C). These observations suggested that the C/N, C/P, and N/P ratios in flowers are significantly affected by developmental stages and organs.

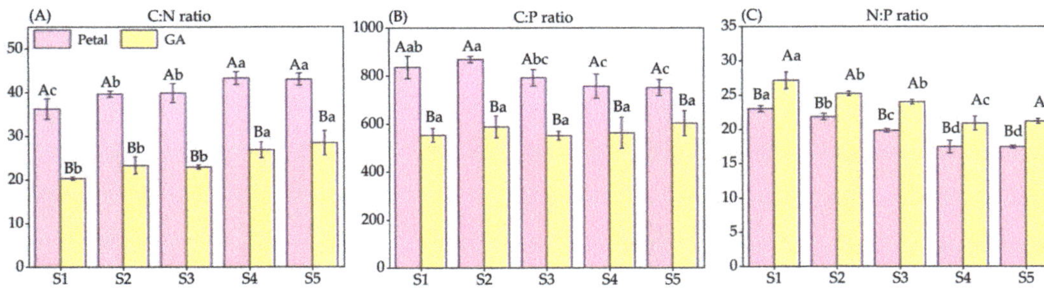

Figure 4. Changes in C/N (**A**), N/P (**B**), and C/P (**C**) ratios at five developmental stages of *M. maudiae* 'Rubicunda' flowers. Each value represents the mean ± standard deviation ($n = 3$). Different lowercase letters above bars indicate significant differences among stages at $p \leq 0.05$. Different uppercase letters above the bars indicate significant differences among organsat $p \leq 0.05$.

3.4. Non-Structural Carbohydrates (NSC) Contents

As shown in Figure 5, significant differences in the NSC contents in the petal and GA were observed at five developmental stages. The glucose contents in the petal ranged from 112.8 to 159.5 mg/g and 106.8 to 127.8 mg/g in the GA, respectively. The glucose contents in the petal had a tendency to increase from S1 to S5, but the GA glucose contents showed a slight decreasing trend from S2 to S3. The values in the petal were significantly higher than those in the GA, except for at S1 (Figure 5A). The fructose contents in the petal and GA ranged from 24.9 to 77.7 mg/g and 17.9 to 50.1 mg/g, respectively. The fructose content in the petal had a tendency to increase from S1 to S5, and the highest and the lowest values were found at S5 (77.7 mg/g DW) and S1 (24.9 mg/g DW), respectively. The GA fructose content also showed aremarkable increasing trend from S1 to S5, reaching its maximum value at S5 (50.1 mg/g DW) and its minimum value at S1(17.9 mg/g DW). The fructose content in the petal was significantly higher than that in the GA at five developmental stages (Figure 5B). The sucrose contents were significantly different at five developmental stages; the values in the petal ranged from 44.3 to 83.2 mg/g and 63.6 to 97.2mg/g in the GA, respectively. In the petal, the maximum and minimum sucrose contents were recorded at S3 and S4, respectively. In the GA, the maximum and minimum sucrose contents were found at S1 and S4, respectively (Figure 5C). The starch contents in the petal and GA had a tendency to continue to decrease fromS1 to S5 and ranged from 34.6 to 77.8 mg/g in the petal and 33.1 to 48.5 mg/g in the GA, respectively (Figure 5D). These observations showed that the glucose, sucrose, fructose, and starch contents in flowers are significantly affected by developmental stages and organs and their interactions.

3.5. Endogenous Hormones Content and Their Specific Values

As shown in Figure 6, developmental stages can significantly affect the contents of ABA, IAA, CTK, and GA3, and significant differences were foundinthe interaction effects of developmental stage and organs ($p \leq 0.05$). The IAA content in the petal gradually decreased from S1 to S5, reaching its minimum level of 1.04 nmol/g (Figure 6A). The ABA, GA3, and CTK contents reached the maximum levels of 1.92 nmol/g (Figure 6B), 0.72 nmol/g (Figure 6C), and 0.94 nmol/g (Figure 6D), respectively. As shown in Figure 7, the GA3/CTK ratio gradually decreased from S1 to S5, and the maximum value was 0.92 at S1. The ratios of GA3/IAA and CTK/IAA ranged from 0.51 to 0.62 and from 0.66 to 0.83 at five developmental stages, respectively. As shown in Figure 8, the ratios of IAA/ABA, GA3/ABA, CTK/ABA, and (IAA + GA3 + CTK)/ABA at S1 and S2 were significantly higher than those at S3, S4, and S5. The highest ratios of IAA/ABA, GA3/ABA, CTK/ABA, and (IAA + GA3 + CTK)/ABA were recorded at S2 (0.81), S1 (0.47), S2 (0.64), and S2 (1.86), respectively. The minimum ratios of IAA/ABA, GA3/ABA, CTK/ABA, and (IAA + GA3 + CTK)/ABA were observed at S3 (0.60), S3 (0.33), S2 (0.41), and S2 (1.34), respectively.

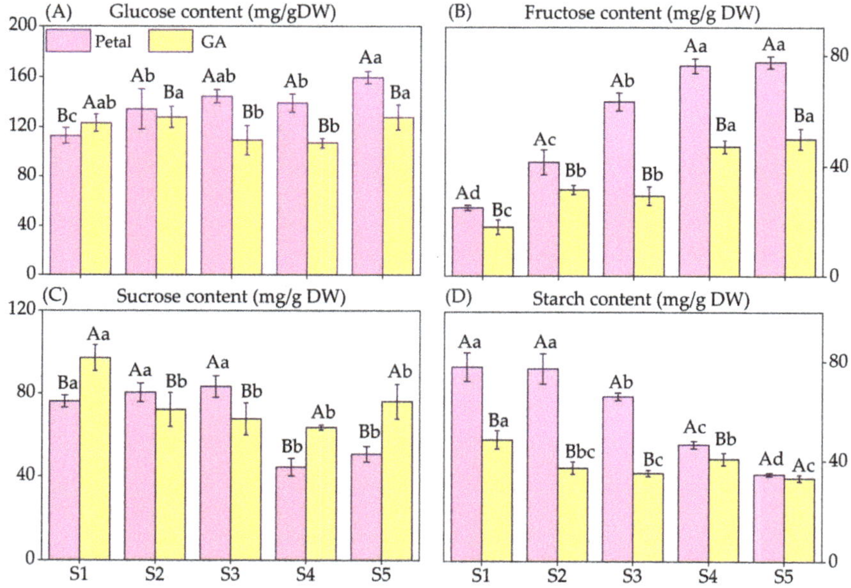

Figure 5. Changes inglucose (**A**), fructose (**B**), sucrose (**C**), and starch (**D**) contents in five developmental stages of *M. maudiae* 'Rubicunda' flowers. Each value representsthe mean ± standard deviation (n = 3). Different lowercase letters above bars denote significant differences among stages at a significance level of $p \leq 0.05$. Different uppercase letters above bars denote significant differences among organs at a significance level of $p \leq 0.05$.

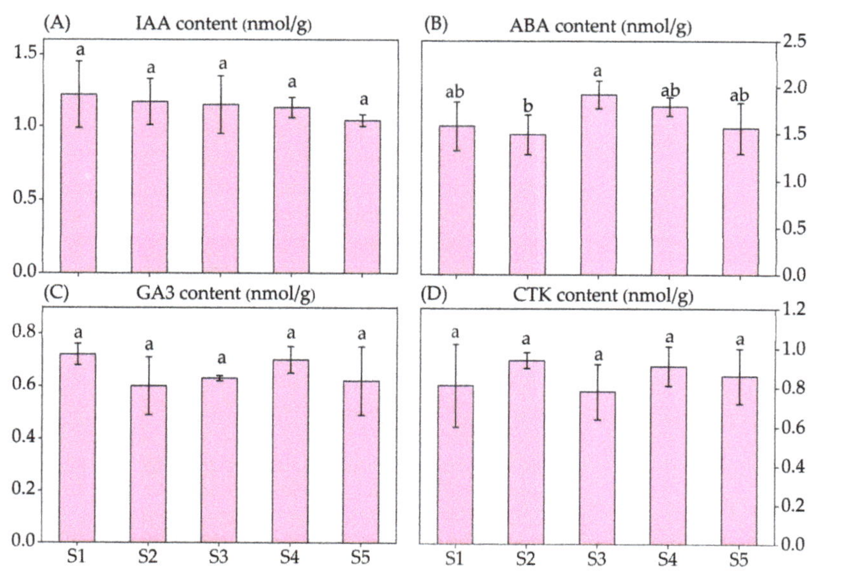

Figure 6. Changes in endogenous hormone contents in the petal during *M. maudiae* 'Rubicunda' flower development at five developmental stages. (**A–D**) represent the acetic acid (IAA), abscisic acid (ABA), gibberellic acid 3 (GA3), and cytokinin (CTK) contents, respectively. Each value represents the mean ± standard deviation (n = 3). Different lowercase letters above bars denote significant differences among stages at a significance level of $p \leq 0.05$.

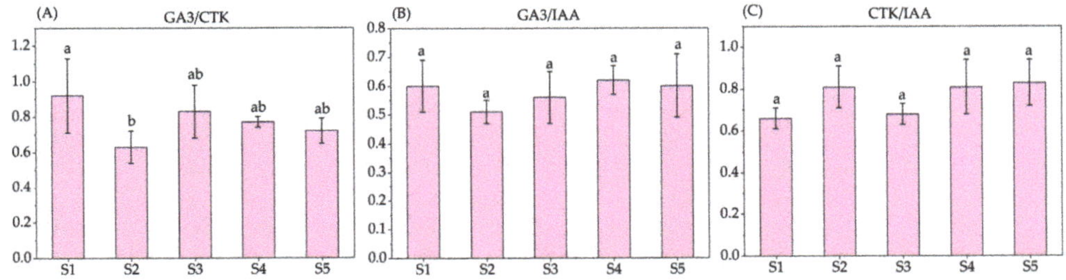

Figure 7. Changes in promoting endogenous hormone ratio in the petal of *M. maudiae* 'Rubicunda' flower at five developmental stages. Value represents the mean ± standard deviation(n = 3). (**A**) the ratios of GA3/CTK. (**B**) the ratios ofGA3/IAA. (**C**) the ratios of CTK/IAA. Different lowercase letters above bars denote significant differences among stages at a significance level of $p \leq 0.05$.

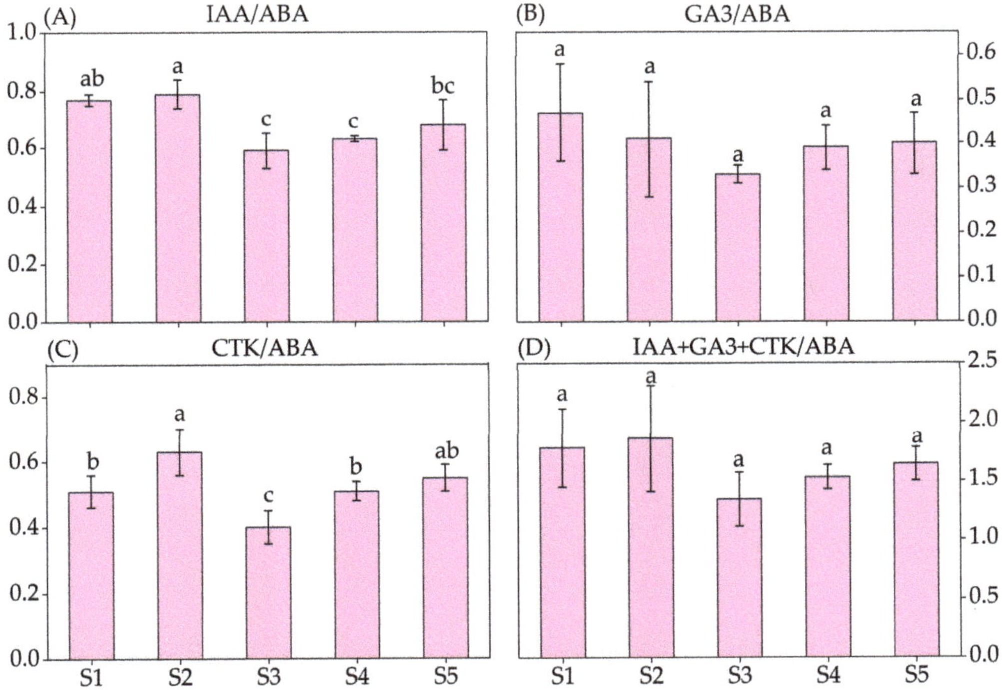

Figure 8. Changes in promoting endogenous hormones to suppress endogenous hormone ratio in the petalof *M. maudiae* 'Rubicunda' flower at five developmental stages. Value represents the mean ± standard deviation(n = 3). (**A**) the ratios of IAA/ABA. (**B**) the ratios of GA3/ABA. (**C**) the ratios of CTK/ABA. (**D**) the ratios of (IAA + GA3 + CTK)/ABA. Different lowercase letters above bars denote significant differences among stages at a significance level of $p \leq 0.05$.

3.6. Principal Component Analysis (PCA)

As shown in Figure 9, PCA results showed that significant differences ($p < 0.05$) are observed in C, N, P, and NSC contents, as well as C/N/P stoichiometric ratios at five developmental stages. Two principal components (PCs) accounted for 100.00% of the total variance, of which principal component 1 (PC1) and principal component 2 (PC2) contributed 55.6% and 23.3%, respectively. The PCA plot revealed that the variables (mineral contents, nutrient ratios, and NSC) were divided into four main clusters. Starch, C/N, and C/P were in the upper right, indicating that theywere significantly positively correlated with five development stages. N/P ratio, fructose, glucose, and sucrose also contributed significantly. N and P showed a negative relationship with other parameters (Figure 9A). These results indicated that mineral contents, C/N/P stoichiometric ratios, and NSC were highly correlated with developmental stages and flower organs. As shown in Figure 9B, the eigenvalues of PC1, PC2, and PC3 are 5.56, 2.33, and 1.16, respectively, and may describe 90.48% of the totalPC. Figure 9C represents the 100% stacked bar, showing the trend of the percentage contribution of each tested parameter over the development stages. The results show that the ratios of C/N and C/P ratiosin PC1 have the highest contribution to flower development, the starch content in PC2 is the key index of flower development, and the C content in PC3 plays an important role during flower development.

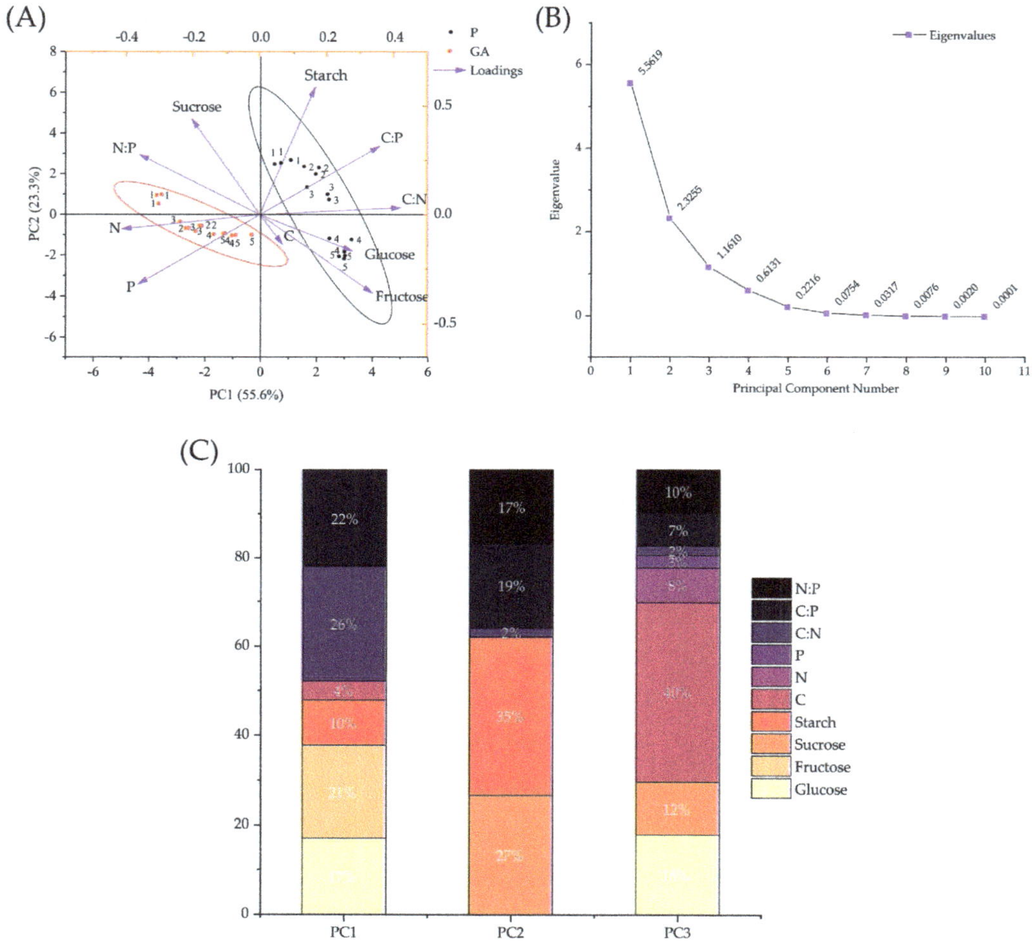

Figure 9. Principal component analysis (PCA) of tested parameters of *M. maudiae* 'Rubicunda' flowersat five developmental stages. (**A**) Distribution plots of PC1 and PC2. (**B**) The histogram from PC1 to PC10. (**C**) The contribution of the tested parameters to PC1, PC2, and PC3. The loadings of the variables in PC1 and PC2 are indicated by the direction and strength of the vector lines; 1, 2, 3, 4, and 5 represent S1, S2, S3, S4, and S5, respectively. The black and red elliptical regions represent petals and GA, respectively. The percentage of variation explained by each component is given next to the axis. The location of the trait in the diagram closest to the intersection of 0 on the *X*−axis (PC1) and *Y*-axis (PC2) shows similarity.

4. Discussion

M. maudiae 'Rubicunda' is one of the most popular ornamental plants due to its long-lasting flowering period and distinctive beauty. In general, the variation in flower bud differentiation time may depend on nutrient level, cultivation, and environmental conditions [24,25]. The RWC of petals is an important factor for petal growth, and the quality of oil-bearing rose petals [10]. In our study, the RWC of petals increased with the developmental stage during flower development. Variations in morphological parameters and RWC of flowers have been reported in *Styrax japonicus* and other ornamental species during flower development [10,31,32]. Moreover, during flower development, a longer period and partial/total overlap between inflorescence development and flowering was observed within individual trees of *M. maudiae* 'Rubicunda' (Figure 2A). Similarly, this long period of inflorescence bud development has also been observed in many species, such as *Pistacia vera*, *Hysalis peruviana*, and *Abelmoschus manihot* [30,33,34]. However, during the flower development of *M. Maudiae* 'Rubicunda', this phenomenon occurred simultaneously for 3–4 months, the reason for which requires further analysis.

N and P are two of the most important mineral nutrientsthatinfluence growth, biomass production, metabolism, energy, and protein synthesisin plants [7]. C, N, and P contents and C/N/P stoichiometry largely reflect thenutrient status and the ability of plants to capture resources, which are strongly coupled in their biochemical functioning and balance [5,6]. Studies have reported the variationof C, N, and P contents during flower development, such as for *Juglans sigillata*, *Cercis chinensis*, olive, etc. [11,12,35]. In the current study, the C contents of the petals and GA showed less variation compared to those of N and P at five developmental stages in *M. maudiae* 'Rubicunda' flowers. Generally, total N and P contents in the GA were significantly higher than those in petals and decreased with developmental times (Figure 3A). One potential explanation is that the petals may have a buffering function to store N and P in flowers, but the GA is an important organ for maintaining flower development in *M. maudiae* 'Rubicunda' [2,4]. The N and P patterns may reflect the general constraints or allocation rules governing the partitioning of nutrients among the organs of flowers. This canalso be explained by the fact that the rapidly growing organ needs relatively more P-rich ribosomal RNA (approx. 9% by mass) to maintain a rapid rate of protein synthesis [5,7,36]. Thus, the differences between N and P content may be due to the uptake, allocation, and transport of these nutrients across compartments based on different physiological needs in different organs of flowers. Moreover, there are great variations of C/N/P stoichiometric ratios in the petal and GA of the *M. maudiae* 'Rubicunda' flower during flower development. The ratios of C/Nand C/P in the petals were remarkably higher than those in the GA at different stages of development, while an opposite trend was observed for the N/P ratio (Figure 4). There are two possible reasons for these differences. One potential reason is an imbalance between C assimilation and N supply for the different efficiencies in the petals and GA. The other potential reason is that the N supply surpasses demand when more N is allocated to the GAand/or petals during flower development in *M. maudiae* 'Rubicunda'. Guo et al. (2022) noted that the C/N ratio of buds was higher at the stage of flower bud differentiation in *Lycium ruthenicum*, which may be due to the fact that carbohydratesmayincrease vascular sap levels and provide key nutrients for flower bud differentiation [23]. N/P ratios may provide a reliable index of N and/or Plimitation in plants, and the optimal range of N/P ratios is between 10 and 20 on a mass basis. As a general rule, N/P ratios < 14 indicate N limitation, while N/P ratios > 16 indicate P limitation [5,7]. The present study showed that N/P ratios in the petal and GA ranged from 17.5 to 23.1 and 20.9 to 27.2 at five developmental stages, respectively, and were significantly higher than 16. This is consistent with the previous finding that flowers are rich in mitochondria, which have higher N and P levels [37]. Based on the above results, the variation in C, N, and P contents, as well as stoichiometric ratios, reflects the nutrientrequirements and dynamic balances between developmental stages and organs, as demonstrated by the strong positive correlations with NSC and hormone contents (Figure 9). However, it remains unclear to what extent the differences in nutrient levels and their stoichiometry in *M. maudiae* 'Rubicunda' affect flower development, and further studies are still needed.

During flower development, NSCs have important contributions as energy sources, osmotic regulators, and precursors of metabolic processes [15,38]. Numerous studies have confirmed that flower development is dependent on carbohydrate metabolism in *Dendrobium crumenatum*, *Oncidium orchid*, and *Creeping bellflower* [13,17,39]. Some reports have also shown that soluble sugars are the main source of energy for flowers, and the lack of sugar leads to undersized petals or arrests flower development [21]. In our research, the higher glucose, fructose, and starch contents in petals were observed compared to those in the GA, and the NSC contents in the petals appeared to be more viable than those in the GA. Moreover, the significant variations in glucose, sucrose, fructose, and starch were corrected with developmental stages and organs (Figure 5). From the above results, it can be concluded thatthe variations of NSC contentsin developing flowers of *M. maudiae* 'Rubicunda' were more sensitive to developmental stages and/or organs. Contradictory to our results, sucrose, and glucose content did not change, but fructose content increased during the development of borage flowers. In cornflowers, the sucrose content did not show any changes, but glucose and fructose content increased as flower development progressed in sweet briar rose [21,40]. These different phenomena may be due to the fact that the variations of NSC contentsnot only depend on flower development stages but also on plant species, nutrient availability, growth conditions, etc. Moreover, highly significant correlations were described between NSC and C/N/P stoichiometric variables at five developmental stages in *M. maudiae* 'Rubicunda'flowers, which are the probably result of the dynamic balances of NSC storage, migration, conversion, and availability.

Endogenous hormone levels and their ratiosplay important rolesduring flower development. Numerous studies have reported that IAA, ABA, GA3,and CTK are the primary factors involved in the regulation of flower development [2,9,41,42]. In *Lycium ruthenicum*, IAA levels decreased significantly from flower bud pre-differentiation to late flower differentiation, which indicated that floral induction needs lower IAA levels. ABA levels significantly increased from flower bud pre-differentiation to late flower differentiation, showing that ABA could promote flower differentiation, flower opening, and senescence in *Lycium ruthenicum* [23]. Moreover, previous studies have also reported that GA3 levelswere significantly changed in the homeotic transformation of flower organs in double-flower loquat, and the higher GA3 can promote the flower differentiation of *Lycium ruthenicum*, olive, and loquat [23,35,43]. These findings showed that endogenous hormones are also controversial, showing that they have positive and/or negative effects during flower development. In the present study, IAA, ABA, GA3,and CTK levels in the petals were dynamicat five developmental stages in *M. maudiae* 'Rubicunda' flowers (Figure 6). During flower development, flower formation is the result of hormone levels, their interaction, and/or their ratio, including promoting endogenous hormone ratio and promoting endogenous hormones to suppress endogenous hormone ratio [23,44]. The present results showed that the ratios of IAA, ABA, GA3,and CTK in the petal varied significantly at five developmental stages in *M. maudiae* 'Rubicunda' flowers (Figures 7 and 8). It is worth noting that the ABA content correlates significantly with the developmental stages and is the main reason for the changes in the ratios between the phytohormone concentrations. However, further studies are needed to clarify the relationship between developmental stages and hormone levels in *M. maudiae* 'Rubicunda' flowers.

5. Conclusions

In summary, the current study provides a detailed description of the morphological features and changes in C, N, P, and NSC contents and hormone content in *M. maudiae* 'Rubicunda' at five developmental stages. The C, N, P, and NSC contents and endogenous hormone content in *M. maudiae* 'Rubicunda' flower exhibited variations at different development stages and organs. In addition, higher contents of ABA, GA3, and CTK and higher ABA/IAA ratio and ABA/GA3 ratio in petals may also be beneficial to the flower bud differentiation of *M. maudiae* 'Rubicunda'. Taken together, the variations of C, N, and P contents were closely related to the NSC and endogenous hormone contents during the flowering process in *M. maudiae* 'Rubicunda'. PCA analysis revealed that C, N, and P contents, C/N/P stoichiometric ratios, and NSC were highly correlated with developmental stages and flower organs in *M. maudiae* 'Rubicunda'. Overall, these tested parameters help to understand the flower development of *M. maudia* 'Rubicunda' and provide valuable information for the effective management of this species. Further studies will focus on the complex regulatory networks of color formation in the petals underlying flower development in *M. maudiae* 'Rubicunda'.

Supplementary Materials: The following supporting information can be downloaded at https://www.mdpi.com/article/10.3390/horticulturae9111198/s1. Table S1: Morphological characteristics of *M. maudiae* 'Rubicunda' flowers at five development stages. Table S2: Correlation coefficients of C, N, P, and NSC contents as well as C/N/P ratio variables of *M. maudiae* 'Rubicunda' flowers at five developmental stages.

Author Contributions: Conceptualization, H.L. and S.G.; methodology, T.Y., Y.Y. and H.W.; software, validation, and formal analysis, T.Y., Y.Y. and H.W.; investigation, T.Y., Y.Y., H.W. and W.Q.; writing, original draft preparation, T.Y., H.L. and S.G.; writing, review and editing, T.Y., Y.Y., H.W., W.Q., Y.H., H.L. and S.G.; visualization and supervision, H.L. and S.G. All authors have read and agreed to the published version of the manuscript.

Funding: This research received no external funding.

Data Availability Statement: The data presented in this study are available on request from the corresponding author.

Acknowledgments: We are grateful to all of the group members and workers for their assistance in the field experiment.

Conflicts of Interest: The authors declare no conflict of interest.

References

1. Pinakin, D.J.; Kumar, V.; Suri, S.; Sharma, R.; Kaushal, M. Nutraceutical potential of tree flowers: A comprehensive review on biochemical profile, health benefits, and utilization. *Food Res. Int.* **2020**, *127*, 108724. [CrossRef] [PubMed]
2. Reid, M.S. Flower development: From bud to bloom. In Proceedings of the ISHS Acta Horticulturae 669: VIII International Symposium on Postharvest Physiology of Ornamental Plants, Doorwerth, The Netherlands, 1 February 2005; ISHS: Korbeek-Lo, Belgium, 2005; Volume 669, p. 12. [CrossRef]
3. Sood, S.; Vyas, D.; Nagar, P.K. Physiological and biochemical studies during flower development in two rose species. *Sci. Hortic.* **2006**, *108*, 390–396. [CrossRef]
4. van Doorn, W.G.; Kamdee, C. Flower opening and closure: An update. *J. Exp. Bot.* **2014**, *65*, 5749–5757. [CrossRef] [PubMed]
5. Agren, G.I. Stoichiometry and nutrition of plant growth in natural communities. *Annu. Rev. Ecol. Evol. Syst.* **2008**, *39*, 153–170. [CrossRef]
6. Güsewell, S. N:P ratios in terrestrial plants: Variation and functional significance. *New Phytol.* **2004**, *164*, 243–266. [CrossRef]
7. Minden, V.; Kleyer, M. Internal and external regulation of plant organ stoichiometry. *Plant Biol.* **2014**, *16*, 897–907. [CrossRef]
8. Jia, S.; Wang, Y.; Hu, J.; Ding, Z.; Liang, Q.; Zhang, Y.; Wang, H. Mineral and metabolic profiles in tea leaves and flowers during flower development. *Plant Physiol. Biochem.* **2016**, *106*, 316–326. [CrossRef]
9. Trivellini, A.; Ferrante, A.; Vernieri, P.; Mensuali-Sodi, A.; Serra, G. Effects of promoters and inhibitors of ethylene and ABA on flower senescence of *Hibiscus rosa-sinensis* L. *J. Plant Growth Regul.* **2011**, *30*, 175–184. [CrossRef]
10. Önder, S.; Tonguç, M.; Erbaş, S.; Önder, D.; Mutlucan, M. Investigation of phenological, primary and secondary metabolites changes during flower developmental of *Rosa damascena*. *Plant Physiol. Biochem.* **2022**, *192*, 20–34. [CrossRef]
11. Zhang, W.; Li, J.; Zhang, W.; Njie, A.; Pan, X. The changes in C/N, carbohydrate, and amino acid content in leaves during female flower bud differentiation of *Juglans sigillata*. *Acta Physiol. Plant.* **2022**, *44*, 19. [CrossRef]
12. Ren, H.Y.; Qian, W.Z.; Yi, L.; Ye, Y.L.; Gu, T.; Gao, S.; Cao, G.X. Nutrient composition and antioxidant activity of *Cercis chinensis* flower in response to different development stages. *Horticulturae* **2023**, *9*, 961. [CrossRef]
13. Liu, Z.; Shi, Y.; Xue, Y.; Wang, X.; Huang, Z.; Xue, J.; Zhang, X. Non-structural carbohydrates coordinate tree peony flowering both as energy substrates and as sugar signaling triggers, with the bracts playing an essential role. *Plant Physiol. Biochem.* **2021**, *159*, 80–88. [CrossRef]
14. Han, H.; He, H.; Wu, Z.; Cong, Y.; Zong, S.; He, J.; Fu, Y.; Liu, K.; Sun, H.; Li, Y.; et al. Non-structural carbohydrate storage strategy explains the spatial distribution of treeline species. *Plants* **2020**, *9*, 384. [CrossRef]

15. Cho, L.H.; Pasriga, R.; Yoon, J.; Jeon, J.S.; An, G. Roles of sugars in controlling flowering time. *J. Plant Biol.* **2018**, *61*, 121–130. [CrossRef]
16. Park, C.H.; Yeo, H.J.; Kim, Y.J.; Nguyen, B.V.; Park, Y.E.; Sathasivam, R.; Kim, J.K.; Park, S.U. Profiles of secondary metabolites (phenolic acids, carotenoids, anthocyanins, and galantamine) and primary metabolites (carbohydrates, amino acids, and organic acids) during flower development in *Lycoris radiata*. *Biomolecules* **2021**, *11*, 248. [CrossRef] [PubMed]
17. Yap, Y.M.; Loh, C.S.; Ong, B.L. Regulation of flower development in *Dendrobium crumenatum* by changes in carbohydrate contents, water status and cell wall metabolism. *Sci. Hortic.* **2008**, *119*, 59–66. [CrossRef]
18. Önder, S.; Tonguç, M.; Önder, D.; Erbaş, S.; Mutlucan, M. Flower color and carbohydrate metabolism changes during the floral development of *Rosa damascena*. *S. Afr. J. Bot.* **2023**, *156*, 234–243. [CrossRef]
19. Roderick, B.; John, E.; Julian, H.; Allan, W. Flower opening in asiatic lily is a rapid process controlled by dark-light cycling. *Ann. Bot.* **2000**, *86*, 1169–1174. [CrossRef]
20. Vergauwen, R.; Ende, W.V.D.; Van Laere, A. The role of fructan in flowering of Campanula rapunculoides. *J. Exp. Bot.* **2000**, *51*, 1261–1266. [CrossRef]
21. Fernandes, L.; Pereira, J.A.; Saraiva, J.A.; Ramalhosa, S.E.; Casal, B. Phytochemical characterization of *Borago officinalis* L. and *Centaurea cyanus* L. during flower development. *Food Res.Int.* **2019**, *123*, 771–778. [CrossRef]
22. Arrom, L.; Munné-Bosch, S. Hormonal changes during flower development in floral tissues of *Lilium*. *Planta* **2012**, *236*, 343–354. [CrossRef] [PubMed]
23. Guo, Y.; An, L.; Yu, H.; Yang, M. Endogenous hormones and biochemical changes during flower development and florescence in the buds and leaves of *Lycium ruthenicum* Murr. *Forests* **2022**, *13*, 763. [CrossRef]
24. Callaghan, C.B.; Png, S.K. Twenty-six additional new combinations in the Magnolia (Magnoliaceae) of China and Vietnam. *PhytoKeys* **2020**, *146*, 1–35. [CrossRef] [PubMed]
25. Lang, X.; Yang, L.; Cui, T.; Zhang, S.; Qi, J. Relationship between diurnal changes of *Micheliamaudiae* var. rubicunda photosynthetic rate and environmental factors. *J. Southwest For. Univ.* **2017**, *37*, 22–27.
26. Lang, X.A.; Yang, L.L.; Zhang, S.Z.; Li, L.F.; Zhang, S.Z. Morphological variation and reproductive application of Micheliamaudiae var. rubicunda. *Subtrop. Plant Sci.* **2019**, *48*, 50–55. [CrossRef]
27. Meier, U.; Bleiholder, H.; Buhr, L.; Feller, C.; Hack, H.; Heß, M.; Lancashire, P.D.; Schnock, U.; Stauß, R.; van denBoom, T. The BBCH system to coding the phenological growth stages of plants, history and publications. *J. Für Kult.* **2009**, *61*, 41–52.
28. Qian, W.; Hu, Y.; Lin, X.; Yu, D.; Jia, S.; Ye, Y.; Mao, Y.; Yi, L.; Gao, S. Phenological growth stages of *Abelmoschus manihot*: Codification and description according to the BBCH Scale. *Agronomy* **2023**, *13*, 1328. [CrossRef]
29. Zhi, X.; Zou, J.; Ma, N.; Liu, T.; Hu, Y.; Yao, Y.; Wang, H.R.; Gao, S. Responses of the growth and nutrient stoichiometry in *Ricinus communis* seedlings on four soil types. *J. Elem.* **2022**, *27*, 223–238. [CrossRef]
30. Liu, Q.; Huang, Z.; Wang, Z.; Chen, Y.; Wen, Z.; Liu, B.; Tigabu, M. Responses of leaf morphology, NSCs contents and C:N:P stoichiometry of *Cunninghamia lanceolata* and *Schima superba* to shading. *BMC Plant Biol.* **2020**, *20*, 354. [CrossRef]
31. Chen, C.; Chen, H.; Ni, M.; Yu, F. A study on petal morphological and physiological characteristics of *Styrax japonicus* during the flowering period. *Agronomy* **2021**, *11*, 1498. [CrossRef]
32. Chumber, M.; Jhanji, S. Morpho-physiological and biochemical characterization of Chrysanthemum varieties for early flowering under heat stress. *S. Afr. J. Bot.* **2022**, *146*, 603–613. [CrossRef]
33. Polito, J.I.H.S. Pistillate and staminate flower development in dioecious *Pistacia vera* (Anacardiaceae). *Amer. J. Bot.* **1996**, *83*, 759–766. [CrossRef]
34. Ramírez, F.; Fischer, G.; Davenport, T.L.; Pinzón, J.C.A.; Ulrichs, C. Cape gooseberry (*Physalis Peruviana* L.) phenology according to the BBCH phenological scale. *Sci. Hortic.* **2013**, *162*, 39–42. [CrossRef]
35. Ulger, S.; Sonmez, S.; Karkacier, M.; Ertoy, N.; Akdesir, O.; Aksu, M. Determination of endogenous hormones, sugars and mineral nutrition levels during the induction, initiation and differentiation stage and their effects on flower formation in olive. *Plant GrowthRegul.* **2004**, *42*, 89–95. [CrossRef]
36. Kerkhoff, A.J.; Fagan, W.F.; Elser, J.J.; Enquist, B.J. Phylogenetic and growth form variation in the scaling of nitrogen and phosphorus in the seed plants. *Am. Nat.* **2006**, *168*, E103–E122. [CrossRef]
37. Handa, H.; Itani, K.; Sato, H. Structural features and expression analysis of a linear mitochondrial plasmid in rapeseed (*Brassica napus* L.). *Mol. Genet. Genom.* **2002**, *267*, 797–805. [CrossRef]
38. Boriboonkaset, T.; Theerawitaya, C.; Yamada, N.; Pichakum, A.; Supaibulwatana, K.; Cha-um, S.; Takabe, T.; Kirdmanee, C. Regulation of some carbohydrate metabolism-related genes, starch and soluble sugar contents, photosynthetic activities and yield attributes of two contrasting rice genotypes subjected to salt stress. *Protoplasma* **2013**, *250*, 1157–1167. [CrossRef]
39. Wang, C.Y.; Chiou, C.Y.; Wang, H.L.; Krishnamurthy, R.; Venkatagiri, S.; Tan, J.; Yeh, K.W. Carbohydrate mobilization and gene regulatory profile in the pseudobulb of Oncidium orchid during the flowering process. *Planta* **2008**, *227*, 1063–1077. [CrossRef]
40. Msaada, K.; Hammami, M.; Marzouk, B.; Salem, N.; Limam, F. Variation in anthocyanin and essential oil composition and their antioxidant potentialities during flower development of Borage (*Borago officinalis* L.). *Plant Biosyst.* **2014**, *148*, 444–459. [CrossRef]
41. Shalit, A.; Rozman, A.; Alvarez, J.P.; Bowman, J.L.; Eshed, Y.; Lifschitz, E. The flowering hormone florigen functions as a general systemic regulator of growth and termination. *Proc. Natl. Acad. Sci. USA* **2009**, *106*, 8392–8397. [CrossRef] [PubMed]

42. Kitamura, Y.; Yamane, H.; Gao-Takai, M.; Tao, R. Changes in plant hormone contents in Japanese apricot flower buds during prolonged chilling exposure. In Proceedings of the ISHS Acta Horticulturae 1208: II Asian Horticultural Congress, Chengdu, China, 30 August 2018; ISHS: Korbeek-Lo, Belgium, 2018; Volume 1208, pp. 251–256. [CrossRef]
43. Chi, Z.; Wang, Y.; Deng, Q.; Zhang, H.; Pan, C.; Yang, Z. Endogenous phytohormones and the expression of flowering genes synergistically induce flowering in loquat. *J. Integrat. Agric.* **2020**, *19*, 2247–2256. [CrossRef]
44. Wu, Y.; Peng, C.; Yu, X.; Shen, Y. Biochemical Changes during fruit and seed development in Nanjing Linden (*Tilia miqueiana* M.). *Forests* **2023**, *14*, 969. [CrossRef]

horticulturae

MDPI

Article

Nutrient Composition and Antioxidant Activity of *Cercis chinensis* Flower in Response to Different Development Stages

Hong-Yu Ren [1], Wen-Zhang Qian [1], Lu Yi [1], Yu-Lin Ye [1], Tao Gu [1], Shun Gao [1,2,]* and Guo-Xing Cao [1,2,]*

[1] Department of Forestry, Faculty of Forestry, Sichuan Agricultural University, Chengdu 611130, China; renhongyu@stu.sicau.edu.cn (H.-Y.R.); 202001247@stu.sicau.edu.cn (W.-Z.Q.); yilu@stu.sicau.edu.cn (L.Y.); yeyulin@stu.sicau.edu.cn (Y.-L.Y.); gutao1@stu.sicau.edu.cn (T.G.)

[2] National Forestry and Grassland Administration Key Laboratory of Forest Resources Conservation and Ecological Safety on the Upper Reaches of the Yangtze River, Sichuan Province Key Laboratory of Ecological Forestry Engineering on the Upper Reaches of the Yangtze River, Sichuan Agricultural University, Chengdu 611130, China

* Correspondence: shungao@sicau.edu.cn (S.G.); cgxing@sicau.edu.cn (G.-X.C.)

Abstract: *Cercis chinensis* Bunge (*C. chinensis*), well known as an ornamental plant widely distributed in China, and its flowers, bark, fruit, etc., have multiple bioactivities. However, reports on the changes in mineral elements, nutrient composition and antioxidant activity in *C. chinensis* flower at different development stages are rare. In this study, the flower samples were collected every 20 days from March 2023 to May 2023. The changes in carbon (C), nitrogen (N), phosphorous (P), soluble protein (SP), amino acid (AA), non-structural carbohydrate (NSC), total phenol (TP) and total flavonoids (TF) content as well as antioxidant activity in *C. chinensis* flower at different development stages were investigated. The results suggested that C, N, and P content, the C:N:P stoichiometric ratio, NSC contents, SP, AA, TP, TF and antioxidant activity of flower showed large variations at three development stages. This study found that C and P contents showed a significant decrease with the development and opening of flowers, while N content showed an opposite trend. The soluble protein content first decreased and then increased. The amino acid content, total polyphenol content, and total flavonoid content all showed a significant downward trend, while the content of NSC increased. Both ABTS and ferric-reducing antioxidant power (FRAP) showed significant decreases at various developmental stages, but DPPH was completely opposite. The highest NSC content and DPPH activity were observed at stage III, but the highest AA, TP, TF, as well ABTS and FRAP activity wereobserved at stage I. These findings will improve understanding of the requirements and dynamic balance among C, N, and P, NSC and nutrient contents as well as antioxidant activity of *C. chinensis* flowers in response to development stages.

Keywords: *Cercis chinensis*; flowerdevelopment; mineral element; phenol; flavonoids; antioxidant activity

Citation: Ren, H.-Y.; Qian, W.-Z.; Yi, L.; Ye, Y.-L.; Gu, T.; Gao, S.; Cao, G.-X. Nutrient Composition and Antioxidant Activity of *Cercis chinensis* Flower in Response to Different Development Stages. *Horticulturae* **2023**, *9*, 961. https://doi.org/10.3390/horticulturae9090961

Academic Editor: Caiyun Wang

Received: 6 August 2023
Revised: 20 August 2023
Accepted: 22 August 2023
Published: 24 August 2023

1. Introduction

Cercis chinensis (*C. chinensis*) Bunge belongs to Fabaceae *Cercis*, which is native to China and widely distributed in China. It not only has high ornamental value, but also has a long history of utilization as a traditional Chinese medicine, of which various parts can be used. For example, the bark and wood of *C. chinensis* can be used to activate blood, restore menstruation, reduce swelling and detoxification, and the fruit can be used to treat cough [1,2]. In the same way, the flower can be used to treat bronchitis, intestinal worms, fungal infections, hepatitis, dysentery, liver diseases, diarrhea, Leprosy, skin diseases, wounds, tumors and bacterial infections [1,3]. The main chemical components of the flower are flavonoids, phenolic acids, toluene, lignin and polysaccharides, which show the highest antioxidant activity in terms of scavenging DPPH free radicals, ABTS free radicals and reducing iron ions [3]. In addition, studies have found that two homogeneous heteropolysaccharides, with an average molecular weight of 17,060 and 8303 Da, can be isolated from *C. chinensis*. They are all effective flocculants that can significantly shorten activated partial thromboplastin time, prothrombin time, and thrombin time, exerting a coagulation promoting effect [4]. Many studies have shown that there is anthocyanin in *C. chinensis* flowers, so it is widely used for the extraction of natural edible red pigment [5].

Edible flowers are rich sources of phytochemistry, containing protein, amino acids, phenols and other nutrients, and have great growth potential in the food industry and medicine [6]. Edible flowers have potential additive effects in preventing chronic diseases, promoting health, and preventing food oxidation. However, as a natural antioxidant, edible flowers need to be further studied in terms of the antioxidant mechanism, anti-tumor, anti-inflammatory and anti-aging activities [7]. Flowering is one of the most important physiological processes in plants, and the changes in primary and secondary metabolite may play a regulatory role in flower development [8–10]. According to the types of flower, the flower can be consumed as a bud, whole flower and/or petals. The transition from closing bud to opening flower is one of the most active growth stages in the entire development process of plants [11]. Every flower, at each developmental stage, exhibits unique ingredients and nutritional value, which can serve as a new source in the food, cosmetics, and pharmaceutical industries [12].

The supply of sugar is all that is needed to drive flower development. The soluble sugar (SS) content also increases, before or during the rapid expansion of flowering [11,13]. Starch is the main storage carbohydrate in flowering plant, which plays an important role in pollen tube growth, ovule and fruit formation, and flower quality determination, one of the nutrients that may also be converted into SS during flower bud differentiation to meet metabolic energy needs [14,15]. Similarly, soluble protein (SP) is also one of the important energy sources of plants and the basis of morphogenesis of flower organs [16]. Amino acids (AA) also play an important role in flowering and are used to synthesize secondary metabolites, including intracellular signaling molecule, structural proteins and enzymes [9,10]. Plant secondary metabolites are essential molecules for plant growth, development, reproduction, and protection, especially phenolic compounds. They have biological effects such as free radical scavenging, antioxidant, antiviral, anti spasmodic, anti-inflammatory, and antibacterial activities, and are beneficial to human health [17]. Flavonoid have various health promoting effects and are important ingredients in many food supplements, medicines and cosmetics [18]. The various elements contents in plants are correlated, and sufficient elemental contents and relatively stable stoichiometric ratios are crucial for the healthy growth and development of plants [7,19]. Carbon (C) is the basic that constitutes cells, tissues and organs, and provides energy for various life activities of plants [20]. Nitrogen (N) and phosphorus (P) are particularly important among various nutrient elements required by plants, which are nutrient limiting indicators for biomass production, and important factors that limit physiological activities such as plant growth and reproduction [21]. They are not only components of many important organic compounds in plants, but also participate in various metabolic processes in plants.

Various cultures around the world consume flowers as food, as part of traditional cuisine or alternative medicine, because their nutritional characteristics, especially their antioxidant compound content, can play an important role in promoting health and preventing different diseases. The nutritional characteristics and antioxidant capacity of flowers vary at different stages.The purpose of this study is to investigate the changes in mineral elements and nutrient contents as well as antioxidant activity of *C. chinensis* flower at different development stages. The results will help to better understand the changes in nutrients and antioxidant activity of *C. chinensis* flower during flower development, and find the best harvest period to develop its functional ingredients in medicine and the food industry, and provide a new plant source for the edible flower market.

2. Materials and Methods

2.1. Study Site

The experiment was conducted in Sichuan Agricultural University, District Wenjiang, Chengdu, China. It has a mid-latitude inland subtropical monsoon climate with a mild climate, four distinct seasons and abundant rainfall. The annual average rainfall is 896.1 mm, and the rainy season is mainly from June to September. The annual average temperature is 16.4 °C, the annual average relative humidity is 84.0%, the annual average sunshine duration was 1104.5 h, and the annual frost-free period is 282 days.

2.2. Chemicals

We used the Milli-Q system (Millipore Corp., Billerica, MA, USA) to produce ultrapure water. Ethanol, Folin-Ciocalteu reagent, gallic acid, bovine serum albumin, glucose, Iron(III) chloride hexahydrate ($FeCl_3 \cdot 6H_2O$), 1,1-diphenyl-2-picrylhydrazine (DPPH) and 2,2′- azido bis (3-ethyl Benzothiazole lin-6-sulfo propionic acid) (ABTS) were purchased from Sigma Aldrich (St. Louis, MO, USA). The other reagents are analytical-grade reagents purchased from China National Pharmaceutical Group Chemical Reagent Co., Ltd. (Shanghai, China).

2.3. Experiment Design and Plant Materials

From March 2023 to May 2023, samples every 20 days were collected from more than three trees at the campus of Sichuan Agricultural University in Wenjiang District, Chengdu, China. The Biologische Bundesantalt, Bundes-sortenamt and Chemische Industrie scales (BBCH) represent a unified coding system for describing phenologically similar growth stages in mono- and dicotyledonous plants [22]. Referring to the BBCH edited by Mishchenko and Rana et al. [23,24], the flower development of *C. chinensis* was coded. Flower development was divided into three stages: stage I—closed bud flower clustered tightly, showing some petal color at the tip of the bud end (Figure 1A); stage II—closed bud flower grown and expanded, inflorescences scattered, showing the color of petals in their entire surface (Figure 1B); stage III—completely opened flower, without symptoms of senescence (Figure 1C). The flower samples were collected from these three stages. Samples were dried at 80 °C for 48 h, and the relative water content was determined. Samples were also used for subsequent experiments. The samples are stored at −60 °C.The transverse diameter, longitudinal diameter and biomass of flower buds in three stages were measured by using the vernier scale. Each measurement consists of three technical repetitions, and then the average of the three values is calculated.

Figure 1. *C. chinensis* flower at different development stages. (**A**) stage I, (**B**) stage II, and (**C**) stage III.

2.4. Determination of C, N and P Contents

The content of total C is determined by the potassium dichromate oxidation method. The contents of N and P was determined by the sulfuric acid–perchloric acid digestion method. The total N content is determined by the Kjeldahl method. The molybdenum antimony colorimetric method was used to determine the total P content [25]. The concentration of C, N and P is expressed in g/kg dry weight (DW), and the C:N, C:P, and N:P ratios are expressed on a mass basis.

2.5. Determination of Protein and Amino Acid Content

Protein content was determined by the Coomassie brilliant blue method [26]. An amount of 0.5 g of sample was put into a mortar and ground into a homogenate with distilled water, centrifuged, part of the supernatant was added to Coomassie brilliant blue, left for 15 min, and then we measured the absorbance value of the sample with a spectrophotometer at 595 nm. The protein content in the sample was obtained according to the standard curve and formula (Table S1). Quantification of soluble protein was performed by following the Coomassie brilliant blue G-250 method using bovine serum albumin as a standard. The results are expressed in mg/g fresh weight (FW).

Amino acid content was determined by the ninhydrin color method [9]. An amount of 0.5 g of fresh leaf was homogenized with 5 mL 10% acetic acid, and the extraction was centrifugated at 12,000 rpm at 4 °C. A volume of 0.5 mL supernatant of per sample was placed in a test tube, and 2 mL of distilled water, 3.0 mL of ninhydrin hydrate and 0.1 mL of ascorbic acid were added. The reaction mixtures were heated in boiling water for 15 min, cooled to room temperature, and then diluted to 20 mL with 60% ethanol. Absorbance was measured at 570 nm. The results are expressed in mg/g FW (Table S1).

2.6. Determination of NSC Content

The NSC content was determined using the anthrone colorimetric method [27].The soluble sugar content was determined using the anthrone colorimetric method. An amount of 0.2 g of samples was mixed with 10 mL of 80% ethanol, and the mixture was incubated in a boiling water bath for 30 min, and then centrifuged at 5000 rpm for 10 min. The supernatant was collected, and this process was performed three times to ensure complete sugar extraction. These supernatants were added to achieve a 50 mL constant volume for measuring soluble sugar content. For glucose measurement, 5 mL of anthrone reagent was added to the 0.1 mL soluble sugar extraction liquid and placed in a 90 °C water bath for 15 min. For sucrose measurement, 0.1 mL of sugar extract and 0.1 mL of 7.6 mol/L KOH solution were mixed, and incubated at 100 °C for 15 min. After cooling, 5 mL of anthrone solution was added, and placed in a 90 °C water bath for 15 min, and then the absorbance at 620 nm was recorded. For fructose measurement, 0.1 mL of sugar extract and 5 mL of anthrone solution were mixed at 25 °C for 90 min. For starch measurement, add 10 mL of 30% perchloric acid to the precipitate after centrifugation of soluble sugars. Let it stand overnight, and then extract accurately in an 80 °C water bath for 10 min. After cooling, centrifuge and collect the supernatant. The measurement steps were the same as above the glucose. These reaction mixtures were measured at 620 nm, and the contents were calculated as the glucose, sucrose, fructose and starch standard curves, respectively (Table S1). The results were expressed as μg per mg of sample. The results are expressed in mg/g DW.

2.7. Determination of Total Phenol (TP) and Total Flavonoid (TF) Contents

The total phenol (TP) content is estimated by the Folin-Ciocalteau method [28]. The content of polyphenols was determined using the Folin phenol method. A sample of 0.2 g was taken and ground into a homogenate with 60% ethanol in a mortar. The mixture was then placed in a centrifuge tube and boiled in a water bath for 30 min. After cooling and centrifugation, the supernatants were mixed with Folin phenol reagent, 7.5% sodium carbonate solution, and distilled water. After avoiding light for 2 h at room temperature, the absorbance value was measured at 765 nm using a spectrophotometer. The calculation result is based on the calibration curve of gallic acid and expressed as the gallic acid equivalent mg/g DW (Table S1).

The total flavonoid (TF) content was determined using Yang's method [29]. A sample of 0.3 g of ground samples was extracted three times using 10 mL of 60% ethanol at 65 °C, and the prepared extraction was centrifuged at 4000 rpm for 10 min at 4 °C. These supernatants were added to achieve a 50 mL constant volume for measuring the TF content. A volume of 1 mL of extract sample was mixed with 5 mL of 60% ethanol, and 300 μL of a 5% $NaNO_2$ solution. After 6 min, 300 μL of 10% $AlCl_3$ solution was added. After 6 min, 4 mL of 1 M NaOH and 400 μL of distilled water were added to prepare the reaction mixture. These solutions were mixed well and the absorbance was read at 510 nm.The results were calculated based on the calibration curve of rutin and expressed as the rutin equivalent mg/g DW (Table S1).

2.8. Assay of DPPH Radical-Scavenging Activity

The DPPH radical-scavenging rate was measured using the Gulcin method [30]. In the preparation method, 0.1 mM DPPH:2 mL of DPPH solution plus 100 μL sample solution, and 1 mL Tris-HCl (pH7.4) were mixed evenly, and stored in the dark for 30 min at room temperature. The absorbance values were recorded at 517 nm. The antioxidant capacity is expressed as the tea polyphenol equivalent (GTP mg/g DW) according to the standard curve of tea polyphenol (GTP) antioxidant activity (Table S1). The rate of DPPH radical-scavenging capability (%) = $(1 - A/A_0) \times 100\%$, where A_0 is the absorbance of the control and A is the absorbance of the sample extract.

2.9. Assay of ABTS Radical-Scavenging Activity

The ABTS radical-scavenging activity was measured using the method described by Srelatha and Padma [31]. The ABTS solution was prepared by mixing 7 mM of ABTS with 2.45 mM of potassium persulfate, and allowed to react at room temperature in the dark for 16 h. The stock solution was diluted with ethanol to an absorbance of 0.70 ± 0.02 at 734 nm. TP extracts (200 μL) were allowed to react with 800 μL of ABTS for 6 min, and then the absorbance was measured at 734 nm. The deionized water was used as the control. The antioxidant capacity is expressed as the tea polyphenol equivalent (GTP mg/g DW) according to the standard curve of tea polyphenol (GTP) antioxidant activity (Table S1). The rate ABTS free radical-scavenging activity (%) = $(1 - A/A_0) \times 100\%$, where A_0 is the absorbance of the control and A is the absorbance of the sample extract.

2.10. Assay of Ferric-Reducing Antioxidant Power (FRAP)

The FRAP was measured using El Karkouri's method [32]. The reaction solutions included 2 mL of sample, 2 mL of 1% potassium ferrocyanide, and 2 mL of 0.2 M phosphate buffer (pH 6.6), and then were mixed evenly, and then place the centrifuge tube in warm water at 50 °C for 20 min. After cooling, the 2 mL of 10% TCA, deionized water and 0.3 mL of 0.1% iron chloride were added, and mixed evenlyfor reaction for 10 min. The absorbance values were recorded at 700 nm. The antioxidant capacity was expressed as the tea polyphenol equivalent (GTP mg/g DW) according to the standard curve of tea polyphenol (GTP) antioxidant activity (Table S1).

2.11. Statistical Analysis

Experiments were carried out in a randomized way with three replicates. The data were analyzed using one-way analysis of variance (one-way ANOVA), and expressed as the means ± SD. One-way analysis of variance (ANOVA) was performed using the Waller–Duncan multi-interval test using SPSS26.0 software (IBM® Corporation, USA). Moreover, the minimum significant difference (LSD) in the ANOVA test was used to analyze the significant difference. Statistical significance was set at a 95% confidence level ($p < 0.05$).

3. Results

3.1. Changes in Morphological Features and Biomass, and Water Content

As shown in Figure 2, the transverse diameter of *C. chinensis* flowers significantly increased at each stage from stage I to stage III, ranging from 3.22 ± 0.32 mm to 13.99 ± 0.30 mm. The longitudinal diameter showed the same trend, but there was no significant change between stage II and stage III, ranging from 8.06 ± 0.50 mm to 13.19 ± 0.67 mm (Figure 2A,B). The biomass of *C. chinensis* flowers also significantly increased with each stage from stage I to stage III, ranging from 27.7 ± 1.53 mg to 68.00 ± 0.40 mg (Figure 2C). The water content of *C. chinensis* flowers showed a significant increase in each period from stage I to stage III (Figure 2D). There are obvious changes in the morphology of *C. chinensis* flowers during three developmental stages. At stage I, the petals are arranged tightly, forming female stamens. At stage II, the length and width of petals increase, the arrangement becomes loose, and the pistils and stamens are in the stage of development and elongation. At stage III, the *C. chinensis* flowers fully open and the pistils and stamens fully develop. In the three developmental stages of *C. chinensis* flowers, from one stage to the next, there is an increase in biomass and water content.

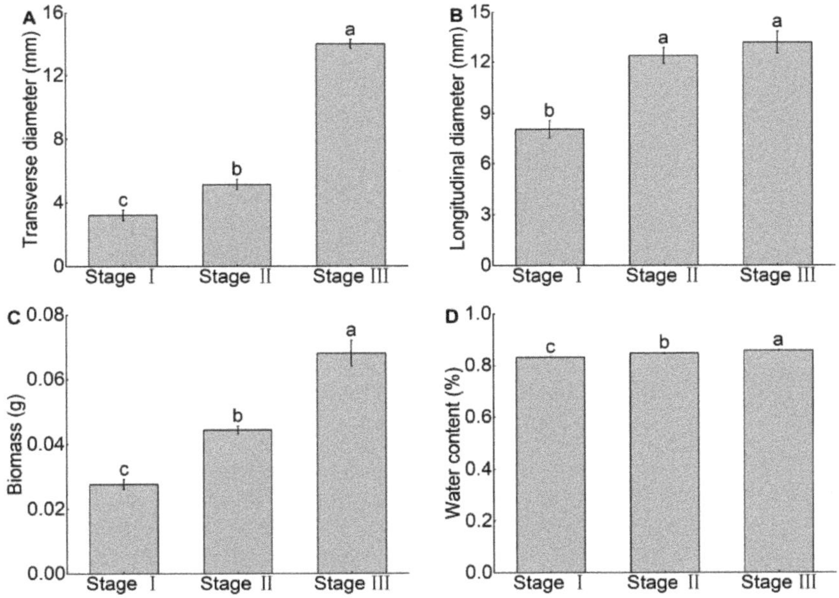

Figure 2. Changes in transverse diameter, longitudinal diameter, biomass and water content of *C. chinensis* flower at different development stages. (**A**) Transverse diameter, (**B**) longitudinal diameter, (**C**) biomass, and (**D**) water content. Data represent the mean ± SEM, n = 3. Different lowercase letters (a–c) represent significant differences among different development stages ($p < 0.05$).

3.2. Changes in C, N, and P Content and the C:N:P Stoichiometric Ratio

As shown in Figure 3A, the C content in *C. chinensis* flowers shows a significant downward trend with flower development and opening (467.46 ± 7.43 g/kg to 410.17 ± 9.95 g/kg), and shows significant differences in all three stages. Similarly, the P content also showed a significant decrease with the development and opening of flowers (1.45 ± 0.11 g/kg to 0.91 ± 0.02 g/kg), and the P content at stage I was significantly higher than that at stages II and III (Figure 3C). In Figure 3B, the range of N content in the three developmental stages of *C. chinensis* flowers follows the floral development stage, but the fluctuation is relatively gentle and the difference is not significant (24.29 ± 0.53 g/kg to 26.26 ± 1.37 g/kg). The C:N ratio showed a continuous decreasing trend during the three developmental stages of *C. chinensis* flowers, but the difference between stages I and II was not significant, with 19.25 ± 0.55 and 17.83 ± 0.14, respectively. However, it significantly decreased to 15.66 ± 1.12 during stage III (Figure 3D). However, the C:P and C:N ratios are indeed the opposite. The C:Pratio showed a continuous upward trend during the three developmental stages. The ratio significantly increased from 322.67 ± 23.36 to 451.09 ± 4.12 between stages I and II, and reached its highest point at 452.54 ± 16.76 during stage III. However, the upward trend between stages II and III was relatively small (Figure 3E). The N:P ratio showed a very significant stepwise increase in all three stages, with values of 16.79 ± 1.63, 25.30 ± 0.19, and 28.96 ± 1.12, respectively (Figure 3F).

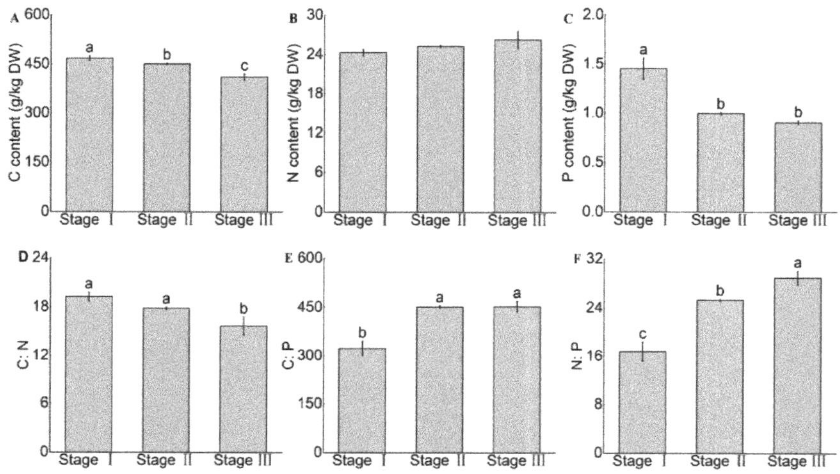

Figure 3. Changes in C, N, and P contents and C:N:P stoichiometry of *C. chinensis* flower at different development stages. (**A**) C contents, (**B**) N contents, (**C**) P contents, (**D**) the C:N ratio, (**E**) the C:P ratio, and (**F**) the N:P ratio. Data represent the mean ± SEM, $n = 3$. Different lowercase letters (a–c) represent significant differences among different development stages ($p < 0.05$).

3.3. Changes in NSC Content

As shown in Figure 4, the content of NSC continued to increase from stage I to stage III, and there was a significant increase at each stage, ranging from 106.74 ± 7.55 mg/g to 169.35 ± 14.77 mg/g (Figure 4A). The soluble sugar showed a decreasing trend, reaching 48.72 ± 2.55 mg/g at stage I, 45.51 ± 1.46 mg/g at stage II, and 63.60 ± 5.59 mg/g at stage III (Figure 4B). Starch was the lowest at stage I (58.01 ± 7.43 mg/g), and significantly increased at stage II (96.01 ± 7.94 mg/g), and reached its highest level at stage III (105.75 ± 13.00 mg/g) (Figure 4C).

3.4. Changes in Soluble Protein, Amino acid, TP and TF Content

As shown in Figure 5A, the soluble protein content was 0.68 ± 0.06 mg/g at stage I, decreased to 0.67 ± 0.11 mg/g at stage II, and then reached its highest level of 0.78 ± 0.01 mg/g at stage III. The soluble protein content showed at first decrease and then increase during the three developmental stages of *C. chinensis*, but there was no significant change between them. In Figure 5B, the decrease in amino acids during each of the three periods is very significant, ranging from 7.21 ± 0.35 μmol/g to 3.19 ± 0.08 μmol/g. According to Figure 5C, it can be seen that the TP content shows a significant decrease trend with the development stage of *C. chinensis* flowers, and shows significant differences in all three stages (55.75 ± 5.79 mg/g to 10.54 ± 3.15 mg/g). The highest TF content at stage I was 9.51 ± 1.00 mg/g, significantly higher than those of 3.73 ± 0.20 mg/g and 2.33 ± 0.64 mg/g levels at stages II and III. Similar to the content of polyphenols, the flavonoid content decreased from the tight bud stage to the fully open stage (Figure 5D).

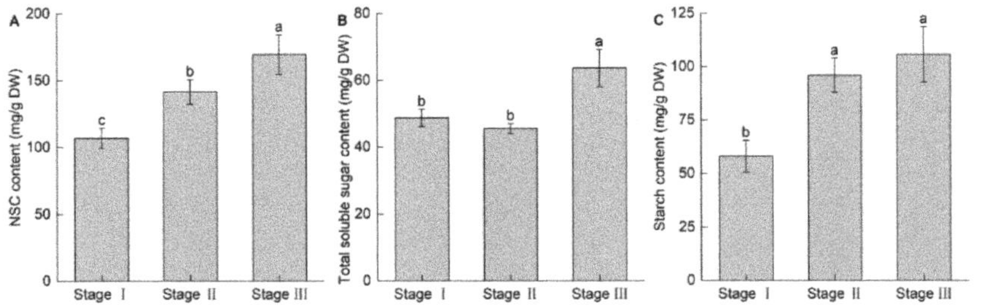

Figure 4. Changes in NSC, total soluble sugar, and starch contents of *C. chinensis* flower at different development stages. (**A**) NSC contents, (**B**) Total soluble sugar contents, (**C**) Starch contents.Data represent the mean ± SEM, $n = 3$. Different lowercase letters (a–c) represent significant differences among different development stages ($p < 0.05$).

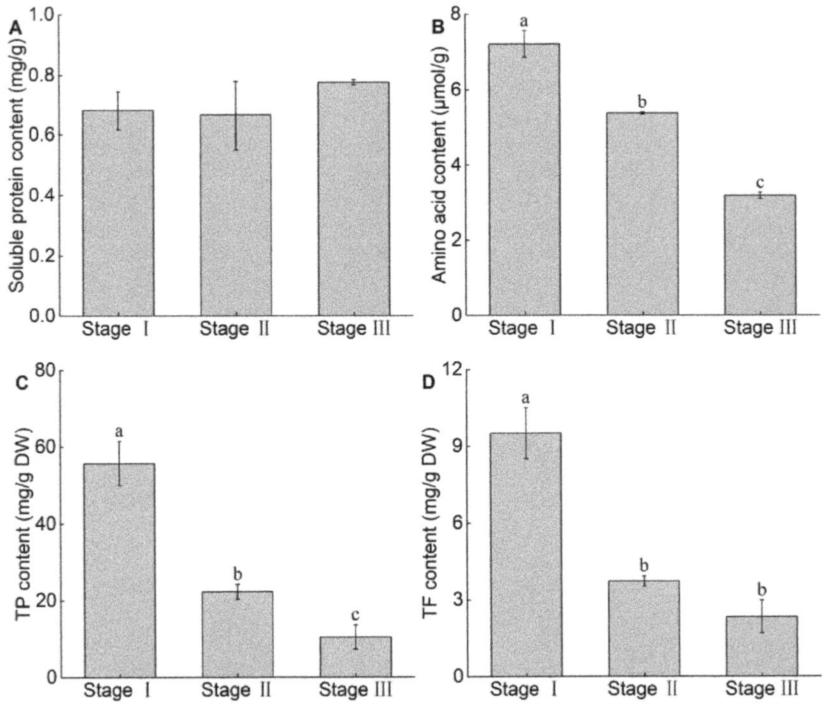

Figure 5. Changes in soluble protein, amino acid, TP and TF contents of *C. chinensis* flower at different development stages. (**A**) Soluble protein contents, (**B**) Amino acid contents, (**C**) TP contents, (**D**) TF contents. Data represent the mean ± SEM, $n = 3$. Different lowercase letters (a–c) represent significant differences among different development stages ($p < 0.05$).

3.5. Changes in Antioxidant Activity

Both ABTS cation radical and FRAP showed significant decreases at various developmental stages of *C. chinensis*, ranging from 18.81 ± 0.97 mg/g to 4.97 ± 1.31 mg/g and 4.67 ± 0.11 mg/g to 1.80 ± 0.40 mg/g, respectively (Figure 6A,C). The difference is that at Figure 6B, the DPPH radical showed a significant increase at stages I to III (9.29 ± 0.03 mg/g to 9.38 ± 0.01 mg/g). Both ABTS and FRAP show significant decreases at various developmental stages of *C. chinensis*. However, DPPH is different, showing a significant increase from stage I to stage III.

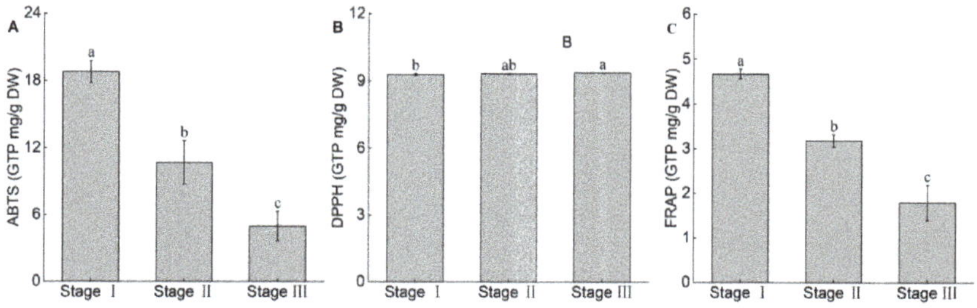

Figure 6. Changes in antioxidant activity of *C. chinensis* flower extraction at different development stages. (**A**) ABTS radical-scavenging activity, (**B**) DPPH radical-scavenging activity, (**C**) FRAP. Data represent the mean ± SEM, $n = 3$. Different lowercase letters (a–c) represent significant differences among different development stages ($p < 0.05$).

3.6. Correlation Analysis

As shown in Figure 7, at three stages of flower development, NSC, C, N, P, the C:N ratio, the N:P ratio, the C:P ratio, total polyphenol, total flavone, amino acid, ABTS, DPPH and FRAP are closely related and have significant effects on each other ($p < 0.05$, $p < 0.01$), but there is no significant relationship between the content of soluble protein and other nutrients, mineral elements and antioxidant activity ($p > 0.05$). C, P content had a significant negative correlation with NSC and DPPH ($p < 0.05$, $p < 0.01$), and a significant positive correlation with total polyphenols, medium brass, amino acids, ABTS, and FRAP ($p < 0.01$). The effect of N content was opposite to that of C and P content, and the degree of influence on various nutrients and oxidative activity was mostly lower than that of CP ($p < 0.05$, $p < 0.01$). The N:P and C:P ratios were significantly positively correlated with NSC and DPPH ($p < 0.05$, $p < 0.01$), and negatively correlated with total polyphenols, total flavonoids, amino acids, ABTS, and FRAP ($p < 0.01$). However, the C:N ratio showed completely opposite behavior ($p < 0.01$). Total polyphenols, flavonoids, and amino acids were significantly positively correlated with ABTS and FRAP ($p < 0.01$), while DPPH was significantly negatively correlated ($p < 0.05$, $p < 0.01$).

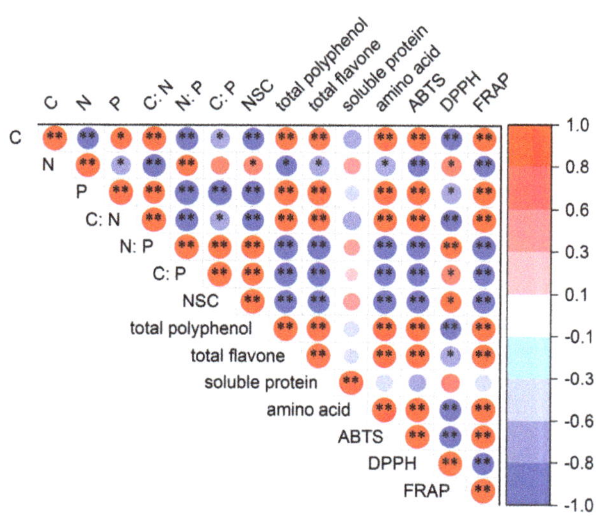

Figure 7. Correlation analysis between different chemical compositions and development stages in *C. chinensis* flower. Correlation matrix showing significant *p*-values (<0.05; <0.01) of different chemical compositions, where its color indicates the correlation slope (red Pearson's correlation coefficient = 1.0 and blue one = −1.0). Asterisks indicate significant differences: * $p < 0.05$, ** $p < 0.01$.

4. Discussion

Flowers are rich sources of phytochemistry, including amino acids, soluble protein, phenols, carbohydrates, and mineral, which havepotentialapplications in the food, cosmetics, and pharmaceutical industries [33,34]. The flower can be consumed as a bud and/or as a whole flower, or only as petals in flowering plants, which exhibits unique ingredients and nutritional value at different developmental stages [35]. *C. chinensis*, well known as an ornamentalplant in China, contained flavonoids, polyphenols, organic acids, lignans, essential oil and pigment, and showed antioxidant, tyrosinase and α-glucosidase inhibition activities [3]. Moreover, its flowers as medicine can treat rheumatism, muscle and bone pain, etc. [1,5]. As they are highly prized objects of beauty with commercial and industrial value, further studies are needed during flower development. The present study showed that *C. chinensis* flowers are described in detail based on the BBCH scale, and showed different morphological and physicochemical properties from mature flower buds to open flowering. These findings will help to enhance the visual quality and vase-life of flowers, and thus increasing their commercial value.

C is a skeleton element and a structural substance of life, and its content not only reflects the assimilation ability of plants, but also one of the response indicators of plants to environmental nutrient status [20]. The effectiveness of N and P affects plant growth and element balance, and the N:P ratio is related to the relative growth rate of plants [36]. Under natural conditions, it is generally believed that C does not limit plant growth, while changes in N and P are the main limiting factors for ecosystem productivity. The content and stoichiometry of elements such as C, N, and P in plants reflect their energy and nutritional status, as well as their interactions with the environment. At different developmental stages, they adjust their stoichiometry to adapt to environmental changes and meet their own metabolic needs [19]. Therefore, through N and P stoichiometric analysis can understand the nutrient utilization and limitation status of plants. Usually, a high C:N ratio and a low N:P ratio both represent nitrogen limitation, and plants have a higher nitrogen protection ability, However, high C:P and N:P ratios indicate phosphorus limitation, and plants may have a higher ability to protect phosphorus [37]. In this experiment, it is evident that the development of *C. chinensis* flowers is limited by P content, which inhibits the growth rate and biomass of *C. chinensis* flowers. The *C. chinensis* flower undergoes a significant increase in the C:P ratio and the N:P ratio during all three developmental stages. In short, during the period of *C. chinensis* flowers from bud to full opening, the limiting effect of P also increases with development and opening. Research has found that when the N:P ratio is greater than 20, the biomass can be increased by applying phosphorus [21]. So, during the development of flowers, appropriate application of P can alleviate the inhibition of the growth rate and biomass.

Sugar, as the fundamental molecule of carbon metabolism, can serve as an important energy substance throughout the entire life cycle of plants. In addition, sugar can interact with other inorganic regulatory networks as a signaling molecule, thus affecting the transformation of plant from vegetative growth to reproductive growth [38]. The proportion of different types of sugars determines the physical and chemical properties of flower buds and flowers, and the metabolic process of sugars affects the amount and form of sugars stored in flower buds and flowers [39]. When the flower bud is about to open, starch and oligosaccharides in the cells are degraded, producing a large amount of soluble sugars such as sucrose, glucose, and fructose, which can cause significant changes in the type and content of carbohydrates in the flower bud [40]. In addition to providing energy for flower development, they also provide precursors for the synthesis of biological molecules such as amino acids, nucleotides and fatty acids, so as to further synthesize other bioactive substances such as phenols and glucosinolate [39]. NSC content in this experiment also increases with flower development stage. The reason may be that a large amount of energy is consumed during the development process of flowers. *C. chinensis* flowers provide metabolic energy requirements for reproductive development through the synthesis of carbohydrates during the development process. The starch content is a pattern of sustained growth. Soluble sugars, as carriers of plant energy and the main form of carbohydrate transport, can be directly utilized by plants [14]. The changes in soluble sugar content were gentle between stages I and II, but significantly increased during stage III, possibly due to the fact that soluble sugar is one of the important factors controlling plant flowering. During stage III, a large amount of soluble sugar is accumulated to achieve the goal of regulating flowering.

Soluble proteins are one of the factors affecting plant flowering [16]. The changes in soluble protein content in *Styrax japonicus* are the same as in this experiment, showing a "decreasing–increasing" trend [14]. The increase in the concentration of soluble protein in *Helleborus orientalis* occurs during the stage of semi opening to full opening [13]. It can be seen that soluble protein remains at a high level during the early stages of flower development, but due to the consumption of many nutrients during flower development, the content of soluble protein decreases. Subsequently, the petals absorb and store soluble protein from nutrient organs, resulting in an increase in content [14]. Amino acids and proteins are the basis of cell proliferation and Morphogenesis, play an important role in the process of flower bud differentiation, and are used to synthesize secondary metabolites, including intracellular signaling molecule, structural proteins and enzymes [9,41]. The synthesis of secondary metabolites is the reason for the stable decrease in total free amino acid content of *C. chinensis* flowers during their development period.

During the same stage of flower development, the total phenolic content exhibits three different trends: first increasing, then decreasing, continuously increasing, and continuously decreasing [8,13,17], and the changes in total flavonoid content are also similar [9,17,42]. In this experiment, both polyphenol content and flavonoid content showed a significant downward trend with the development stage of *C. chinensis* flowers. Both ABTS and FRAP show significant decreases at various developmental stages of *C. chinensis*. However, DPPH is completely the opposite. Polyphenols are the most important secondary metabolites and pigment products in plants. They include anthocyanidin, flavanols, xanthones, bicarbonate and phenolic acids [43]. Flower color presentation is closely related to the composition and accumulation of anthocyanidin and flavonol [44]. Anthocyanidin is the main component of plant petal coloring, which has strong antioxidant activity [17,45]. During the development of *C. chinensis* flower, the petal area continues to expand, and various physiological changes take place inside the petal. During the flowering process, the petal area expands faster than the speed of pigment synthesis, which leads to the decline of pigment content in the petal per unit area, The higher the water content of the petals, the lighter the color [46], resulting in varying degrees of fading of the flower color. Studies have shown that as flowers fade, the content of phenolic compounds and antioxidant activity significantly decrease [43].

5. Conclusions

In summary, the variations in minerals and nutrient contents, and antioxidant activity of flowers extraction in *C. chinensis* flowers were recorded, indicating that they were affected by different harvest stages. The results indicate that different developmental stages have significant effects on the C, N, and P content, the stoichiometric ratio, NSC, amino acid and polyphenol, and flavonoid contents, as well as antioxidant activity of flower extraction. The highest NSC content and DPPH activity were observed at stage III, but the highest amino acid, polyphenol, and flavonoid contents, as well ABTS and FRAP activity were observed at stage I. These findings will help to provide a broader range of options for developing the practical applications of different bioactive component at the optimal harvest stages of *C. chinensis* flowers. However, the relationship between development stages and chemical compositions is very complex in *C. chinensis* flowers, and more detailed studies will be needed to further explore its high-value utilization based on extraction and isolation, identified from aqueous extracts of *C. chinensis* flowers.

Supplementary Materials: The following supporting information can be downloaded at: https://www.mdpi.com/article/10.3390/horticulturae9090961/s1, Table S1 Standard curves of test parameters in this study.

Author Contributions: Conceptualization, G.-X.C. and S.G.; methodology, H.-Y.R. and W.-Z.Q.; software, H.-Y.R. and W.-Z.Q.; validation, H.-Y.R. and W.-Z.Q.; formal analysis, H.-Y.R. and W.-Z.Q.; investigation, H.-Y.R. and W.-Z.Q. and L.Y.; writing—original draft preparation, H.-Y.R., G.-X.C. and S.G.; writing—review and editing, H.-Y.R., W.-Z.Q., L.Y., Y.-L.Y., T.G., S.G. and G.-X.C.; visualization, S.G. and G.-X.C.; supervision, S.G. and G.-X.C. All authors have read and agreed to the published version of the manuscript.

Funding: This work was financially supported by the National Undergraduate Training Program for Innovation and Entrepreneurship (No. 202310626038) and the Cultivation of Scientific Research Interest Project for Undergraduate of Sichuan Agricultural University (No. 2023297, No. 2023301).

Data Availability Statement: The data presented in this study are available on request from the corresponding author.

Acknowledgments: We are grateful to all of the group members and workers for their assistance in the field experiment.

Conflicts of Interest: The authors declare no conflict of interest.

References

1. Li, Y.; Zhang, D.; Yu, S. A new stilbene from *Cercis chinensis* Bunge. *J. Integr. Plant Biol.* **2005**, *47*, 1021–1024. [CrossRef]
2. Mu, L.; Li, J.; Yang, J.; Zhang, D. New dibenz[b, f]oxepins from *Cercis chinensis* Bunge. *J. Asian Nat. Prod. Res.* **2007**, *9*, 649–653. [CrossRef]
3. Zhang, J.; Zhou, L.; Cui, L.; Liu, Z.; Wei, J.; Kang, W. Antioxidant and α-glucosidase inhibitiory activity of *Cercis chinensis* flowers. *Food Sci. Hum. Wellness* **2020**, *9*, 313–319. [CrossRef]
4. Yin, Z.; Zhang, J.; Guo, Q.; Sun, K.; Chen, L.; Zhang, W.; Yang, B.; Kang, W. Two novel heteroglycan with coagulant activity from flowers of *Cercis chinensis Bunge*. *J. Mol. Struct.* **2021**, *1243*, 130756. [CrossRef]
5. Hong, Y.; Liao, X.; Chen, Z. Determination of bioactive components in the fruits of *Cercis chinensis* Bunge by HPLC-MS/MS and quality evaluation by principal components and hierarchical cluster analyses. *J. Pharm. Anal.* **2021**, *11*, 465–471. [CrossRef]
6. Pinakin, D.J.; Kumar, V.; Suri, S.; Sharma, R.; Kaushal, M. Nutraceutical potential of tree flowers: A comprehensive review on biochemical profile, health benefits, and utilization. *Food Res. Int.* **2020**, *127*, 108724. [CrossRef]
7. Monaci, F.; Ancora, S.; Paoli, L.; Loppi, S.; Franzaring, J. Differential elemental stoichiometry of two Mediterranean evergreen woody plants over a geochemically heterogeneous area. *Perspect. Plant Ecol. Evol. Syst.* **2022**, *55*, 125672. [CrossRef]
8. Shamsi, S.R.H.A.; Rabert, G.A.; Kurup, S.S.; Alyafei, M.A.M.; Jaleel, A. Biochemical changes and antioxidant variations in date palm (*Phoenix dactylifera* L.) varieties during flower induction and development. *Plants* **2021**, *10*, 2550. [CrossRef]
9. Önder, S.; Tonguç, M.; Erbaş, S.; önder, D.; Mutlucan, M. Investigation of phenological, primary and secondary metabolites changes during flower developmental of *Rosa damascena*. *Plant Physiol. Biochem.* **2022**, *192*, 20–34. [CrossRef]
10. Anwesha, C.; Rim, C.; Smritikana, D.; Mridushree, B.; Sonali, D.; Anton, R.S.; Malay, D. Role of metabolites in flower development and discovery of compounds controlling flowering time. *Plant Physiol. Biochem.* **2022**, *190*, 109–118. [CrossRef]
11. Reid, M.S. Flower development: From bud to bloom. *Acta Hortic.* **2005**, *669*, 105–110. [CrossRef]
12. Fernandes, L.; Ramalhosa, E.; Baptista, P.; Pereira, J.A.; Saraiva, J.A.; Casal, S.I.P. Nutritional and nutraceutical composition of pansies (*Viola* × *wittrockiana*) during flowering. *J. Food Sci.* **2019**, *84*, 490–498. [CrossRef]
13. Shahri, W.; Tahir, I.; Islam, S.T.; Bhat, M.A. Physiological and biochemical changes associated with flower development and senescence in so far unexplored *Helleborus orientalis* Lam. cv. Olympicus. *Physiol. Mol. Biol. Plants* **2011**, *17*, 33–39. [CrossRef]
14. Chen, C.; Chen, H.; Ni, M.; Yu, F. A Study on petal morphological and physiological characteristics of *Styrax japonicus* during the flowering period. *Agronomy* **2021**, *11*, 1498. [CrossRef]
15. Çetinbaş-Genç, A.; Yanık, F.; Vardar, F. Histochemical and biochemical alterations in the stigma of *Hibiscus syriacus* (Malvaceae) during flower development. *Caryologia* **2019**, *72*, 3–13. [CrossRef]
16. Yu, P.; Pu, D.; Zhou, W. Analysis of soluble proteins during flowering of *Petunia hybrida*. *J. Plant Physiol. Mol. Biol.* **2004**, *30*, 179–182. [CrossRef]
17. Yeo, H.J.; Park, C.H.; Park, Y.E.; Hyeon, H.; Kim, J.K.; Lee, S.Y.; Park, S.U. Metabolic profiling and antioxidant activity during flower development in *Agastache rugosa*. *Physiol. Mol. Biol. Plants* **2021**, *27*, 445–455. [CrossRef]
18. Panche, A.N.; Diwan, A.D.; Chandra, S.R. Flavonoids: An overview. *J. Nutr. Sci.* **2016**, *5*, e47. [CrossRef]
19. Minden, V.; Kleyer, M. Internal and external regulation of plant organ stoichiometry. *Plant Biol. (Stuttg.)* **2014**, *16*, 897–907. [CrossRef]
20. Agren, G.I. Stoichiometry and nutrition of plant growth in natural communities. *Annu. Rev. Ecol. Evol. Syst.* **2008**, *39*, 153–170. [CrossRef]
21. Güsewell, S. N:P ratios in terrestrial plants: Variation and functional significance. *New Phytol.* **2004**, *164*, 243–266. [CrossRef]
22. Qian, W.; Hu, Y.; Lin, X.; Yu, D.; Jia, S.; Ye, Y.; Mao, Y.; Yi, L.; Gao, S. Phenological growth stages of *Abelmoschus manihot*: Codification and description according to the BBCH Scale. *Agronomy* **2023**, *13*, 1328. [CrossRef]

23. Mishchenko, S.; Mokher, J.; Laiko, I.; Burbulis, N.; Kyrychenko, H.; Dudukova, S. Phenological growth stages of hemp (*Cannabis sativa* L.): Codification and description according to the BBCH scale. *Žemės Ūkio Mokslai* **2017**, *24*, 31–36. [CrossRef]
24. Rana, N.; Singh, S.; Dhakad, A.K.; Dhatt, K.K. Coding phenological growth stages of yellow bells (*Tecoma stans* (L.) Juss. ex Kunth) based on BBCH scale and its implications for urban greening. *Curr. Plant Biol.* **2023**, *34*, 100284. [CrossRef]
25. Zhi, X.; Gao, S.; Zhi, X.; Zou, J.; Ma, N.; Liu, T.; Hu, Y.; Yao, Y.; Hong Rui, W. Responses of the growth and nutrient stoichiometry in *Ricinus communis* seedlings on four soil types. *J. Elem.* **2022**, *27*, 223–238. [CrossRef]
26. Jhanji, S.; Kaur, G.; Kaur, R.; Dhatt, U.K. Physiological and biochemical changes during flower development and senescence in *Chrysanthemum* and *Gladiolus*. *Acta Physiol. Plant.* **2023**, *45*, 14. [CrossRef]
27. Liu, Q.; Huang, Z.; Wang, Z.; Chen, Y.; Wen, Z.; Liu, B.; Tigabu, M. Responses of leaf morphology, NSCs contents and C:N:P stoichiometry of *Cunninghamia lanceolata* and *Schima superba* to shading. *BMC Plant Biol.* **2020**, *20*, 354. [CrossRef]
28. Flávia de Abreu Pinheiro, F.; Ferreira Elias, L.; de Jesus Filho, M.; Uliana Modolo, M.; Gomes Rocha, J.D.C.; Fumiere Lemos, M.; Scherer, R.; Soares Cardoso, W. Arabica and Conilon coffee flowers: Bioactive compounds and antioxidant capacity under different processes. *Food Chem.* **2021**, *336*, 127701. [CrossRef]
29. Zhang, J.; Lee, T.G. Optimization of phenolics and flavonoids extraction from the fruit of *Empetrum nigrum* var. *japonicum* from Jeju Island in South Korea. *J. Ind. Eng. Chem.* **2021**, *98*, 350–357. [CrossRef]
30. Gulcin, İ. Antioxidants and antioxidant methods: An updated overview. *Arch. Toxicol.* **2020**, *94*, 651–715. [CrossRef]
31. Sreelatha, S.; Padma, P.R. Antioxidant activity and total phenolic content of *Moringa oleifera* leaves in two stages of Maturity. *Plant Food. Hum. Nutr.* **2009**, *64*, 303–311. [CrossRef] [PubMed]
32. El Karkouri, J.; Kchibale, A.; Chroho, M.; Eddamsyry, B.; Touijer, H.; El Makhoukhi, F.; Handaq, N.; Eto, B.; Salamatullah, A.; Bourhia, M.; et al. Phytochemical profile, antioxidant activity, anti-hyperglycemic effect and toxicity assessment of *Ridolfia segetum* (L.) moris extract. *Life* **2023**, *13*, 44. [CrossRef] [PubMed]
33. Motti, R.; Paura, B.; Cozzolino, A.; Falco, B.D. Edible flowers used in some countries of the Mediterranean basin: An ethnobotanical overview. *Plants* **2022**, *11*, 3272. [CrossRef]
34. Dujmović, M.; Radman, S.; Opačić, N.; Fabek Uher, S.; Mikuličin, V.; Voća, S.; šic žlabur, J. Edible flower species as a promising source of specialized metabolites. *Plants* **2022**, *11*, 2529. [CrossRef] [PubMed]
35. Fernandes, L.; Pereira, J.A.; Saraiva, J.A.; Ramalhosa, E.; Casal, S. Phytochemical characterization of *Borago officinalis* L. and *Centaurea cyanus* L. during flower development. *Food Res. Int.* **2019**, *123*, 771–778. [CrossRef] [PubMed]
36. Yan, Z.; Kim, N.; Han, W.; Guo, Y.; Han, T.; Du, E.; Fang, J. Effects of nitrogen and phosphorus supply on growth rate, leaf stoichiometry, and nutrient resorption of *Arabidopsis thaliana*. *Plant Soil* **2015**, *388*, 147–155. [CrossRef]
37. Huang, J.; Wang, P.; Niu, Y.; Yu, H.; Ma, F.; Xiao, G.; Xu, X. Changes in C:N:P stoichiometry modify N and P conservation strategies of a desert steppe species *Glycyrrhiza uralensis*. *Sci. Rep.-UK* **2018**, *8*, 12668. [CrossRef]
38. Xuan, L.; Wang, Q.; Liu, Z.; Xu, B.; Cheng, S.; Zhang, Y.; Lu, D.; Dong, B.; Zhang, D.; Zhang, L.; et al. Metabolic analysis of the regulatory mechanism of sugars on secondary flowering in Magnolia. *BMC Mol. Cell Biol.* **2022**, *23*, 56. [CrossRef]
39. Li, X.; Cui, J.; Qin, X.; Wang, J.; Wang, M. Metabolic and transcriptional regulatory mechanisms of differential carbohydrates formation from flower buds to flowers of *Hemerocallis citrina*. *Sci. Hortic.* **2023**, *308*, 111553. [CrossRef]
40. Le, T.N.; Luong, H.Q.; Li, H.; Chiu, C.; Hsieh, P. Broccoli (*Brassica oleracea* L. var. *italica*) sprouts as the potential food source for bioactive properties: Acomprehensive study on in vitro disease models. *Foods* **2019**, *8*, 532. [CrossRef]
41. Zhang, W.; Li, J.; Zhang, W.; Njie, A.; Pan, X. The changes in C/N, carbohydrate, and amino acid content in leaves during female flower bud differentiation of *Juglans sigillata*. *Acta Physiol. Plant.* **2022**, *44*, 19. [CrossRef]
42. Li, J.; Ye, G.; Liu, H.; Wang, Z. New insights on *Abelmoschusmanihot* flower development: Dynamic changes of flavonoids based on a metabolomic approach. *J. Plant Biochem. Biot.* **2022**, *31*, 351–360. [CrossRef]
43. Han, M.; Zhao, Y.; Meng, J.; Yin, J.; Li, H. Analysis of physicochemical and antioxidant properties of *Malus* spp. petals reveals factors involved in flower color change and market value. *Sci. Hortic.* **2023**, *310*, 111688. [CrossRef]
44. Guo, C.; Liu, K.; Li, E.; Chen, Y.; He, J.; Li, W.; Dong, W.; Suo, Z. Maternal donor and genetic variation of *Lagerstroemia indica* cultivars. *Int. J. Mol. Sci.* **2023**, *24*, 3606. [CrossRef] [PubMed]
45. Kozicka, M.; Hallmann, E. Identification and quantification of bioactive compounds in organic and conventional edible pansy flowers (*Viola* × *wittrockiana*) and their antioxidant activity. *Plants* **2023**, *12*, 1264. [CrossRef] [PubMed]
46. Zhang, D.; Xie, A.; Yang, X.; Yang, L.; Shi, Y.; Dong, L.; Lei, F.; Sun, L.; Bao, M.; Sun, X. Analysis of physiological and biochemical factors affecting flower color of herbaceous peony in different flowering periods. *Horticulturae* **2023**, *9*, 502. [CrossRef]

horticulturae

MDPI

Article

Combined Analysis of Volatile Compounds and Extraction of Floral Fragrance Genes in Two *Dendrobium* Species

Yanni Yang [1], Ke Xia [1], Qiaofen Wu [1], Xi Lu [1], Shunjiao Lu [2], Zhiguo Zhao [1] and Shuo Qiu [1,*]

1 Guangxi Key Laboratory of Plant Functional Phytochemicals and Sustainable Utilization, Guangxi Institute of Botany, Guangxi Zhuang Autonomous Region and Chinese Academy of Sciences, Guilin 541006, China; yangyanni219@126.com (Y.Y.); xiake4502@163.com (K.X.); wuqfmj@163.com (Q.W.); luluxi_97@163.com (X.L.); 13788588632@139.com (Z.Z.)

2 Tropical Crops Genetic Resources Institute, Chinese Academy of Tropical Agricultural Sciences, Haikou 571101, China; lushunjiao@catas.cn

* Correspondence: qiushuo001@163.com; Tel.: +86-773-3550103; Fax: +86-773-3550067

Abstract: Many species of the *Dendrobium genus* are traditional Chinese herbal medicine and ornamental plants. Flower fragrance is one of the most important horticultural ornamental characters and plays a crucial role in the ecology, economy, and aesthetics of plants. However, the volatile constituents and key regulatory genes related to floral biosynthesis are poorly understood. In this experiment, the flowers from two species of *Dendrobium* with high-scent smells, *Dendrobium moniliforme* (L.) Sw. (*D. moniliforme*), and light-scent smells, *Dendrobium nobile* "H1" (*D.* "H1"), were selected. The aim of this study was to explore the key gene expression profiles of floral biosynthesis by combining volatile constituent determination and transcriptome analysis in two different *Dendrobium* species. Physiological determination results showed that 60 volatile compounds were identified in *D. moniliforme* and 52 volatile compounds were identified in *D.* 'H1' flowers in four flowering stages, and the full bloom stage was the most complicated stage because there were 41 and 33 volatile compounds, respectively. These compounds belong to terpenes, aromatics, fatty acids, nitrogenous compounds, ketones, alcohols, and alkanes, respectively. The components identified in the gynandrium and petals revealed that the petals were probably the most important sites affecting the release of volatiles. The relative content of terpene compounds was the highest, with 77.25% (*D. moniliforme*) and 50.38% (*D.* "H1"), respectively. Transcriptome analysis showed that differentially expressed genes (DEGs) were highly enriched in terpenoid backbone biosynthesis and that four linalool synthetase (LIS) genes were up-regulated in high-scent smell species. This study is helpful to explore the key genes of flower fragrance and provides a theoretical basis for further understanding of the regulatory molecular functions of floral synthesis and release, as well as for the cultivation of new aromatic species.

Keywords: *Dendrobium*; volatile compounds; transcriptome; floral metabolites; terpenoid

Citation: Yang, Y.; Xia, K.; Wu, Q.; Lu, X.; Lu, S.; Zhao, Z.; Qiu, S. Combined Analysis of Volatile Compounds and Extraction of Floral Fragrance Genes in Two *Dendrobium* Species. *Horticulturae* **2023**, *9*, 745. https://doi.org/10.3390/horticulturae9070745

Academic Editor: Jiri Gruz

Received: 29 May 2023
Revised: 20 June 2023
Accepted: 21 June 2023
Published: 26 June 2023

1. Introduction

Dendrobium genus belongs to the orchidaceae herbs, which make up one of the largest families of flowering plants with abundant ecological habitats and extensive economic impacts for their ornamental and medicinal values in the world [1,2]. There are more than 74 species and 2 variant *Dendrobium* species in China [3], of which only a few can release their flower fragrance [4]. Floral fragrances not only attract insects and other pollinators to help plants reproduce [5], driving away natural enemies for self-protection and resisting abiotic stress [6], but are also one of the most important horticultural ornamental characters of *Dendrobium* [7].

The main volatile compounds of *Dendrobium* included terpenoids, aldehydes, esters, and alcohols [8]. Among them, terpenoids are one of the most abundant and diverse aromatic compounds in flowers [9,10]. In recent years, more than 80,000 terpenoids have been identified, and the high volatility of terpenoid compounds promotes the fragrance of *Dendrobium* [11–15]. Terpenoids have diverse biological functions in nature and play an essential role in plant-animal [16,17], plant-pathogen [9,18], and plant–plant interactions and the regulation of antagonism between organisms [19]. There are two main metabolic pathways for forming terpenoids: the mevalonate pathway (MVA pathway) and the 2C-methyl-D-erythritol-4-phosphate pathway (MEP pathway). Both pathways (MVA and MEP) are responsible for forming isoprene, which is the building block of various types of terpenes [20]. All-trans-farnesyl diphosphate (FPP) is transformed into a sesquiterpene skeleton by the MVA pathway, while geranyl diphosphate (GPP) and all-trans-geranylgeranyl diphosphate (GGPP) are produced as monoterpene and diterpene skeletons by the MEP, respectively [21]. TPS is the final enzyme converting the precursors FPP, GPP, and GGPP to different kinds of sesquiterpene, monoterpenes, and diterpene. In addition, TPS harbors conserved motif structures such as DDxxD (an aspartate-rich motif that interacts with divalent metal ions involved in positioning the substrate for catalysis) in plants [22], but their catalytic activity and products vary greatly. Structural differences may lead to new catalytic activities, resulting in the formation of many different species of volatile terpenoids from the same substrate [23]. Monoterpenes and sesquiterpenes are the majority of volatile compounds [18]. TPSs are highly differentiated gene families that have been found in *Arabidopsis thaliana*, *Solanum lycopersicum*, and *Vitis vinifera* [24]. In recent years, genome-wide analyses of TPSs have identified them in plants such as *Gossypium hirsutum* [25], *Brachypodium distachyon* [26], and *Liriodendron chinense* [27]. Bioinformatics techniques were used to compare plant TPS family members, which can be divided into 8 subfamilies: TPSa (sesquiterpenes), TPSb (cyclic monoterpenes and hemiterpenes), TPSc (copalyl diphospate synthases), TPSd (gymnosperm-specific), TPSe (ent-kaurene synthases), TPSf (other diterpene synthases), TPSg (acyclic monoterpenes), and TPSh (lycopod-specific) [28].

The volatile compounds of orchids contain various terpenes. Monoterpenes, such as linalool, pinene, ylangene, and limonene, are major floral scent compounds in Orchid flowers [20,29–31]. Among the monoterpene synthetase genes, linalool synthetase (LIS), geraniol synthetase (GES), and β-Basil synthetase (OS) have been studied more, which belong to the TPSg family [22,32]. To date, only a few TPS genes have been identified in orchids. A total of 34 TPS genes were identified In *Dendrobium officinale*, and these genes were mainly expressed in the flowers, followed by the roots and stems [24]. PbTPS5 and PbTPS10 genes are involved in monoterpene biosynthesis in *Phalaenopsis bellina* [33]. In the flowers of the *Freesia hybrid*, a total of 8 TPS genes were identified. Among them, FhTPS1 catalyzed the formation of linalool, whereas FhTPS4, FhTPS6, and FhTPS7 were poly-product enzymes that could recognize both substrates, GPP and FPP [34].

There are many species of *Dendrobium*, but few have flowers that are both fragrant and beautiful. In the experiment, we selected two species of *Dendrobium*: *Dendrobium moniliforme* (L.) Sw. (*D. moniliforme*) and *Dendrobium nobile* "H1" (*D.* "H1"). *D. moniliforme* is a wild germplasm widely distributed in the south of China, the Korea Peninsula, northeastern India, and Japan [35]. It grows on tree trunks, valley rock walls, or even cliffs. It has the characteristics of thin stems, small flowers, and a high-scent fragrance of flowers. Unlike *D. moniliforme*, *D.* "H1" is a cultivated line with large and bright flowers but only a faint fragrance. By combining volatile constituent determination and transcriptome sequencing analysis of these two different *Dendrobium* species, the characteristics of volatile constituents in different florescences and organs were analyzed, and the key genes involved in terpenoid synthesis were explored. Furthermore, real-time quantitative PCR (qRT-PCR) was used for the detection of the correlation between transcriptome sequencing and gene expression levels.

2. Materials and Methods

2.1. Plant Materials

In this experiment, the fresh flowers of light-scent *D.* "H1" and high-scent *D. moniliforme* were used as materials, both of which were obtained from the laboratory of Professor

Shuo Qiu of the Guangxi Institute of Botany, complying with Chinese legislation (Figure 1). *D.* "H1" was bought from Zhejiang Senhe Seed Co., Ltd. (Hangzhou City, Zhejiang Province, China), and *D. moniliforme* came from Baise City, Guangxi Province, China. Twelve flower samples were collected from each variety in different florescence stages and different organs, with three replicates for each of the two species, to determine the natural volatile constituents. The petals at the full bloom stage were chosen for transcriptome analysis. Using solid phase microextraction (SPME) and gas chromatography coupled with mass spectrometry (GC-MS) to determine the natural volatile constituents of these two *Dendrobium* species in different florescence stages (bud stage, first flowering stage, full bloom stage, and declining stage) and flower parts (gynandrium and petal) [36,37]. To investigate the temporal and spatial correlation between volatile compound-related genes, according to the results of physiological determination, we collected petals at full bloom stage, including two species of flowers, for RNA extraction. All samples were immediately frozen in liquid nitrogen and stored at −80 °C until needed.

Figure 1. Two species of *Dendrobium* flowers. (**A**) *D.* "H1". (**B**) *D. moniliforme*.

2.2. Apparatus

Manual solid phase micro extraction injector and 50/30 µm PDMS/CAR/DVB extractor (SUPELCO, Inc., Bellefonte, PA, USA), 6890N-5975B Gas chromatography-mass spectrometer (GC-MS) apparatus (Agilent Technologies, Santa Clara, CA, USA), 40 mL brown headspace sampling bottle, and water bath (Shanghai Jingxue Scientific Instruments Co., Ltd., Shanghai, China).

2.3. GC-MS Analysis

For the analysis of natural volatile compounds, *D. moniliforme* and *D.* "H1" fresh flowers were enclosed, sampled in brown headspace sampling bottles, and repeated three times. The extractor was inserted into the GC-MS injector for 30 min at 250 °C. Fiber heads were inserted in brown headspace sampling bottles containing 12 flowers, and headspace extraction was performed at 40 °C for 30 min. After extraction, the fiber head was removed and inserted into the GC-MS injection port. After analysis for 5 min, the sample was injected for further analysis [38].

The flow rate of the HP5-MS quartz capillary column (30 m × 0.25 mm × 0.25 μm) was 0.8 $mL \cdot min^{-1}$. The carrier gas was high-purity helium (99. 999%), and splitless mode was selected. Program temperature setting: the initial column temperature was at 40 °C for 3 min, then increased at a rate of 3 °C min^{-1} to 73 °C for 3 min, and then the temperature was heated to 220 °C at 5 $°C \cdot min^{-1}$ for 2 min. The temperature of the electron ionization (EI) ion source was maintained at 230 °C, and the electron energy was 70 eV. The temperature of the GC-MS transmission line was 250 °C, and the scanning range was 40–450 amu [39].

According to the total ionization chromatography of GC-MS and to determine the volatile compounds detected during the experiment, the National Institute of Standards and Technology, https://www.nist.gov/ (accessed on 18 March 2023) was searched and compared with the standard spectrum of the Eight Peaks Index and the EPA/NIH mass spectral data base, respectively [40]. Using the Xcalibur1.2 software, quantitative analysis was carried out according to the peak area normalization method, and the phase pair contents of each chemical component were obtained, respectively.

2.4. Transcriptome Analysis

The TRIzol Reagent (Life Technologies, Carlsbad, CA, USA) was used for the total RNA extraction of each sample. Using the Agilent 2100 Bioanalyzer (Agilent Technologies, Santa Clara, CA, USA), checked RNA integrity and concentration. The mRNA was isolated by the NEBNext Poly (A) mRNA Magnetic Isolation Module (NEB, E7490). The Illumina NEBNext Ultra RNA Library Prep Kit (NEB, E7530) and NEBNext Multiplex Oligos (NEB, E7500) were used in the cDNA library construction [41]. Finally, the constructed cDNA libraries were sequenced on a flow cell on the Illumina HiSeq™ sequencing platform.

Low quality and shorter reads were removed by the Perl script. The clean reads that were filtered from the raw reads were mapped to the *Dendrobium* genome (OGSv3.2) using Tophat2 software. The transcriptome data have been deposited into the NCBI Sequence Read Archive with the identifier PRJNA976822. Then gene expression levels were calculated using FPKM values (fragments per kilobase of exon per million fragments mapped) by the Cufflinks software. DESeq and Q-value were employed and used to evaluate differential expression genes (DEGs) between *D.* 'H1' and *D. moniliforme*. After that, gene abundance differences between those samples were calculated based on the ratio of the FPKM values [42].

To compute the significance of the differences, only genes with a *p*-value of log2 ratio ≥ 2 and a false discovery rate (FDR) significance score < 0.01 were used for subsequent analysis. With a cut-off E-value of 10^{-5}, genes were compared against various protein databases by BLASTX, using the National Center for Biotechnology Information (NCBI) non-redundant protein (Nr) database and Swiss-Prot database [43]. Furthermore, the Nr BLAST results were imported into the Blast2 GO program to obtain the gene ontology (GO) term annotation genes [44]. Then, a Perl script was used to plot GO functional classification, and KEGG pathways were assigned to the assembled sequences for the unigenes [45]. The resulting comments were enriched and refined using TopGo (the R package) [46]. Gene sequences were also compared with the Cluster Homologous Group (COG) database to predict and classify function [47].

2.5. RT-PCR Analysis

According to the physiological determination, combined transcriptome sequencing analysis were performed to identify the differentially expressed genes related to floral biosynthesis and their expression levels were verified. Fresh *D.* "H1" and *D. moniliforme* flowers in the full bloom stage of petals were taken for qRT-PCR analysis. HUAYUEYANG RNA extraction kit (HUAYUEYANG Biotechnology, Beijing, China) was used to isolate total RNA from the samples. According to the manufacturer's protocol, 1 μg total RNA was used to synthesize cDNA using the Takara cDNA Synthesis Kit (biological engineering company, Dalian, China), and fluorescent quantitative primers were designed using primer 5.0 (Supplementary Table S5). Using the Roche Lightcycler 480 Real-time PCR System (Roche, Switzerland) and the SYBR Green PCR Reaction Master Mix Kit (Vazyme, Nanjing, China), the expression level of genes was detected by qRT-PCR. Procedure: 95 °C pre-denaturation for 3 min, then 95 °C reaction for 10 s, and finally 60 °C reaction for 30 s, 45 cycles. Each sample was repeated three times, and the relative expression level of each gene was calculated using $2^{-\Delta\Delta CT}$.

3. Results

3.1. Analysis the Characteristics of Volatile Constituents in Different Florescences of Two Dendrobium Species

According to the SPME and GC-MS analysis of volatile compounds from *D. moniliforme* and *D.* "H1", a total of seven main chemical groups, including terpenes, aromatics, fatty acids, nitrogenous compounds, ketone, alcohols, and alkanes, were collected. However, there were significant differences in the compositions and content of volatile compounds in the two *Dendrobium* species. As shown in Figure 2, terpenoids are the main components of the aromatic compounds in the flowers at the four flowering stages. The contents of terpenoids in *D. moniliforme* were 71.78% at the bud stage, 89.83% at the first flowering stage, 76.69% at the full bloom stage, and 64.33% at the declining stage. Otherwise, the terpenoids contents of *D.* "H1" in the four stages are 53.27%, 89.74%, 79.09%, and 35.78%, respectively. Among the 60 volatile compounds identified in *D. moniliforme*, 1R-α-pinene and Linalool were found in all four flowering stages. The relative contents of 1R-α-pinene and Linalool accounted for more than 40% of all compounds, which was the highest among all compounds in the four flowering stages (Supplementary Table S1). On the other hand, we found 52 volatile compounds in the flowers of *D.* "H1", and the relative contents of 1R-α-pinene were more than 21.11%, which was detected in four flowering stages. Except for the bud stage, the content of cis-β-ocimene was also relatively high (10.36–54%) (Supplementary Table S2). These results indicated the composition of volatile compounds becomes more complex as the flowers open, with the most complicated in the full bloom stage, which has 41 (*D. moniliforme*) and 33 (*D.* "H1") volatile compounds, respectively. In particular, the emissions of Linalool (one of the terpenoids) were higher in *D. moniliform* throughout all the flowering stages, which accounted for 10.21–27.84% of the total volatiles in *D. moniliform* but for only 0–1.37% in *D.* "H1". The difference in linalool content may be one of the main volatile components that distinguish floral fragrance release between these two *Dendrobium* species.

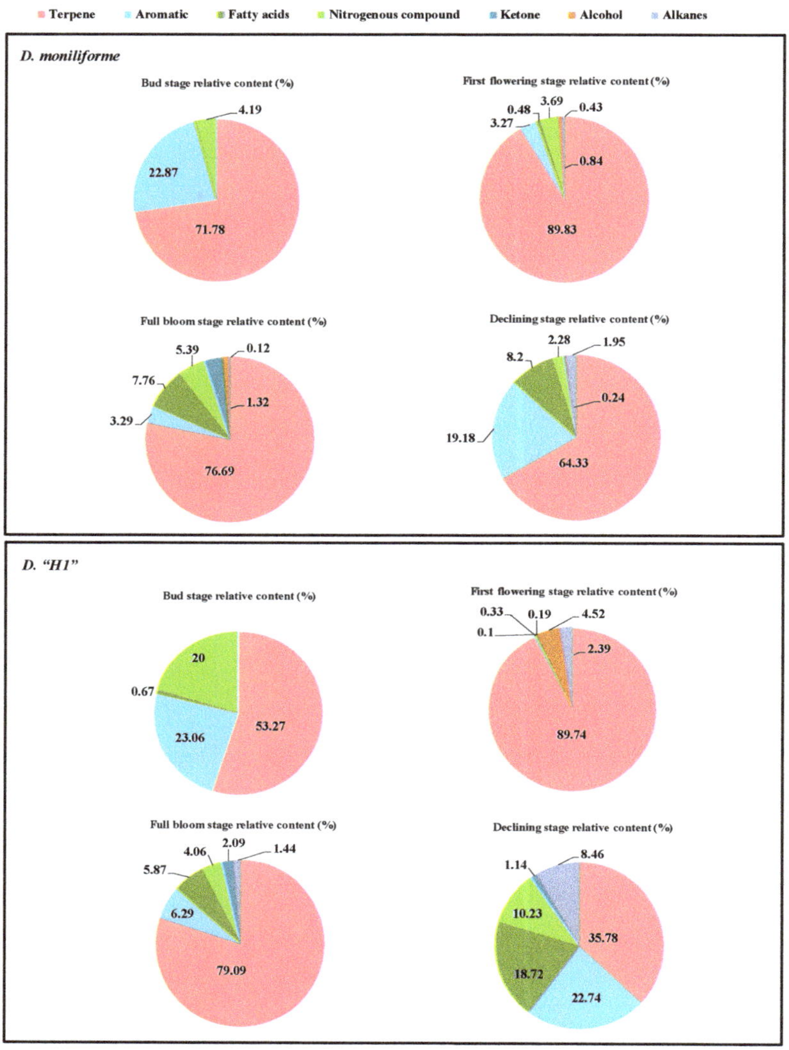

Figure 2. Volatile constituents in different florescences of two *Dendrobium* species.

3.2. Analysis of the Characteristics of Volatile Constituents in Different Organs of Two Dendrobium Species

By GC/MS analysis, the components and relative contents of the same compounds in different flower parts of the two *Dendrobium* species at full bloom stage were measured (Table 1). The highest relative content in the petals was still the terpenoids contents, and there were 15 kinds of terpenoids identified in *D. moniliforme* with a relative total content of 77.25%. *D.* 'H1' contained 13 kinds of terpenoids with a relative total content of 50.38%. Although there were only 7 kinds of terpenoids in the gynandrium of the two species, the relative total contents reached 81.75% (*D. moniliforme*) and 63.31% (*D.* 'H1'), respectively.

Table 1. Classification of scent compositions of different fluorescence in two flower parts of two *Dendoubium* species.

Cultivar and Flower Part	Compound Name	Variety	Relative Content (%)
D. moniliforme Petals	Terpene	15	77.25
	Aromatic	2	2.12
	Fatty acids	2	9.99
	Nitrogenous compound	6	4.94
	Ketone	0	0
	Alcohol	1	1.94
	Alkanes	0	0
	Total	26	96.24
D. moniliforme Gynandrium	Terpene	7	81.75
	Aromatic	3	5.8
	Fatty acids	0	0
	Nitrogenous compound	4	3.03
	Ketone	3	6.19
	Alcohol	0	0
	Alkanes	0	0
	Total	17	96.77
D. H1 Petals	Terpene	13	50.38
	Aromatic	7	18.16
	Fatty acids	4	11.24
	Nitrogenous compound	7	13.1
	Ketone	1	1.5
	Alcohol	1	0.58
	Alkanes	3	2.88
	Total	36	97.84
D. 'H1' Gynandrium	Terpene	7	63.31
	Aromatic	4	18.12
	Fatty acids	1	3.36
	Nitrogenous compound	4	7.58
	Ketone	1	1.29
	Alcohol	1	1.67
	Alkanes	1	2.21
	Total	19	97.54

The results in Tables S1 and S2 showed that at the full flowering stage, the volatile compounds of *D. moniliforme* and *D.* "H1" in petals were 26 and 36, respectively, and 17 and 19 volatile compounds in gynandrium, respectively. Thus, the volatile compounds of the petal are more complex than those of the gynandrium. These results suggested that the petals may be the most important part affecting the release of volatiles.

3.3. Identification and Analysis of the DEGs

Based on the determination of physiological composition, petals in full bloom were selected for RNA-seq analysis of differentially expressed genes (DEGs) in the two *Dendrobium* species. In three independent experiments, 24,199 genes were matched to the *Dendrobium* genome, of which 16,735 were expressed. Of these expressed genes, 841 genes were unique to *D.* 'H1', 1912 genes were unique to *D. moniliforme,* and 13,982 genes were found in both species (Figure 3A). DESeqR and Q-value were applied to determine if the genes in *D.* 'H1' and *D. moniliforme* species were significantly different based on 16,735 expressed genes with a 1% false discovery rate (FDR). According to fold change (FC) > 1 and $p < 0.01$, 5790 differentially expressed genes (DEGs) showed dynamic changes between *D.* "H1" and *D. moniliforme.* If compared with *D.* "H1", 3210 DEGs were up-regulated, and 2580 DEGs were down-regulated in *D. moniliforme* (Figure 3B).

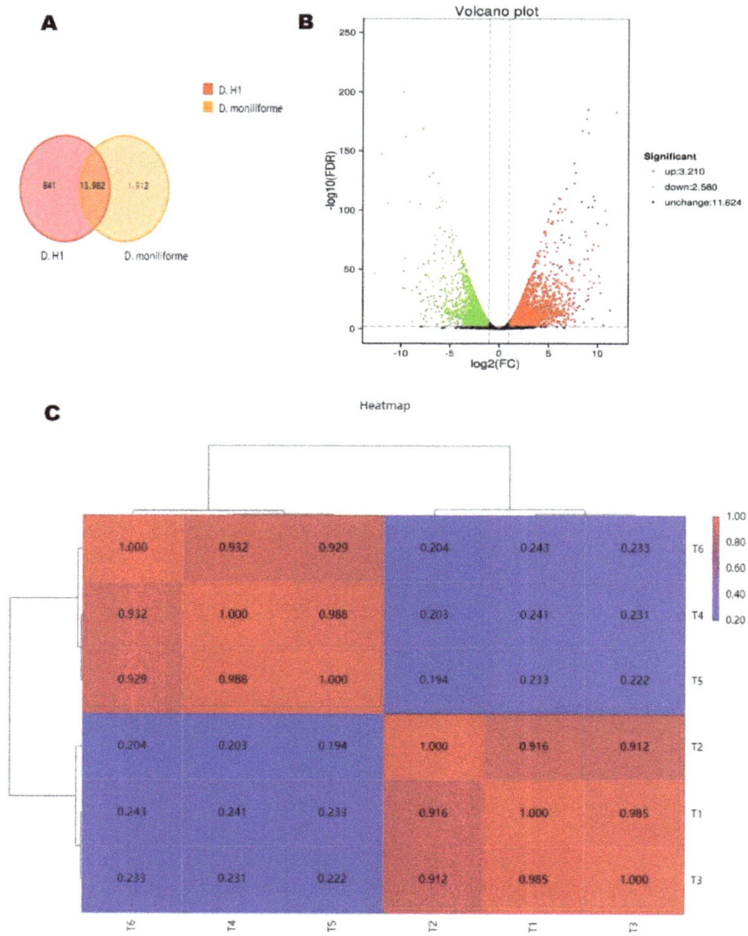

Figure 3. General information about the identified genes. (**A**) The distribution of expressed genes. The red circle were unique to *D.* "H1" and the yellow circle were unique to *D. moniliforme*, with genes found in both species shown at the intersection of the circles; (**B**) Venn diagram showing quantitative distribution of differentially expressed genes of the *D.* "H1" and *D. moniliforme*. The red dots represent upregulated differentially expressed genes, the green dots represent downregulated differentially expressed genes; (**C**) Hierarchical clustering analysis of the *D.* "H1" and *D. moniliforme*. T1–T3 represented the three replicates of *D.* "H1", and T4–T6 represented the three replicates of *D. moniliforme*.

Using the pHEATMAP package in R, we performed hierarchical clustering to show a comprehensive overview of the two species (Figure 3C). These results indicated that the three replicates of each sample from each species belong to the same respective clades, demonstrating the reproducibility and reliability of our RNA-Seq data.

3.4. KEGG Functional Classifications and Enrichment Analysis of the DEGs

The Kyoto Encyclopedia of Genes and Genomes (KEGG) functional classification analysis showed that the DEGs were mainly divided into five pathways: cellular processes, environmental information processing, genetic information processing, metabolism, and organismal systems. Among them, 932 unigenes were involved in the metabolism pathway, which was the most annotated gene in all pathways (Figure 4A). To further reveal the metabolic pathways and functions of the identified DEGs, KEGG pathway enrichment analysis was performed (Figure 4B,C). The up-regulated genes were significantly enriched in plant hormone signal transduction, phenylpropanoid biosynthesis, and terpenoids backbone biosynthesis. The down-regulated proteins were significantly enriched in Glycolysis/Gluconeogenesis, fatty acid metabolism, and fatty acid biosynthesis. Based on the number of genes, these DEGs from *D.* "H1" and *D. moniliforme* were highly associated with phenylpropanoid biosynthesis (49 unigenes involved), terpenoids backbone biosynthesis (20 unigenes involved), and fatty acid biosynthesis (26 unigenes involved).

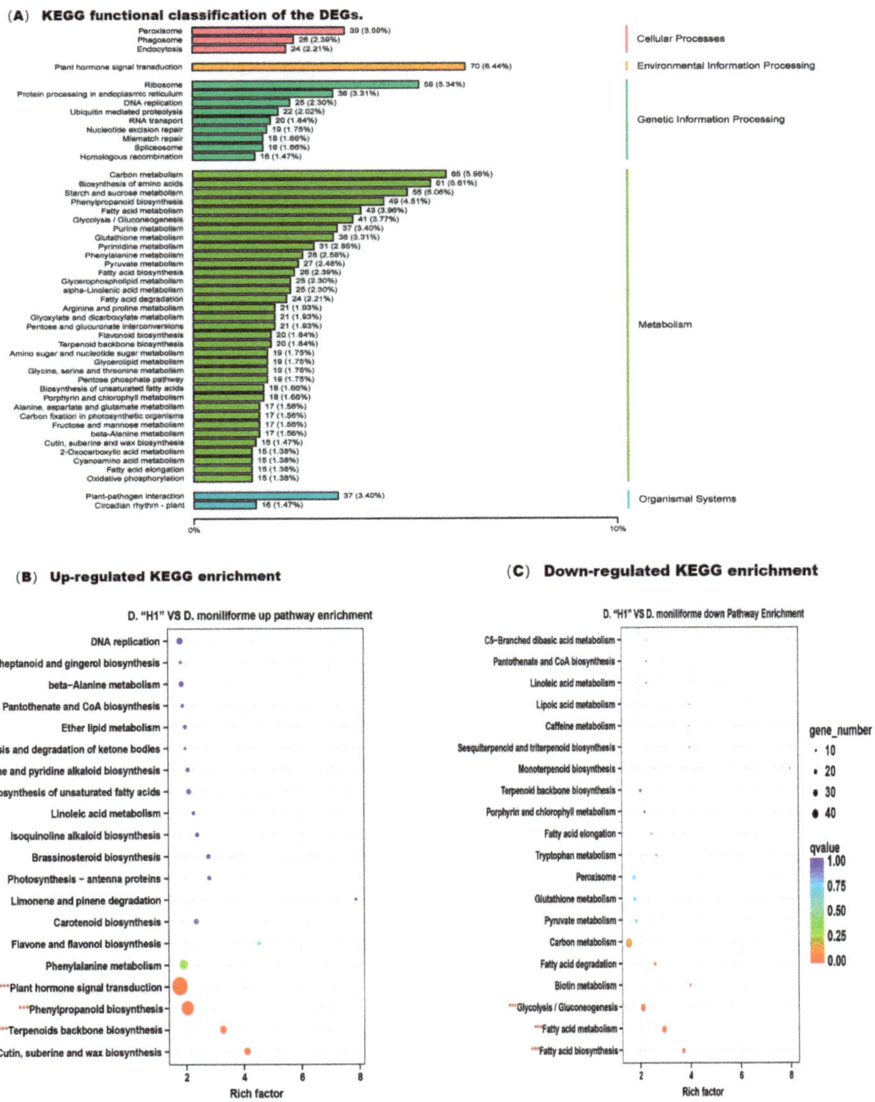

Figure 4. KEGG functional classifications and enrichment analysis of the DEGs. The *** in the figure indicate the statistical significance ($p < 0.01$) of the two *Dendoubium* species.

3.5. Protein-Protein Interaction (PPI) Network Analysis of the DEGs

PPI network analysis of a total of unigenes 95 DEGs co-regulated by phenylpropanoid biosynthesis, terpenoids backbone biosynthesis, and the fatty acid biosynthesis pathway was performed based on the STRING database. The results showed that proteins of 15 DEGs interacted with each other, as shown in Figure 5, and four of the up-regulated genes were LIS genes. Detailed information and nucleotide sequences for these 15 genes are listed in Supplementary Table S3. Combined with the results of physiological volatile compounds, the content of LIS might be critical for fragrance formation.

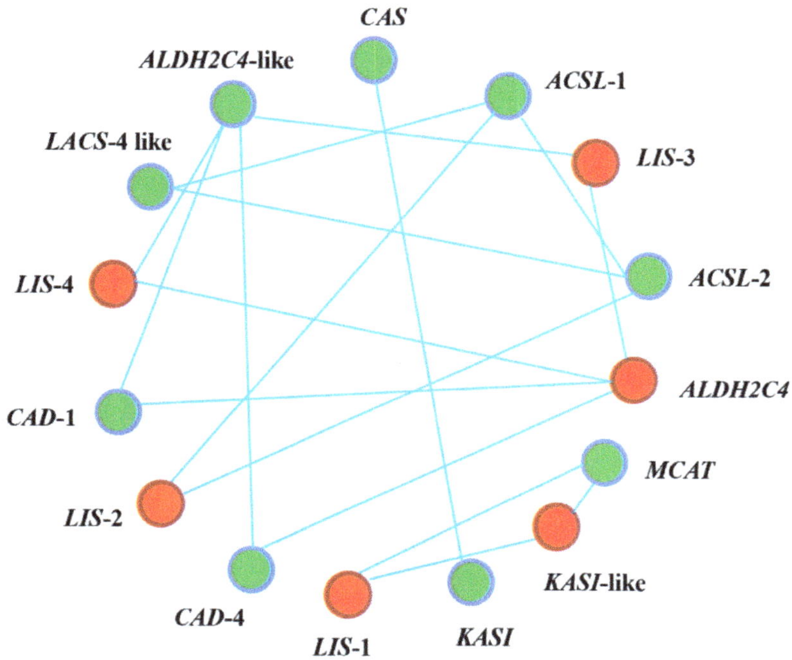

Figure 5. Protein-protein interaction network analysis of DEGs. Each node in the figure represents a protein of the gene, and the size of the nodes indicates the degree of correlation between them. Each line indicates the interaction between proteins. The green and red dots represent down-regulated and up-regulated, respectively.

3.6. Transcription Factor Prediction in Two Dendrobium Species

A total of 1563 transcription factors (TF) covering 66 TF families were identified from the transcriptome data of the two *Dendrobium* species. Family information on transcription factors is listed in Supplementary Table S4. C2H2, MYB, AP2/ERF-ERF, bHLH, and NAC families were significantly enriched, among which C2H2, MYB, and AP2/ERF-ERF families were dominant, with 117, 115, and 104 differentially expressed transcription factors, respectively (Figure 6). Our results suggested that these enriched TFs activated or inhibited downstream functional genes by regulating specific transcription factors during floral fragrance formation in *Dendrobium*.

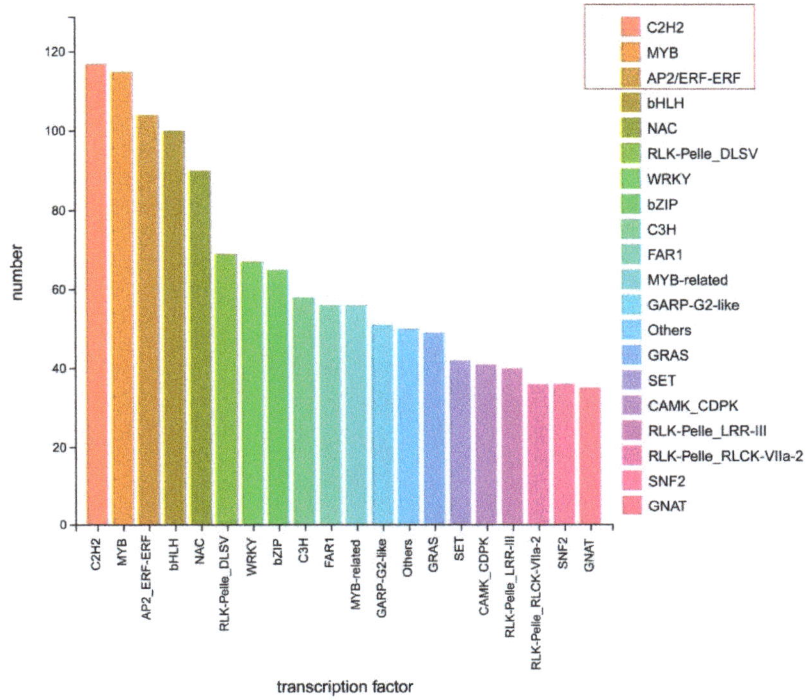

Figure 6. Transcription factor predictions.

3.7. Validation of the DEGs Expression

Fifteen DEGs associated with phenylpropanoid biosynthesis, terpenoids backbone biosynthesis, and fatty acid biosynthesis were selected for qRT-PCR analysis to verify the transcriptome data. These genes information as well as primers are shown in Supplementary Table S5. The qRT-PCR results showed the expression profiles of the selected genes were in agreement with the transcriptome data (Figure 7). The consistent expression profiles of qRT-PCR and transcriptome indicated the reliability of the transcriptome data. Interestingly, four of them were Linalool synthase (LIS) genes belonging to the terpene synthase family, which are associated with the monoterpene biosynthesis pathway and showed up-regulation in transcriptome and qRT-PCR expression. These results suggest that the expression level of LIS genes may play a crucial role in the genetic regulation mechanism of floral formation and synthesis.

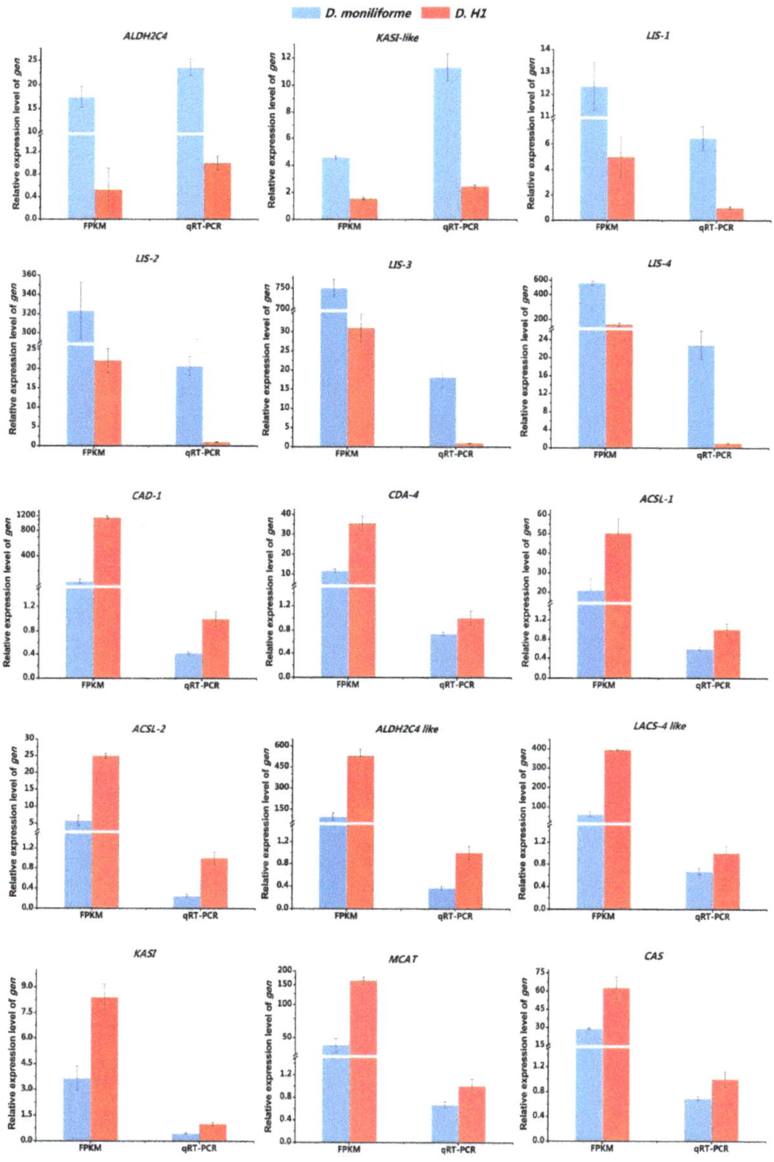

Figure 7. Relative expressions of 15 genes at the transcriptome and mRNA levels.

4. Discussion

There were significant differences in the volatile components and relative contents of different *Dendrobium* species, and the main volatile components played a decisive role in their fragrance. For instance, esters are the aroma volatile components of *D. lohohense*, while in *D. densiflorum* are mainly alkanes. The volatile content of linoleic acid and linolenic acid accounted for more than 60% of the total content in *D. thyrsiflorum* [48]. Geraniol and linalool are major floral scent compounds in *D. hancockii* [49]. The main volatile terpenoids in *D. chrysanthum* flowers were α-phellandrene, α-pinene and α-thujene [50]. In *D. officinale*, different cultivars with different smells had different contents of volatile compounds such as α-thujene, linalool, and α-terpineol [29,51]. In this experiment, SPME and GC-MS were used to extract and detect the volatile components of flowers. Although there were certain components and content differences in different florescences and parts, 60 and 52 volatile compounds were collected from *D. moniliforme* and *D.* "H1", respectively. These compounds, according to their main chemical group, were classified as terpenes, aromatics, fatty acids, nitrogenous compounds, ketone, alcohols, and alkanes. The results showed that terpenoids were the main components of aromatic compounds in the two species, and their relative contents were 76.69% (*D. moniliforme*) and 79.09% (*D.* "H1") at full bloom stage, respectively. In petals, the contents of terpenoids in *D. moniliforme* and *D.* "H1" accounted for 77.25% and 50.38%, respectively. Among terpenoids, α-Pinene is the most abundant compound in the four florescences and different parts of the two species flowers. At the same time, we found that the contents of volatile terpenoids in two *Dendrobium* species were significantly different. For example, linalool was detected in four flowering stages in *D. moniliforme*, with the content ranging from 10.21% to 27.84% and reaching 30.86% in the petals. In contrast, *D.* "H1" can only be detected in three flowering stages, with the content ranging from 0.44% to 1.37% and only 2.73% in the petals. We hypothesized that LIS content is one of the reasons for the difference in scent between the two species.

Plant volatile terpenes play critical roles in the formation of orchid floral scents, which include monoterpenes, sesquiterpenes, and diterpenes [52]. In the current study, up-regulated KEGG enrichment pathway analysis showed that the terpenoids backbone biosynthesis pathway was highly enriched in *D. moniliforme*. The findings reinforce that terpenoids biosynthesis increased in high-scent *Dendrobium*. This study suggests a consistent relationship between the *Dendrobium* of floral scents and terpenoids biosynthetic pathways. It is urgent to study the differential gene functions of these two species.

In addition to terpenoids, the floral scents of *Dendrobium* contain a series of secondary metabolites such as benzenes/phenylpropanes, fatty acid derivatives, sulfur, or nitrogen compounds released by flowers [53]. Thus, a total of 95 genes related to these pathways in KEGG were selected for PPI network analysis to clarify the functional network that controls Aroma formation. The results showed that there were 15 closely connected genes, among which 4 were LIS genes, and their expressions were up-regulated. Real time PCR analysis of the 15 closely related genes showed that their expressions were completely consistent with the transcriptome data, indicating that the transcriptome data used in this study had strong credibility.

As an important member of the monoterpene family, linalool has a unique fragrance and various biological properties, which show important application value in attracting predators and repelling pests [54,55]. For example, Arabidopsis plants release linalool by transferring the TPS gene to repel aphids [56]. Additionally, spraying inducible (3S)-linalool on OsLIS-silenced rice plants could repel infestation by the rice brown planthopper Nilaparvata lugens [57]. TPS family genes that catalyze the production of linalool have also been found in *freesia hybrids* [58], *Malus domestica* [59], and *Vitis vinifera* [60].

TPS genes are mainly responsible for the production of volatile terpenes in plants [52]. In the present study, the LIS genes we found in *D. moniliforme* and *D.* "H1" that synthesize linalool belong to the TPS gene family. TPS is the primary enzyme that catalyzes the formation of linalool from the substrate GPP. In orchid flowers, the high volatility of terpenoids promotes floral fragrances. It was confirmed that linalool and geraniol were the main floral compounds in *Phalaenopsis bellina* [61]. In in vitro experiments, the TPS10 gene uniquely converted GPP to linalool in *D. officinale* [24]. In our present research, combined analysis of volatile compounds and transcriptomes in two *Dendrobium* species revealed that four LIS genes belonging to the TPS family, which were selected from the terpenoid biosynthesis, were involved in the positive regulation of floral formation in high-scent species. We speculate that inducing specific LIS gene expression can increase the production of linalool, which promotes the formation of flower fragrance. These results deepen our understanding of the molecular mechanism of *Dendrobium* flowers. It is valuable to further study the function of these genes, and these findings will provide a valuable reference about the terpene biosynthetic pathway in orchids.

5. Conclusions

There are many species of *Dendrobium*, but few have flowers that are both fragrant and beautiful. By combining volatile constituent determination and transcriptome sequencing analysis, we compared floral components and genes of the high-scent species *Dendrobium moniliforme* (*D. moniliforme*) and light-scent species *Dendrobium* "H1" (*D.* "H1"). Physiological results showed that the terpenoids content (especially linalool content) in different species is closely related to floral fragrance, and the petals may be the most important part affecting the release of volatiles. In the transcriptome analysis results, 5790 DEGs were identified in *D.* "H1" and *D. moniliforme*. These DEGs were highly enriched in phenylpropanoid biosynthesis (49 genes), fatty acid biosynthesis (26 genes), and terpenoid backbone biosynthesis (20 genes). Using them to construct the PPI network, four LIS (linalool) genes in the terpenoids biosynthesis pathway were closely connected, and all their expressions were shown to be up-regulated. The results indicated that the linalool biosynthesis in terpenoids was closely related to the formation of floral fragrances in *D. moniliforme* species. The key DEGs related to LIS metabolism screened in this study suggested that they played important regulatory roles in floral synthesis and release. The future work should focus on confirming the function of these genes and providing ideas for the cultivation of new aromatic species.

Supplementary Materials: The following supporting information can be downloaded at: https://www.mdpi.com/article/10.3390/horticulturae9070745/s1, Table S1: Changes of the main scent compositions and relative content in *Dendrobium moniliforme*; Table S2: Changes of the main scent compositions and relative content in *Dendrobium* "H1"; Table S3: Detailed information and nucleotide sequences of PPI network genes; Table S4: Family information of transcription factors; Table S5: Primers for qRT-PCR.

Author Contributions: S.Q. initiated, designed the experiment and revised the manuscript. Y.Y., K.X., Q.W., X.L., S.L. and Z.Z. performed the experiments and collected the data. Y.Y. analyzed the data and wrote the manuscript. All authors have read and agreed to the published version of the manuscript.

Funding: This work was funded by the National Natural Science Foundation of China (31560567), the Hainan Natural Science Fund Project (320RC722), the National Natural Science Foundation of Guangxi (2020GXNSFAA297260), the Start-up Fund of Innovation Team of Guangxi Academy of Sciences for Innovation and Utilization of Germplasm in Horticultural Crops (CQZ-E-1919), the Guilin Innovation Platform and Talent Plan (20210102-3), the Fundamental Research Fund of Guangxi Institute of Botany (23011), and the fund of the Guangxi Key Laboratory of Plant Functional Phytochemicals and Sustainable Utilization (ZRJJ2023-1).

Data Availability Statement: All sequencing data are available through the NCBI Sequence Read Archive under the accession number PRJNA976822.

Conflicts of Interest: The authors declare no conflict of interest.

References

1. Ge, D.; Zhang, D.; Ding, X.; Zhou, Q.; Zhang, W.; Li, X. Genetic variation and conservation of the endangered Chinese endemic herb *Dendrobium* officinale based on SRAP analysis. *Plant Syst. Evol.* **2008**, *276*, 149–156.
2. Zhang, K.H.; Wang, M.Q.; Wei, L.L. Investigation of the Effects and Mechanisms of *Dendrobium* loddigesii Rolfe Extract on the Treatment of Gout. *Evid.-Based Complement. Altern. Med.* **2020**, *2020*, 4367347. [CrossRef]
3. Liu, W.M.; Chen, R.; Wei, R.F. Advances in development and utilization of *Dendrobium*. *Subtrop. Agric. Res.* **2011**, *7*, 87–91.
4. Ai, Y.M. *Annals of Medicinal Plants in China*; Peking University Medical Press: Beijing, China, 2013.
5. Sun, Z.; Song, Y.; Zeng, R. Advances in studies on intraspecific and interspecific relationships mediated by plant volatiles. *J. South China Agric. Univ.* **2019**, *40*, 166–174.
6. Wu, J.N.; Liu, Z.Y.; Wu, M.Q. Analysis of Volatile Components in the Flowers of Seven Species of *Dendrobium*. *Mol. Plant Breed.* **2023**, *3*, 1–16.
7. Kai, L. The Scent of Orchids. Olfactory and chemical investigations. *Nord. J. Bot.* **2010**, *14*, 302.
8. Robustelli della Cuna, F.S.; Calevo, J.; Bari, E.; Giovannini, A.; Boselli, C.; Tava, A. Characterization and Antioxidant Activity of Essential Oil of Four Sympatric Orchid Species. *Molecules* **2019**, *24*, 3878. [CrossRef]
9. Abbas, F.; Ke, Y.G.; Yu, R.C.; Yue, Y.; Amanullah, S.; Jahangir, M.M.; Fan, Y. Volatile terpenoids: Multiple functions, biosynthesis, modulation and manipulation by genetic engineering. *Planta* **2017**, *246*, 803–816. [CrossRef]

10. Yu, F.N.; Utsumi, R. Diversity, regulation, and genetic manipulation of plant mono- and sesquiterpenoid biosynthesis. *Cell. Mol. Life Sci.* **2009**, *66*, 3043–3052. [CrossRef]
11. Qiao, Z.L.; Hu, H.Z.; Shi, S.B.; Yuan, X.; Yan, B.; Chen, L. An Update on the Function, Biosynthesis and Regulation of Floral Volatile Terpenoids. *Horticulturae* **2021**, *7*, 451. [CrossRef]
12. Ramya, M.; Jang, S.; An, H.R.; Lee, S.Y.; Park, P.M.; Park, P.H. Volatile Organic Compounds from Orchids: From Synthesis and Function to Gene Regulation. *Int. J. Mol. Sci.* **2020**, *21*, 1160. [CrossRef]
13. Dinesh, A.; Priyanka, G. Advances in biosynthesis, regulation, and metabolic engineering of plant specialized terpenoids. *Plant Sci.* **2020**, *294*, 457–475.
14. Dudareva, N.; Pichersky, E.; Gershenzon, J. Biochemistry of plant volatiles. *Plant Physiol.* **2004**, *135*, 1893–1902. [CrossRef]
15. DeerInWater, K.M. *Plant Volatiles as Information Sources Influencing Herbivore Preference and Performance*; University of California: Davis, CA, USA, 2015.
16. Paschold, A.; Halitschke, R.; Baldwin, I.T. Using 'mute' plants to translate volatile signals. *Plant J.* **2010**, *45*, 275–291. [CrossRef]
17. Ali, J.G.; Alborn, H.T.; Campos-Herrera, R.; Kaplan, F.; Duncan, L.W.; Rodriguez-Saona, C.; Koppenhöfer, A.M.; Stelinski, L.L. Subterranean, Herbivore-Induced Plant Volatile Increases Biological Control Activity of Multiple Beneficial Nematode Species in Distinct Habitats. *PLoS ONE* **2012**, *7*, 38146. [CrossRef]
18. Cheng, A.X.; Lou, Y.G.; Mao, Y.B.; Lu, S.; Wang, L.-J.; Chen, X.-Y. Plant Terpenoids: Biosynthesis and Ecological Functions. *Acta Bot.* **2007**, *49*, 8. [CrossRef]
19. Dudareva, N.; Andersson, S.; Orlova, I.; Gatto, N.; Reichelt, M.; Rhodes, D.; Boland, W.; Gershenzon, J. The nonmevalonate pathway supports both monoterpene and sesquiterpene formation in snapdragon flowers. *Proc. Natl. Acad. Sci. USA* **2005**, *102*, 933–938. [CrossRef]
20. Dong, X.X.; Gao, L.; Liu, L.; Chen, L.; Liu, Q.; Zhang, D. Function-specific volatiles and volatilization characteristics of *Dendrobium* officinale. *J. King Saud Univ.-Sci.* **2020**, *32*, 2020–2028. [CrossRef]
21. Irmisch, S.; Krause, S.T.; Kunert, G.; Gershenzon, J.; Degenhardt, J.; Köllner, T.G. The organ-specific expression of terpene synthase genes contributes to the terpene hydrocarbon composition of chamomile essential oils. *BMC Plant Biol.* **2012**, *12*, 84–97. [CrossRef]
22. Zeng, X.L.; Liu, C.; Zheng, R.R.; Cai, X.; Luo, J.; Zou, J.; Wang, C. Emission and Accumulation of Monoterpene and the Key Terpene Synthase (TPS) Associated with Monoterpene Biosynthesis in Osmanthus fragrans Lour. *Front. Plant Sci.* **2015**, *6*, 1232. [CrossRef]
23. Zhou, F.; Pichersky, E. The complete functional characterisation of the terpene synthase family in tomato. *New Phytol.* **2020**, *226*, 1341–1360. [CrossRef]
24. Yu, Z.M.; Zhao, C.H.; GZhang, G.H. Genome-Wide Identification and Expression Profile of TPS Gene Family in *Dendrobium* officinale and the Role of DoTPS10 in Linalool Biosynthesis. *Int. J. Mol. Sci.* **2020**, *21*, 5419. [CrossRef]
25. Zhang, C.P.; Zhang, J.L.; Sun, Z.R.; Liu, X.Y.; Shu, L.Z.; Wu, H.; Song, Y.; He, D.H. Genome-wide identification and characterization of terpene synthase genes in Gossypium hirsutum. *Gene* **2022**, *828*, 146462. [CrossRef]
26. Song, W.; Kai, O.Y.; Kai, W. Genome-Wide Identification, Evolution, and Expression Analysis of TPS and TPP Gene Families in Brachypodium distachyon. *Plants* **2019**, *8*, 362.
27. Cao, Z.Q.; Ma, Y.; Weng, J.; Shi, J.; Chen, J.; Hao, Z. Genome-Wide Identification and Expression Analysis of TPS Gene Family in Liriodendron chinense. *Genes* **2023**, *14*, 770. [CrossRef]
28. Chen, F.; Tholl, D.; Bohlmann, J.; Pichersky, E. The family of terpene synthases in plants: A mid-size family of genes for specialized metabolism that is highly diversified throughout the kingdom. *Plant J.* **2011**, *66*, 212–229. [CrossRef]
29. Li, N.Y.; Dong, M.; Lv, L.; Qian, L.; Sun, X.; Liu, L.; Cai, Y.; Fan, H. Combined Analysis of Volatile Terpenoid Metabolism and Transcriptome Reveals Transcription Factors Related to Terpene Synthase in Two Cultivars of *Dendrobium* officinale Flowers. *Front. Genet.* **2021**, *12*, 661296. [CrossRef]
30. Robert, A.; Flath, K.O. Volatile Components of the Orchid *Dendrobium* superbum Rchb. f. *J. Agric. Food Chem.* **1982**, *30*, 842–845.
31. Baek, Y.S.; Ramya, M.; An, H.R.; Park, P.M.; Lee, S.Y.; Baek, N.I.; Park, P.H. Volatiles Profile of the Floral Organs of a New Hybrid Cymbidium, 'Sunny Bell' Using Headspace Solid-Phase Microextraction Gas Chromatography-Mass Spectrometry Analysis. *Plants* **2019**, *8*, 251. [CrossRef]
32. Bohlmann, J.; Meyer-Gauen, G.; Croteau, R. Plant terpenoid synthases: Molecular biology and phylogenetic analysis. *Proc. Natl. Acad. Sci. USA* **1998**, *95*, 4126–4133. [CrossRef]
33. Chuang, Y.C.; Hung, Y.C.; Tsai, W.C.; Chen, W.H.; Chen, H.H. PbbHLH4 regulates floral monoterpene biosynthesis in *Phalaenopsis* orchids. *J. Exp. Bot.* **2018**, *69*, 4363–4377. [CrossRef]
34. Gao, F.; Liu, B.; Li, M.; Gao, X.; Fang, Q.; Liu, C.; Ding, H.; Wang, L.; Gao, X. Identification and Characterization of Terpene Synthase Genes Accounting for the Volatile Terpene Emissions in Flowers of *Freesia hybrida*. *J. Exp. Bot.* **2018**, *69*, 4249–4265. [CrossRef]
35. Sheng, C.Z.; Wei, M.Z. A New Bibenzyl Glycoside from *Dendrobium* moniliforme. *Chin. Chem. Express* **2003**, *14*, 276–277.
36. Li, N.; Mao, Y.; Zhang, X. Separation and identification of volatile constituents in Artemisia argyi flowers by GC-MS with SPME and steam distillation. *J. Chromatogr. Sci.* **2008**, *46*, 401–405. [CrossRef]
37. Medina, A.; Lucero, M.; Estell, R.; O'CONNELL, M. A comparison of steam distillation and solid- phase microextraction (SPME) for isolation of the anemopsis californica volatiles. *Soc. Range Manag. New Mex. Sect. Meet.* **2002**, *8*. Available online: https://www.ars.usda.gov/research/publications/publication/?seqNo115=130925 (accessed on 5 April 2023).

38. Wei, S.L.; Liang, X.; Meng, Y.; Li, Y.; Gao, F.; Liu, X.; Wang, S.; Gao, X.; Wang, L. Biochemical and Molecular Characterization of a Flavonoid 3-O-glycosyltransferase Responsible for Anthocyanins and Flavonols Biosynthesis in *Freesia hybrida*. *Front. Plant Sci.* **2016**, *7*, 410.
39. Feng, L.G.; Chen, C.; Sheng, L.X.; Liu, P.; Tao, J.; Su, J.L.; Zhao, L.Y. Comparative analysis of headspace volatiles of Chinese *Rosa rugosa*. *Molecules* **2010**, *15*, 8390–8399. [CrossRef]
40. Heller, S.R.; Milne, G.W. EPA/NIH Mass Spectral Data Base. Supplement 1. 1980. U. S. Governwent Printing Office. 1980. Available online: https://www.researchgate.net/publication/235096355_EPANIH_Mass_Spectral_Data_Base_Supplement_1_1980 (accessed on 3 February 2023).
41. Zhang, H.; Chen, M.; Wang, X.; Dai, J.; Zhang, X.; Zhang, Z.; Zhang, X.; Tang, M.; Tang, J.; Gong, J.; et al. Transcriptome Analysis of *Rhododendron liliiflorum* H. Lév. Flower Colour Differences. *Horticulturae* **2023**, *9*, 82. [CrossRef]
42. Bray NL, H.; Pimentel, P.; Melsted, L. Near-optimal RNA-Seq quantification. *Comput. Sci.* **2015**, *11*, 48550.
43. Li, T.; Cai, Z.L.; Wu, H.J. Transcriptome analysis of the Japanese pine sawyer beetle, *Monochamus alternatus* (Coleoptera: Cerambycidae) by high-throughput Illumina sequencing. *J. Asia-Pac. Entomol.* **2015**, *18*, 439–445.
44. Gaur, R.; Bhatia, S.; Gupta, M. Generation of expressed sequence tags under cadmium stress for gene discovery and development of molecular markers in chickpea. *Protoplasma* **2014**, *251*, 955–972. [CrossRef]
45. Kanehisa, M.; Furumichi, M.; Tanabe, M.; Sato, Y.; Morishima, K. KEGG: New perspectives on genomes, pathways, diseases and drugs. *Nucleic Acids Res.* **2017**, *45*, 353–361. [CrossRef]
46. Martin IR, E.; Vigne, V.; Komar, A.; Amandine, V.; Schmitt-Keichinger, C. Transcriptomic analysis of grapevine cv. Gewurztraminer infected by two strains of grapevine fanleaf virus with distinct phenotypes. In Proceedings of the 15th Meeting of the International Council for the Study of Virus and Virus-Like Diseases of the Grapevine (ICVG), Santiago de Chile, Chile, 3–7 April 2018.
47. Han, X.F.; Ling, Q.F.; Li, C.J.; Wang, G.; Xu, Z.; Lu, G. Characterization of pikeperch (*Sander lucioperca*) transcriptome and development of SSR markers. *Biochem. Syst. Ecol.* **2016**, *66*, 188–195. [CrossRef]
48. Zhang, G.N.; Zhang, C.F.; Wang, Z.T.; Xu, L.S. Studies on chemical constituents of *Dendrobium* thyrsiflorum rchb. (I). *Chin. J. Nat. Med.* **2004**, *2*, 78–81.
49. Li, C.H.; Huang, M.Z.; Huang, S.H.; Yin, J. Volatile Components in Flowers of Four *Dendrobium* Species. *J. Trop. Subtrop. Bot.* **2015**, *23*, 454–462.
50. Wen, X.; Wang, Y.; Chen, L.; Zhang, K.; Zhang, J.; Li, Y. Chemical Constituents from the Flowers of *Dendrobium* chrysanthum. *Agric. Sci. Technol.* **2014**, *15*, 1629–1633.
51. Hua, L.S.; Meng, X.U.; Zhang, X.F.; Liu, J.J.; Si, J.P. Studies on Volatile Constituents of 11 Families of *Dendrobium* officinale Flowers. *Chin. J. Exp. Tradit. Med. Formulae* **2016**, *22*, 52–57.
52. Huang, X.Z.; Xiao, Y.T.; Köllner, T.G.; Jing, W.X.; Kou, J.F.; Chen, J.Y.; Liu, D.F.; Gu, S.H.; Wu, J.X.; Zhang, Y.J.; et al. The terpene synthase gene family in *Gossypium hirsutum* harbors a linalool synthase GhTPS12 implicated in direct defence responses against herbivores. *Plant Cell Environ.* **2017**, *41*, 261–274. [CrossRef]
53. Dudareva, N. Biochemical and Molecular Genetic Aspects of Floral Scents. *Plant Physiol.* **2000**, *122*, 627–633. [CrossRef]
54. Zhou, P.; Du, Y.; Xu, N.; Yue, C.; Ye, L. Improved linalool production in Saccharomyces cerevisiae by combining directed evolution of linalool synthase and overexpression of the complete mevalonate pathway. *Biochem. Eng. J.* **2020**, *161*, 107655. [CrossRef]
55. Zhou, P.; Du, Y.; Fang, X.; Xu, N.; Yue, C.; Ye, L. Combinatorial Modulation of Linalool Synthase and Farnesyl Diphosphate Synthase for Linalool Overproduction in Saccharomyces cerevisiae. *J. Agric. Food Chem.* **2021**, *69*, 1003–1010. [CrossRef]
56. Aharoni, A. Terpenoid metabolism in wild-type and transgenic Arabidopsis plants. *Plant Cell* **2003**, *12*, 107655. [CrossRef]
57. Xiao, Y.; Wang, Q.; Erb, M.; Turlings, T.C.J.; Ge, L.; Hu, L.; Li, J.; Han, X.; Zhang, T.; Lu, J.; et al. Specific herbivore-induced volatiles defend plants and determine insect community composition in the field. *Ecol. Lett.* **2012**, *15*, 1130–1139. [CrossRef]
58. Yang, Z.Z.; Li, Y.Q.; Gao, F.Z.; Jin, W.; Li, S.; Kimani, S.; Yang, S.; Bao, T.; Gao, X.; Wang, L. MYB21 interacts with MYC2 to control the expression of terpene synthase genes in flowers of *Freesia hybrida* and *Arabidopsis thaliana*. *J. Exp. Bot.* **2020**, *71*, 4140–4158. [CrossRef]
59. Nieuwenhuizen, N.J.; Green, S.A.; Chen, X.; Bailleul, E.J.; Matich, A.J.; Wang, M.Y.; Atkinson, R. Functional genomics reveals that a compact terpene synthase gene family can account for terpene volatile production in apple. *Plant Physiol.* **2013**, *161*, 787–804. [CrossRef]
60. Martin, D.M.; Aubourg, S.; Schouwey, M.B.; Daviet, L.; Schalk, M.; Toub, O.; Lund, S.T.; Bohlmann, J. Functional Annotation, Genome Organization and Phylogeny of the Grapevine (*Vitis vinifera*) Terpene Synthase Gene Family Based on Genome Assembly, FLcDNA Cloning, and Enzyme Assays. *BMC Plant Biol.* **2010**, *10*, 226. [CrossRef]
61. Hsiao, Y.Y.; Pan, Z.J.; Hsu, C.C.; Yang, Y.P.; Hsu, Y.C.; Chuang, Y.C.; Shih, H.H.; Chen, W.H.; Tsai, W.C.; Chen, H.H. Research on Orchid Biology and Biotechnology. *Plant Cell Physiol.* **2011**, *52*, 1467–1486. [CrossRef]

horticulturae

MDPI

Article

Characterization, Evolutionary Analysis, and Expression Pattern Analysis of the Heat Shock Transcription Factors and Drought Stress Response in *Heimia myrtifolia*

Guozhe Zhang [1,2,3,†], Cuihua Gu [1,2,3,†], Yacheng Ye [1,2,3,†], Yu Zhao [1,2,3], Linxue Shang [1,2,3], Weili Shao [1,2,3], Sidan Hong [1,2,3,*] and Jin Ma [1,2,3,*]

1 College of Landscape and Architecture, Zhejiang Agriculture & Forestry University, Hangzhou 311300, China; zhangguozhe@stu.zafu.edu.cn (G.Z.); zhaoyu@stu.zafu.edu.cn (Y.Z.); vswl@163.com (W.S.)
2 Zhejiang Provincial Key Laboratory of Germplasm Innovation and Utilization for Garden Plants, Zhejiang Agriculture & Forestry University, Hangzhou 311300, China
3 Key Laboratory of National Forestry and Grassland Administration on Germplasm Innovation and Utilization for Southern Garden Plants, Zhejiang Agriculture & Forestry University, Hangzhou 311300, China
* Correspondence: jude1998@stu.zafu.edu.cn (S.H.); majinzjl@163.com (J.M.)
† These authors contributed equally to this work.

Abstract: Heat shock transcription factors (HSFs) are among the most important regulators of plant responses to abiotic stimuli. They play a key role in numerous transcriptional regulatory processes. However, the specific characteristics of HSF gene family members and their expression patterns in different tissues and under drought stress have not been precisely investigated in *Heimia myrtifolia*. This study analyzed transcriptome data from *H. myrtifolia* and identified 15 members of the HSF family. Using a phylogenetic tree, these members were classified into three major classes and fifteen groups. The amino acid physicochemical properties of these members were also investigated. The results showed that all *HmHSF* genes are located in the nucleus, and multiple sequence alignment analysis revealed that all HmHSF proteins have the most conserved DBD structural domains. Interestingly, a special HmHSF15 protein was found in the three-dimensional structure of the protein, which has a conserved structural domain that performs a function in addition to the unique structural domain of HSF proteins, resulting in a three-dimensional structure for HmHSF15 that is different from other HmHSF proteins. GO enrichment analysis shows that most *HmHSFA*-like genes are part of various biological processes associated with abiotic stresses. Finally, this study analyzed the tissue specificity of *HmHSF* genes in different parts of *H. myrtifolia* by qRT-PCR and found that *HmHSF* genes were more abundantly expressed in roots than in other tissues, and *HmHSF05*, *HmHSF12*, and *HmHSF14* genes were different from other *HSF* genes, which could be further analyzed to verify their functionality. The results provide a basis for analyzing the functions of *HmHSF* genes in *H. myrtifolia* and help to explore the molecular regulatory mechanism of *HmHSF* in response to drought stress.

Keywords: heat shock transcription factors; drought stress; *Heimia myrtifolia*

Citation: Zhang, G.; Gu, C.; Ye, Y.; Zhao, Y.; Shang, L.; Shao, W.; Hong, S.; Ma, J. Characterization, Evolutionary Analysis, and Expression Pattern Analysis of the Heat Shock Transcription Factors and Drought Stress Response in *Heimia myrtifolia*. *Horticulturae* **2023**, *9*, 588. https://doi.org/10.3390/horticulturae9050588

Academic Editor: Sumei Chen

Received: 24 April 2023
Revised: 10 May 2023
Accepted: 12 May 2023
Published: 16 May 2023

1. Introduction

As the global climate gradually deteriorates, localized weather extremes are becoming more frequent, with drought being one of the most obvious climatic features [1]. Extensive research has demonstrated that drought is a significant abiotic stress factor that impacts plant growth and development [2,3]. It frequently causes cell water loss, disconnection of the serum wall and enzyme systems, metabolic imbalances, and ultimately changes the direction of plant growth [4–6]. When plants are subjected to external stress stimuli, they produce a number of signals in the living organism that cause a variety of signaling pathways to be turned on and the activation of subsequent proteins, such as those from HSFs, DREB, and bZIP [7]. The stimulation of plant-specific resistance genes and associated defense mechanisms by transcription factors alters the ability of plants to adapt to their environment [8].

Transcription factors are a class of proteins with specific structures and regulatory functions that can play a significant role in the growth and behavior of plants in the external environment by activating or repressing gene transcription [9]. The heat shock transcription factor class is an important class of plant stress resistance factors [6,10]. It has a major impact on heat, cold, dehydration, drought, salt, and disease resistance processes in plants, and plays an extremely important role in controlling plant adaptation to stress as well as plant development [11–13]. The highly conserved HSF protein in plants is made up of a structurally and functionally conserved DNA-binding domain at the N-terminus (DBD), an adjacent dichotomous domain for oligomers, a nuclear localization signal, a nuclear export signal, and a deterrent domain [14,15]. The DBD structural domain is the one that has been retained the best in HSF out of all of these conserved structural domains [16]. According to related studies, the HSF family members and their activities were exhaustively examined in *Arabidopsis thaliana* [17], *Ziziphus jujuba* [18], and other model plants [15,19–21]. The findings clearly indicate that HSF plays a remarkably valuable role in the resistance of these plants to abiotic stresses [22].

Heimia myrtifolia loves the sun and has few pests and diseases, making it a rare small shrub that flowers in both summer and autumn [23]. When in bloom, the golden-yellow flowers adorn the branches and add great amenity and accent to the landscape. For this reason, it is widely used in landscaping [24]. However, *H. myrtifolia* has a higher water requirement for its habitat. In the event of a water shortage during its lifetime, the leaves will curl, and, if they are not rehydrated in time, the whole plant will promptly die. Therefore, it is essential to conduct research on drought tolerance in *H. myrtifolia* and to progressively select drought-tolerant *H. myrtifolia* germplasm resources. The group determined the transcriptome of *H. myrtifolia* drought stress in the early stage and carried out a series of analyses. Analytical studies revealed that most *HSF* genes play an important role in the response of *H. myrtifolia* to drought stress. However, the study did not investigate the key *HSF* genes. It is known that *HSF* genes are essential for receiving and transmitting signals, recognizing heat shock components, and controlling downstream genes [25–27]. Therefore, this study first identified family members of the *H. myrtifolia HSF* using a bioinformatic approach based on transcriptome data from *H. myrtifolia*. Subsequently, this study investigated the amino acid physicochemical properties, multiple sequence alignment, GO enrichment, and tissue expression pattern analysis of HSF family members under simulated drought conditions. These findings will help us to better understand the evolutionary links and the functional differentiation of *HmHSF* genes in *H. myrtifolia* and provide corresponding theoretical support for future studies on the molecular biology of *H. myrtifolia HSF* genes.

2. Materials and Methods

2.1. HSF Gene Characterization and Sequencing Analysis in H. myrtifolia

This study retrieved the sequences of the HSF family of *A. thaliana* from the TAIR database (https://www.arabidopsis.org/ (accessed on 3 March 2023)). A BLASTP search was then performed on the transcriptome data of *H. myrtifolia*. In the next step, the built-in HMMER 3.0 program was used to download the HSF family members of *H. myrtifolia* by using the Hidden Markov Model of the HSF protein (PF0047) in combination with the Pafm database (http://pfam.xfam.org/ (accessed on 15 March 2023)). This study summarized the potential genes for *HmHSF* identified by these two search techniques. Finally, the potential HmHSF proteins were validated for HSF-conserved structural domains by Web CD-search and SMART, thus ensuring the accuracy of the screened *HmHSF* members. This study finally identified 15 *HmHSF* genes and renamed them in the order of the original assembly ID of the transcriptome.

2.2. Prediction of the Physicochemical Properties and Protein Structure of HmHSF

This study predicted the subcellular localization of HmHSF proteins using Cell-PLoc 2.0, and, through the ExPasy website, the physicochemical properties of all HmHSF proteins.

The secondary structures of the proteins were analyzed by the online site SOPMA, and then the data were collated using Excel. The prediction of the tertiary structure of proteins was presented by the online tool SWISS-MODEL.

2.3. Phylogenetic Tree Construction and Sequence Comparison

Using the NCBI database (https://ngdc.cncb.ac.cn/ (accessed on 26 March 2023)), this study downloaded all *HSF* genes from *Oryza sativa* and *A. thaliana*. Multiple-sequence protein alignment using the 21 AtHSF proteins, 25 OsHSF sequences, and 15 HmHSF protein sequences was performed using MEGA 6.0 to generate NJ trees for HSF proteins. The phylogenetic tree was post-landscaped and processed using iTOL. This study also used Jalview software to perform a multiple sequence alignment analysis on HSF proteins.

2.4. GO Database Annotation and Enrichment Analysis

This study performed GO functional annotation of the *HmHSF* gene via the online tool eggNOG-mapper. The annotation result files were then imported into the eggNOG-mapper program of TBtools for analysis and aggregation. Finally, this study used the online website HIP-LOT to graph and embellish the data results generated by TBtools.

2.5. Plant Material, Methods, and qRT-PCR Validation

This study grew annual cuttings of *H. myrtifolia* in an artificial climate chamber at Zhejiang Agriculture and Forestry University. The soil moisture content in the cuttings' basins was maintained at 33.5% for one week prior to the drought treatment and replenished each morning at 8 a.m. using a TDR100 portable soil moisture meter. For the drought treatment, one control group (CK) and four treatment groups (T1, T2, T3, and T4) were established, each with three replicates. The relative soil moisture contents were 65–75% (CK), 45–60% (T1), 30–45% (T2), 15–30% (T3), and 5–15% (T4). The field water content was measured each morning at 8 a.m. to maintain it within the control range. After 10 days of drought treatment, roots, stems, leaves, and flowers were collected from the control group and leaves from the different treatment groups. The samples were rapidly frozen in liquid nitrogen and stored at −80 °C.

This study extracted total RNA from the leaves of the control and treatment groups using the FastPure Plant Total RNA Isolation Kit from Vazyme. The first strand cDNA was then reverse-transcribed using HiScriptIII All-in-One RT Perfect Hypermix. Specific primers for 15 pairs of genes were designed using Primer 5 software, with *HmGAPDH* as the internal reference gene (Table S1). qRT-PCR amplification was performed using SYBR Premix Ex TaqTM in 10 μL volumes on an ABI 7300 real-time PCR instrument. Three replicates were performed for each selected gene and the relative expression levels of *HmHSF* genes were analyzed using the $2^{-\Delta\Delta Ct}$ method. The experimental data were analyzed and organized using Excel 2010 and SPSS 20.0.

This study performed K-means clustering analysis of 15 HmHSF genes through the online website HIP-LOT, which was then further embellished using Photoshop 2022.

3. Results

3.1. Identification of HSF Genes in H. myrtifolia

This study searched for 15 *HmHSF* genes in the *H. myrtifolia* transcriptome database (transcriptome data available from NCBI under number PRJNA804698) using BLAST and HMMER search methods. For subsequent analysis, all members were named *HmHSF01*-*HmHSF15* in the order of their assembly ID (Table 1).

Table 1. Physicochemical property analysis of HmHSF proteins.

Gene Name	Gene ID	Size (aa)	MW (kDa)	pI	Stability	A.I	GRAVY	Predicted Location
HmHSF01	Gene.80304	446	50501.11	5.83	U	65.88	−0.869	Nuclear
HmHSF02	Gene.33060	363	40958.06	4.81	U	81.87	−0.515	Nuclear
HmHSF03	Gene.82041	290	32889.93	6.35	U	59.55	−0.808	Nuclear
HmHSF04	Gene.67227	391	44075.58	5.11	U	61.10	−0.931	Nuclear
HmHSF05	Gene.83400	339	38308.34	8.23	U	70.44	−0.522	Nuclear
HmHSF06	Gene.78293	325	36158.11	4.64	U	65.38	−0.677	Nuclear
HmHSF07	Gene.92484	372	41725.87	9.07	U	71.05	−0.756	Nuclear
HmHSF08	Gene.77897	303	34137.49	6.19	U	68.58	−0.549	Nuclear
HmHSF09	Gene.93333	253	28667.25	9.33	U	61.70	−0.836	Nuclear
HmHSF10	Gene.82043	295	32617.38	8.31	U	66.88	−0.69	Nuclear
HmHSF11	Gene.98726	427	48851.44	5.16	U	69.13	−0.784	Nuclear
HmHSF12	Gene.96475	392	44531.09	6.04	U	77.83	−0.663	Nuclear
HmHSF13	Gene.44218	320	34804.22	5.33	U	72.28	−0.46	Nuclear
HmHSF14	Gene.51133	279	30884.61	7.02	U	62.26	−0.785	Nuclear
HmHSF15	Gene.86874	492	53592.58	4.85	U	61.20	−0.643	Nuclear

The 15 *HmHSF* genes are anticipated to encode proteins with 253 (*HmHSF09*) to 492 (*MsHSF15*) amino acids based on the projected physicochemical features of the amino acid sequences. The molecular weights (MWs), with an average of 39,513.54 Da, varied from 28667.25 (*MsHSF09*) to 53592.58 Da (*MsHSF15*). The anticipated isoelectric point (pI), with a mean value of 6.42, varied from 4.64 (*MsHSF06*) to 9.33 (*MsHSF09*). The stability index indicates that every HSF protein is erratic in vitro. The aliphatic amino acid index (A.I.) also varied between 59.55 (*MsHSF03*) and 81.87 (*MsHSF02*), indicating a minor difference in thermal stability. The hydrophilic (GRAVY) rankings of each HSF protein have a negative overall mean, demonstrating that all of them are hydrophilic proteins. The final discovery was that all HSF proteins were found in the atomic nucleus, as predicted by subcellular localization predictions.

3.2. HmHSF Protein Sequence Analysis

To further investigate the existence and location of the conserved protein structural domain, in this study, we performed a multiple sequence alignment analysis of the 15 HmHSF proteins using Jalview software and found that all 15 HmHSF proteins contain a highly conserved structural domain, the DBD conserved structural domain (Figure 1B), which consists of approximately 100 amino acids (Figure 1D). Subsequently, to further reveal the structural diversity of the HmHSF protein sequences, this study analyzed the 10 most conserved motifs of the HmHSF proteins using MEME software and named them Motif 1–10. The conserved amino acid patterns range in length from 9 to 50. Motif 1 is the longest in this study with 50 conserved amino acids, while motif 9 is the shortest with 9 conserved amino acids. This study visualized the conserved structural domains of HmHSF using TBtools and found that HmHSF proteins in the same broad class have similar conserved motifs. Interestingly, Motif 5 was only found in HSFA-like proteins. These results suggest that specific motifs may be associated with the functions of different subgroups.

3.3. Phylogenetic Analysis of HmHSF Protein

To investigate the evolutionary features among *H. myrtifolia HSF* genes, in this study, we constructed an NJ phylogenetic tree of the amino acid sequences of 15 *H. myrtifolia* HSF proteins, 21 *A. thaliana* HSF proteins, and 25 *O. sativa* HSF proteins using MEGA 6.0 (Figure 2). According to our previous family classification of *A. thaliana* HSF proteins, in this study, we divided the 15 *H. myrtifolia* HSF proteins into three groups, namely HSFA (green), HSFB (yellow), and HSFC (blue). Of these, HSFA was divided into nine subgroups (A1–A9) with seven members, including HmHSF01, HmHSF02, HmHSF04, HmHSF07, HmHSF11, HmHSF12, and HmHSF15, where the *H. myrtifolia* HSF proteins did not aggregate into five subgroups: A3, A6, A7, A8, and A9. HSFB (B1–B4) has four subgroups within it, consisting of seven members, namely HmHSF03, HmHSF05, HmHSF06, HmHSF09, HmHSF10, HmHSF13, and HmHSF14. *H. myrtifolia* HSFC is the smallest group, containing only one member, HmHSF08, distributed in subgroup C1.

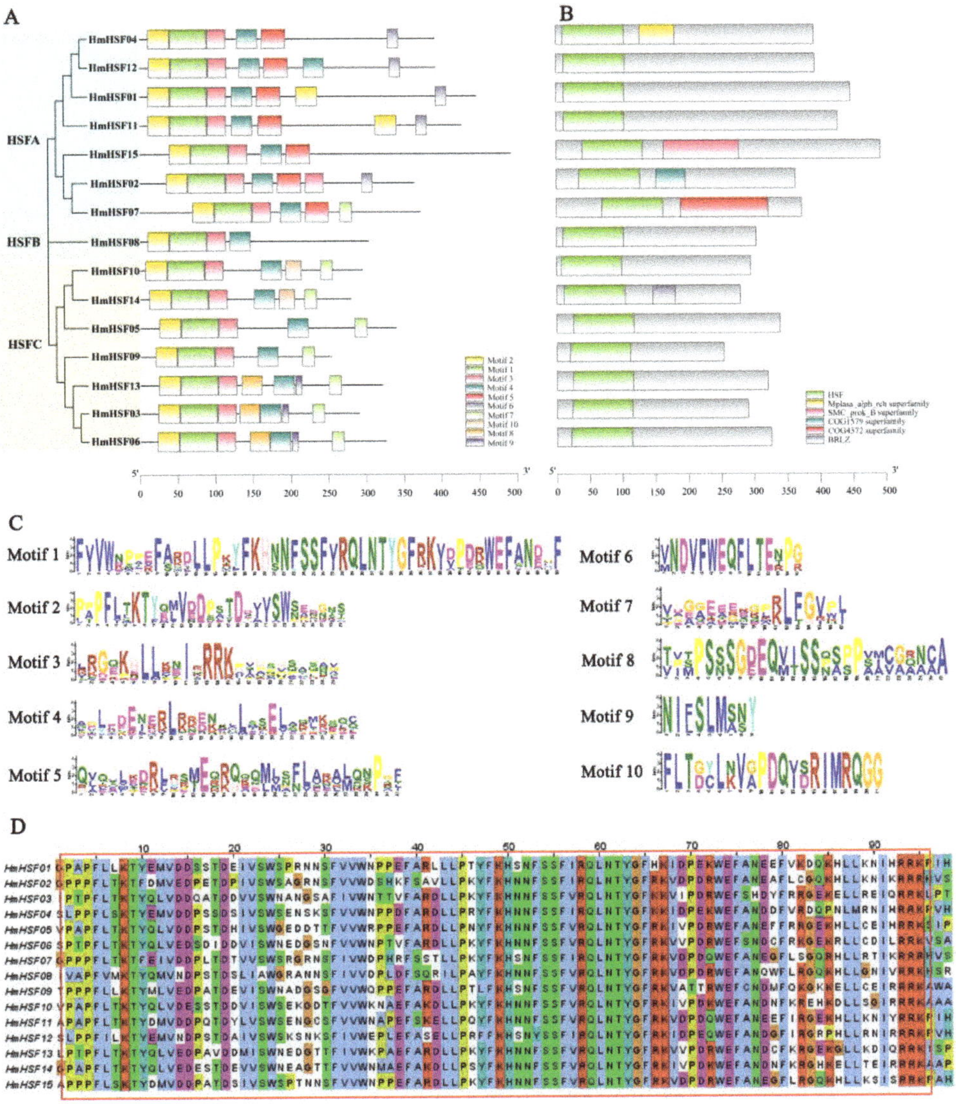

Figure 1. Characterization of *HmHSF* genes. (**A**) Phylogenetic tree of the *HmHSF* gene family. Genes in the *HSFA*, *HSFB*, and *HSFC* groups are shown in blue, green, and yellow, respectively. Different colored boxes represent different conserved motifs. (**B**) Conserved protein structural domains. The green color is the most conserved DBD structural domain of the HSF protein. (**C**) The sequence identity of all identified HmHSF patterns was analyzed. (**D**) DBD structural domain of HmHSF protein multiple sequence alignment. The red boxed part is the most conservative structural domain of HSF, DBD structure.

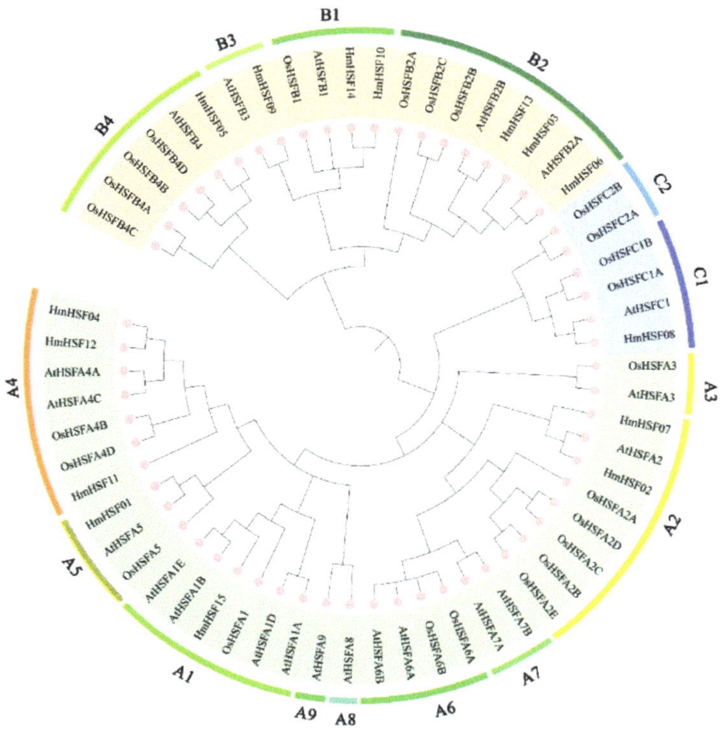

Figure 2. NJ phylogenetic tree of HSF proteins of *H. myrtifolia*, *A. thaliana*, and *O. sativa* constructed using default parameters. The evolutionary tree was divided into three broad categories of different colors: HSFA (green), HSFB (yellow), and HSFC (blue). According to the previous subfamily classification of AtHSF proteins, HSFA contains nine subgroups (A1–A9), HSFB contains four subgroups (B1–B4), and HSFC contains two subgroups (C1–C2).

3.4. Structural and Interaction Network Analysis of the HmHSF Protein

To better understand the interactions between HmHSF proteins, this study used the STRING database to predict their possible interactions (Figure 3). The results showed that the HmHSF15 protein is a central node in the protein interaction network graph and is closely related to other connecting nodes, and, subsequently, this study investigated the HmHSF secondary structure and tertiary structure to further analyze the HmHSF protein (Table S2). The secondary structure analysis showed that the secondary structure of HmHSF02 and HmHSF07 proteins is mainly α-helix, and then randomly becomes a coiled coil; then an extended chain; and lastly, it exhibits β-rotation. The secondary structures of the other 13 HmHSF proteins were mainly random coiling, followed by α-helix, followed by the extension chain, and β-rotation was the least common. Our prediction of protein tertiary structure showed that random coiling and α-helix contribute significantly to the stability of the tertiary structure, while β-rotation plays a crucial role in modification. Interestingly, the tertiary structure models of all HmHSF proteins were highly similar except for HmHSF15 (Figure 4). The difference between the structure of the HmHSF15 protein and the rest of the proteins can be analyzed with the help of the domains in Figure 1B, where it can be clearly seen that HmHSF15 protein has other structural domains besides the conserved structural domain of HSF, which performs other functions and causes changes in the three-dimensional structure of the HmHSF15 protein at the same time.

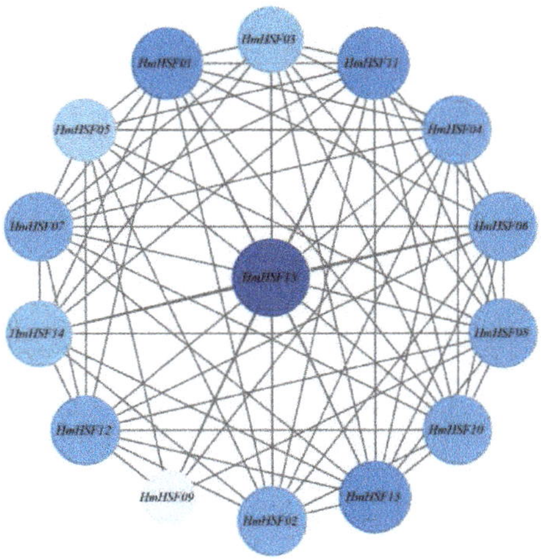

Figure 3. Interaction network diagram of HmHSF proteins. Nodes represent proteins; darker blue indicates stronger connectivity of the protein, and grey lines indicate interactions between nodes.

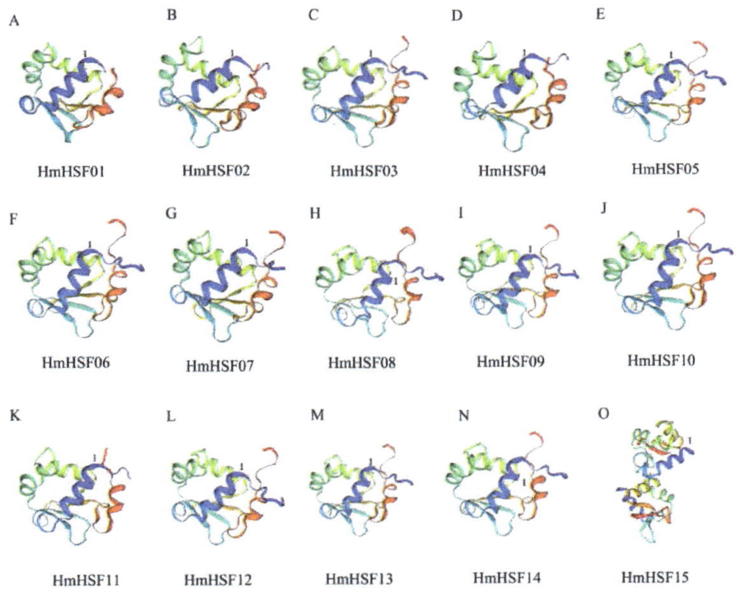

Figure 4. 3D structure model of HmHSF protein. (**A–O**) represents HmHSF01-HmHSF15, respectively. Number 1 represents the DBD structure of HmHSF protein.

3.5. Analysis of the GO Annotation and Enrichment

To understand the biological activities carried out by HmHSF proteins, in this study, we performed a GO annotation analysis of 15 *HmHSF* genes, which showed that 13 genes are involved in the biological processes (BP), cellular components (CC), and molecular functions (MF). This study then performed a GO enrichment analysis on 12 *HmHSF* genes (Figure 5), which revealed that *HmHSF* genes are involved in biological functions related to drought stress, such as in the response to abiotic factors, response to temperature stimuli, and response to xenobiotic stimuli. Interestingly, GO enrichment showed that most of the biological processes responding to stress and other related functions were enriched in *HmHSFA*-like genes; *HmHSFB*-like genes were enriched very rarely, and *HmHSFC*-like genes were not enriched at all. Therefore, *HmHSFA*-like genes play an irreplaceable role in the response to drought stress.

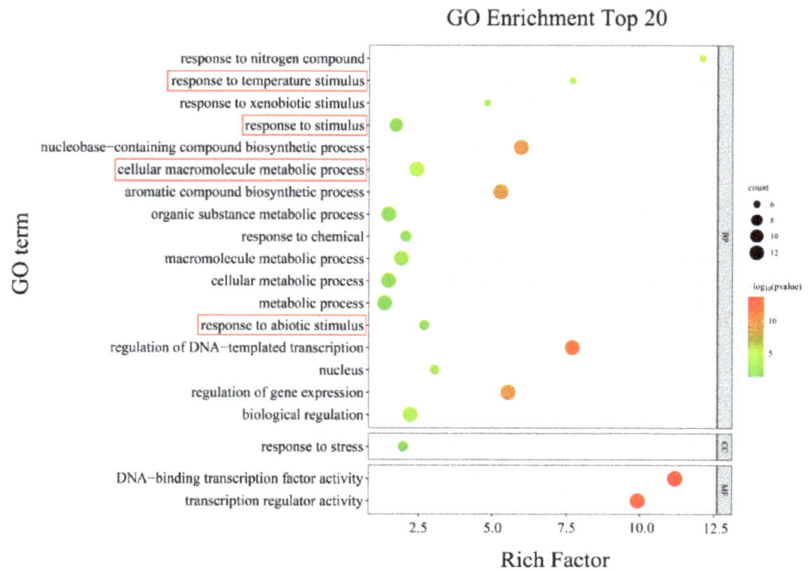

Figure 5. Analysis of the HmHSF proteins′ GO enrichment in comparison to the GO database. The red squares represent is the biological processes associated with drought stress.

3.6. Expression and K-Means Clustering Analysis of HmHSF Gene

To investigate the potential biological functions of *HmHSF* genes in the growth and development of *H. myrtifolia*, this study studied the expression of *HmHSF* genes in roots, stems, leaves, and flowers using real-time quantitative PCR to determine the expression patterns of specific *HmHSF* genes in different tissues. The results were also presented via TBtools software in the form of heat maps (Figure 6A). It was found that *HmHSF* genes were significantly expressed in roots except for *HmHSF08* and *HmHSF09*, suggesting that these genes may be required for root tissue development. Notably, all *HmHSF* genes were barely expressed in flowers and stems except for *HmHSF03*, *HmHSF08*, and *HmHSF09*, which were highly expressed in leaves. These findings imply that there is a clear tissue specificity in the expression of *HmHSF* genes in different tissues.

In addition, this study obtained the expression profile of *HmHSF* genes by transcriptome data extraction and screening (Figure 6B), and found that the genes that exerted the earliest response in the initial stage of drought stress were *HmHSF05*, *HmHSF09*, and *HmHSF12*; with the increase in stress duration, *HmHSF02*, *HmHSF06*, *HmHSF07*, and *HmHSF13* were highly expressed, resisting the damage caused by drought stress; finally, the genes that were not expressed in the first two stages during the continuous increase in drought stress also appeared to be highly expressed to resist the stress. In this study, we also performed a K-means clustering analysis of 15 *HmHSF* genes (Figure 6C), and the analysis showed that 15 *HmHSF* genes were clustered into 10 clusters that functioned in the process of drought stress in the form of regulation without passage, respectively. These results suggest that genes of the HmHSF family may play important roles in different stages of drought stress. Notably, *HmHSF05*, *HmHSF12*, and *HmHSF14* played roles in two of the three stages of drought stress, which subsequently warranted further investigation into their functionality in drought stress.

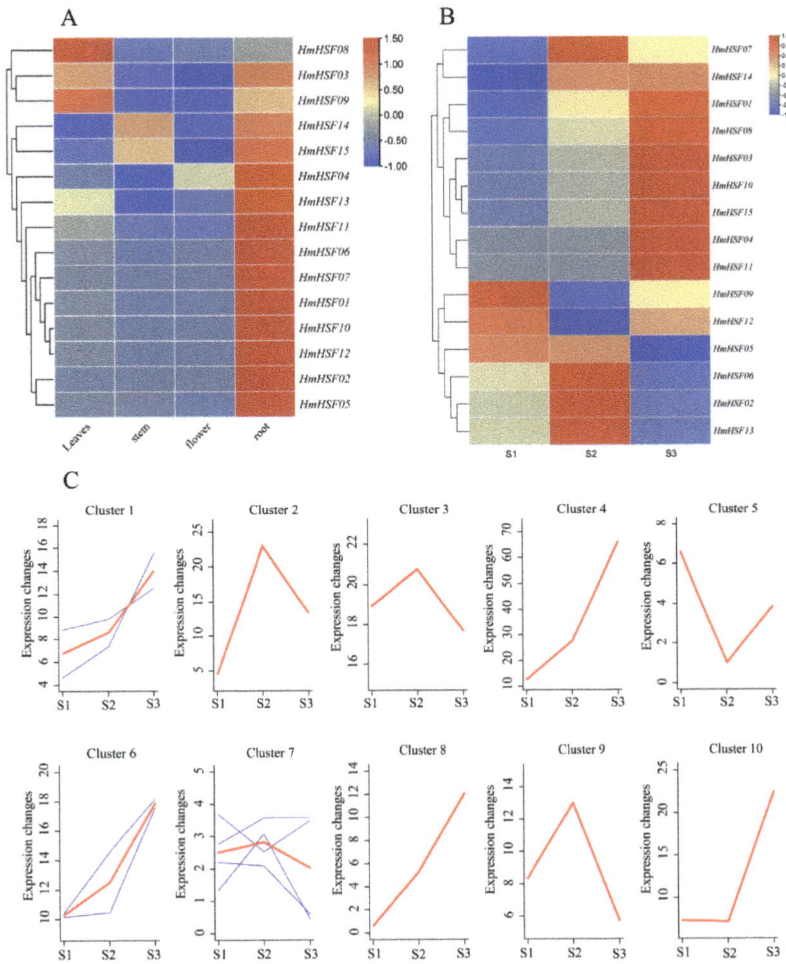

Figure 6. (**A**) The expression of *HmHSF* genes in different tissues using a heatmap, where red represents high expression and blue represents low expression. (**B**) The expression of *HmHSF* genes under drought stress in transcriptome data, with soil water content ranging from 65–75%, 30–45%, and 5–15%. The expression values were visualized in a heatmap, with red representing high expression and blue representing low expression. (**C**) K-means clustering analysis of *HmHSF* gene expression. The red line in the figure indicates the trend of such genes, and the blue line indicates the *HmHSF* genes enriched in this trend.

3.7. Expression of HSF Genes under Drought Stress

The genes *HmHSF05*, *HmHSF12*, and *HmHSF14* were identified by tissue specificity analysis and K-means clustering analysis as being different from other *HmHSF* genes in that they were highly expressed in two tissues (Figure 7). This study further investigated the expression of these genes under drought stress using q-PCR. The analysis showed that the expression trend of the three genes increased and then decreased, with *HmHSF05* and *HmHSF14* being negatively regulated and *HmHSF12* being positively regulated. These results were consistent with the transcriptome data. Therefore, it is hypothesized that *HmHSF* genes may be closely related to the response mechanism of drought stress.

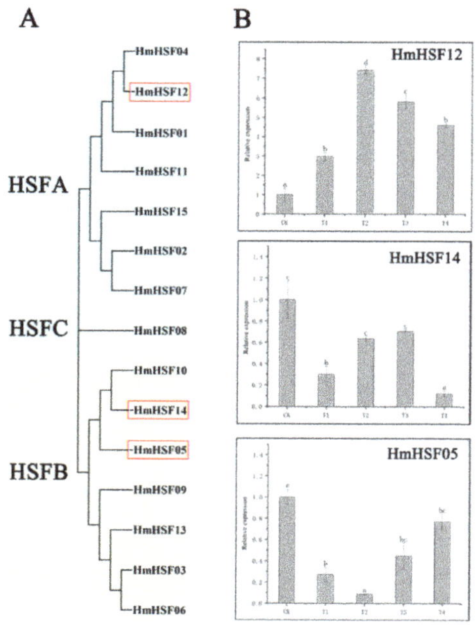

Figure 7. (**A**) The phylogenetic tree of the *HmHSF* gene family is shown. Genes in the *HSFA*, *HSFB*, and *HSFC* groups are shown in blue, green, and yellow, respectively. (**B**) The expression analysis of three *HmHSF* genes under simulated drought stress is presented. Data represent the mean ± SD of three biological replicates. Means indicated by the same letter are not significantly different at $p < 0.05$, as determined by Duncan's multiple range test.

4. Discussion

Knowledge on plant stress under drought stress is important in order to enhance the ornamental value of landscape plants under adversity. HSFs belong to a relatively special category among all transcription factors. Such a class of TFs is essential for plants so that they can withstand diverse biotic and abiotic stresses [28].

Based on the transcriptome data of *H. myrtifolia*, 15 complete HmHSF genes with open reading frames were discovered in this study, and their fundamental characteristics were investigated. *H. myrtifolia* has fewer *HSF* genes than plants such as *A. thaliana* (21), *S. lycopersicum* (25), and *Zea mays* (30); this may be due to the loss of some *HSF* genes during evolution [29–31]. The analysis of the physicochemical properties of *H. myrtifolia* HSF proteins showed that the molecular weight and amino acid content of the proteins did not differ much from each other. Subcellular localization analysis showed that all HSF proteins were localized in the nucleus. This suggests that no differential evolution has occurred during the evolution of HmHSF proteins. Early studies found that the *N*-terminal end of the highly conserved plant HSF DBD is capable of locating and identifying the target gene's heat stress element (HSE) prokinetic region [32,33]. Comparative multiple sequence analysis and protein secondary structure prediction revealed that all HmHSF proteins have DBD structures, which consist of three α-helices and four β-folds. It is noteworthy that HmHSF has other structural domains besides the DBD structural domain; therefore, further experiments are necessary to analyze whether this structural domain affects the function of the HSF gene family. In this study, we also constructed phylogenetic trees for *H. myrtifolia*, *A. thaliana*, and *O. sativa*. The *HmHSF* genes were further grouped into three groups, *HmHSFA*, *HmHSFB*, and *HmHSFC*, using homology matching and multi-species matching. The numbers of genes in the groups *HmHSFA* and *HmHSFB* were the highest, and the numbers of genes in the group *HmHSFC* were the lowest, and this clustering was the same as that of the *A. thaliana* HSF gene family members [34]. A related study found that the expression of class A *HSF* increased most significantly when *A. thaliana* was subjected to heat stress, and excessive manifestation of the *AtHSFA2* gene significantly increased the basal heat resistance of plants [35]. Although class B *HSF* can also recognize and bind to the thermal initiation element HSE, it is prevented from doing so by the carboxy-terminal characteristic structural domain (-LFGV-), which prevents plants from regulating either their ability to tolerate heat or their ability to initiate normal thermal response gene expression [36,37]. Therefore, whether there are differences in the protein function and mode of action of different groups of *H. myrtifolia HmHSF* needs further study. Notable is the absence of the subgroups A3, A6, A7, A8, and A9 in the *H. myrtifolia* HSF proteins; this finding shows that despite having a common ancestor, HSF family proteins developed independently in various species. Nearly all the *H. myrtifolia* HSF proteins are comparable to those of *A. thaliana* but not *Oryza sativa*, indicating that HmHSF and AtHSF have a closer evolutionary connection [38–40]. In addition, GO annotation and enrichment analysis of these 15 predicted *HmHSF* genes showed that most of them are part of biological processes in response to abiotic stresses, such as the responses to abiotic factors, responses to temperature stimuli, and responses to xenobiotic stimuli.

This study observed that most of these *HSF* genes showed a progressive upward trend in expression with increasing drought levels. About 13% of HSF genes were upregulated and 67% were downregulated, suggesting that they may play a particular role in drought conditions. The majority of the regulated genes belong to the HmHSFA group. This result is interesting because *HSFA* genes often perform irreplaceable roles. These genes include *HSFA1* and *HSFA2*, which have been identified as responsive to drought stimulation in *A. thaliana* and *Solanum lycopersicum* [41]. There are also some related studies showing that some *HSFB*-like genes are implicated in osmotic pressure tolerance in *Arachis hypogaea* [18,42]. These results suggest that *HSF* members have species-specific and stress-specific features when it comes to regulating the roles of genes involved in plant stress responses. In addition, in this study, we found that the expression levels of up-regulated genes reduced rapidly while the expression levels of down-regulated genes increased in the reaction to drought pressure in *H. myrtifolia*. These findings suggest a more efficient function of the down-regulated genes, as they function more effectively in response to drought stress. However, the process by which genes exert their regulatory effects is a complex one, and further research on *HSF* genes is therefore essential. It is noteworthy that HmHSF15 protein has a different three-dimensional structure to other HmHSF proteins. Combined with the analysis of the conserved structural domains, it can be concluded that HmHSF15 protein has other structures besides the specific structural domains of the HSF family, and therefore the HmHSF15 protein has more regulatory functions that deserve attention in the follow-up study. In addition, tissue specificity analysis and K-means clustering analysis showed that the *HmHSF* gene has significant tissue specificity in the growth and development of *H. myrtifolia*. Notably, the *HmHSF05*, *HmHSF12*, and *HmHSF14* genes were highly exo-compressed in two tissues, unlike other *HmHSF* genes, and their specific functions could be demonstrated by subsequent experiments. It can be speculated that the above genes may exhibit enhanced functions in response to stress. In addition, a study on the analysis of promoter cis-acting elements of *A. thaliana HSF* genes revealed that *HSF* genes not only play a role in abiotic stresses but also in biological processes involved in growth and development, biosynthesis, and hormone response.

Therefore, in the subsequent bioinformatics study, the promoter cis-acting elements of the key genes can be continually observed in order to further investigate the response of *HSF* genes in drought and in the growth development of *H. myrtifolia*. Based on bioinformatics, the screening of the key *HSF* genes can be further performed for full-length cloning and genetic transformation, and the response mechanism of the key *HSF* in drought stress can be studied in depth to provide a theoretical basis for the subsequent breeding of a new cold-tolerant *H. myrtifolia* species.

5. Conclusions

This study analyzed and identified 15 *HmHSF* genes. Amino acid sequence alignment revealed that all *HmHSF* genes contain a conserved DBD structure. HmHSF proteins were classified into three groups and fifteen sub-groups based on evolutionary relationships. Proteins within the same group were mostly similar, but differed significantly between subgroups. Additionally, it was found that the HmHSF15 protein has a conserved structural domain in addition to the DBD structure. The specific function of this domain needs to be verified by further experiments. The GO analysis revealed that most *HmHSF* genes are a part of various biological processes associated with abiotic stresses. Finally, it was found that *HmHSF* genes showed very obvious tissue specificity in different tissues of *H. myrtifolia*, especially the three genes *HmHSF05*, *HmHSF12*, and *HmHSF14*, which belonged to the same class of genes in the K-means clustering analysis. Further experiments can be arranged to elucidate the molecular mechanism of the key *HSF* genes against drought stress in *H. myrtifolia*.

Supplementary Materials: The following supporting information can be downloaded at https://www.mdpi.com/article/10.3390/horticulturae9050588/s1. Table S1: qRT-PCR primers for the *HmHSF* genes; Table S2: Secondary structure prediction of HmHSF.

Author Contributions: Conceptualization, G.Z., C.G., J.M. and W.S.; methodology, G.Z. and Y.Y; software, G.Z.; validation, Y.Y. and Y.Z.; investigation, J.M. and S.H.; data curation, G.Z. and C.G.; writing original draft preparation, G.Z. and C.G.; writing—review and editing, J.M., W.S., G.Z., S.H. and L.S.; funding acquisition, J.M. All authors have read and agreed to the published version of the manuscript.

Funding: Zhejiang Provincial Natural Science Foundation of China (No. LY21C160001), National Natural Science Foundation of China (31272494), and Zhejiang Provincial Natural Science Foundation of China (No. LY16C170003) all provided financial support for this work.

Institutional Review Board Statement: Not applicable.

Informed Consent Statement: Not applicable.

Data Availability Statement: All data in this study can be found in the manuscript or in the Supplementary Materials.

Conflicts of Interest: The authors declare no conflict of interest.

References

1. Fang, Y.; Xiong, L. General mechanisms of drought response and their application in drought resistance improvement in plants. *Cell Mol. Life Sci.* **2015**, *72*, 673–689. [CrossRef] [PubMed]
2. Joshi, R.; Wani, S.H.; Singh, B.; Bohra, A.; Dar, Z.A.; Lone, A.A.; Pareek, A.; Singla-Pareek, S.L. Transcription Factors and Plants Response to Drought Stress: Current Understanding and Future Directions. *Front. Plant Sci.* **2016**, *7*, 1029. [CrossRef] [PubMed]
3. Ilyas, M.; Nisar, M.; Khan, N.; Hazrat, A.; Khan, A.H.; Hayat, K.; Fahad, S.; Khan, A.; Ullah, A. Drought Tolerance Strategies in Plants: A Mechanistic Approach. *J. Plant Growth Regul.* **2021**, *40*, 926–944. [CrossRef]
4. Batool, M.; El-Badri, A.M.; Hassan, M.U.; Haiyun, Y.; Chunyun, W.; Zhenkun, Y.; Jie, K.; Wang, B.; Zhou, G. Drought Stress in *Brassica napus*: Effects, Tolerance Mechanisms, and Management Strategies. *J. Plant Growth Regul.* **2022**, *42*, 21–45. [CrossRef]
5. Zhang, X.; Lei, L.; Lai, J.; Zhao, H.; Song, W. Effects of drought stress and water recovery on physiological responses and gene expression in maize seedlings. *BMC Plant Biol.* **2018**, *18*, 68. [CrossRef] [PubMed]
6. Nakashima, K.; Yamaguchi-Shinozaki, K.; Shinozaki, K. The transcriptional regulatory network in the drought response and its crosstalk in abiotic stress responses including drought, cold, and heat. *Front. Plant Sci.* **2014**, *5*, 170. [CrossRef]
7. Abdelraheem, A.; Esmaeili, N.; O'Connell, M.; Zhang, J. Progress and perspective on drought and salt stress tolerance in cotton. *Ind. Crop. Prod.* **2019**, *130*, 118–129. [CrossRef]
8. Leng, P.; Zhao, J. Transcription factors as molecular switches to regulate drought adaptation in maize. *Theor. Appl. Genet.* **2020**, *133*, 1455–1465. [CrossRef]
9. Zhang, D.-F.; Zeng, T.-R.; Liu, X.-Y.; Gao, C.-X.; Li, Y.-X.; Li, C.-H.; Song, Y.-C.; Shi, Y.-S.; Wang, T.-Y.; Li, Y. Transcriptomic profiling of sorghum leaves and roots responsive to drought stress at the seedling stage. *J. Integr. Agr.* **2019**, *18*, 1980–1995. [CrossRef]
10. Yao, T.; Zhang, J.; Xie, M.; Yuan, G.; Tschaplinski, T.J.; Muchero, W.; Chen, J.-G. Transcriptional Regulation of Drought Response in Arabidopsis and Woody Plants. *Fron. Plant Sci.* **2021**, *11*, 572137. [CrossRef]

11. Liang, Y.; Wang, J.; Zheng, J.; Gong, Z.; Li, Z.; Ai, X.; Li, X.; Chen, Q. Genome-Wide Comparative Analysis of Heat Shock Transcription Factors Provides Novel Insights for Evolutionary History and Expression Characterization in Cotton Diploid and Tetraploid Genomes. *Front. Genet.* **2021**, *12*, 658847. [CrossRef] [PubMed]
12. Zhou, M.; Zheng, S.; Liu, R.; Lu, J.; Lu, L.; Zhang, C.; Liu, Z.; Luo, C.; Zhang, L.; Yant, L.; et al. Genome-wide identification, phylogenetic and expression analysis of the heat shock transcription factor family in bread wheat (*Triticum aestivum* L.). *BMC Genomics.* **2019**, *20*, 505. [CrossRef] [PubMed]
13. Yu, Y.; Yu, M.; Zhang, S.; Song, T.; Zhang, M.; Zhou, H.; Wang, Y.; Xiang, J.; Zhang, X. Transcriptomic Identification of Wheat AP2/ERF Transcription Factors and Functional Characterization of TaERF-6-3A in Response to Drought and Salinity Stresses. *Int. J. Mol. Sci.* **2022**, *23*, 3272. [CrossRef] [PubMed]
14. Li, W.; Wan, X.-L.; Yu, J.-Y.; Wang, K.-L.; Zhang, J. Genome-Wide Identification, Classification, and Expression Analysis of the Hsf Gene Family in Carnation (*Dianthus caryophyllus*). *Int. J. Mol. Sci.* **2019**, *20*, 5233. [CrossRef]
15. Liu, X.; Meng, P.; Yang, G.; Zhang, M.; Peng, S.; Zhai, M.Z. Genome-wide identification and transcript profiles of walnut heat stress transcription factor involved in abiotic stress. *BMC Genom.* **2020**, *21*, 474. [CrossRef]
16. Chen, X.; Wang, Z.; Tang, R.; Wang, L.; Chen, C.; Ren, Z. Genome-wide identification and expression analysis of Hsf and Hsp gene families in cucumber (*Cucumis sativus* L.). *Plant Growth Regul.* **2021**, *95*, 223–239. [CrossRef]
17. Zhang, Y.; Wang, C.; Wang, C.; Yun, L.; Song, L.; Idrees, M.; Liu, H.; Zhang, Q.; Yang, J.; Zheng, X.; et al. OsHsfB4b Confers Enhanced Drought Tolerance in Transgenic Arabidopsis and Rice. *Int. J. Mol. Sci.* **2022**, *23*, 10830. [CrossRef]
18. Song, X.; Liu, G.; Duan, W.; Liu, T.; Huang, Z.; Ren, J.; Li, Y.; Hou, X. Genome-wide identification, classification and expression analysis of the heat shock transcription factor family in Chinese cabbage. *Mol. Genet. Genom.* **2014**, *289*, 541–551. [CrossRef]
19. Lal, S.V.; Brahma, B.; Gohain, M.; Mohanta, D.; De, B.C.; Chopra, M.; Dass, G.; Vats, A.; Upadhyay, R.C.; Datta, T.K.; et al. Splice variants and seasonal expression of buffalo HSF genes. *Cell Stress Chaperones* **2015**, *20*, 545–554. [CrossRef]
20. Chen, J.; Gao, T.; Wan, S.; Zhang, Y.; Yang, J.; Yu, Y.; Wang, W. Genome-Wide Identification, Classification and Expression Analysis of the HSP Gene Superfamily in Tea Plant (*Camellia sinensis*). *Int. J. Mol. Sci.* **2018**, *19*, 2633. [CrossRef]
21. Jiang, L.; Hu, W.; Qian, Y.; Ren, Q.; Zhang, J. Genome-wide identification, classification and expression analysis of the Hsf and Hsp70 gene families in maize. *Gene* **2021**, *770*, 145348. [CrossRef] [PubMed]
22. Andrási, N.; Pettkó-Szandtner, A.; Szabados, L. Diversity of plant heat shock factors: Regulation, interactions, and functions. *J. Exp. Bot.* **2020**, *72*, 1558–1575. [CrossRef] [PubMed]
23. Jahan, K.; Nie, H.; Yan, X. Revealing the potential regulatory relationship between *HSP70*, *HSP90* and *HSF* genes under temperature stress. *Fish Shellfish Immunol.* **2023**, *134*, 108607. [CrossRef] [PubMed]
24. Lin, L.; Wang, J.; Wang, Q.; Ji, M.; Hong, S.; Shang, L.; Zhang, G.; Zhao, Y.; Ma, Q.; Gu, C. Transcriptome Approach Reveals the Response Mechanism of *Heimia myrtifolia* (Lythraceae, Myrtales) to Drought Stress. *Fron. Plant Sci.* **2022**, *13*, 877913. [CrossRef]
25. Giorno, F.; Guerriero, G.; Baric, S.; Mariani, C. Heat shock transcriptional factors in Malus domestica: Identification, classification and expression analysis. *BMC Genom.* **2012**, *13*, 639. [CrossRef]
26. Huang, B.; Huang, Z.; Ma, R.; Chen, J.; Zhang, Z.; Yrjälä, K. Genome-wide identification and analysis of the heat shock transcription factor family in moso bamboo (*Phyllostachys edulis*). *Sci. Rep.* **2021**, *11*, 16492. [CrossRef]
27. Zhang, H.; Li, G.; Fu, C.; Duan, S.; Hu, D.; Guo, X. Genome-wide identification, transcriptome analysis and alternative splicing events of Hsf family genes in maize. *Sci. Rep.* **2020**, *10*, 8073. [CrossRef]
28. Liu, G.; Chai, F.; Wang, Y.; Jiang, J.; Duan, W.; Wang, Y.; Wang, F.; Li, S.; Wang, L. Genome-wide Identification and Classification of HSF Family in Grape, and Their Transcriptional Analysis under Heat Acclimation and Heat Stress. *Hortic. Plant J.* **2018**, *4*, 133–143. [CrossRef]
29. MMallick, B.; Kumari, M.; Pradhan, S.K.; Acharya, G.C.; Naresh, P.; Das, B.; Shashankar, P. Genome-wide analysis and characterization of heat shock transcription factors (Hsfs) in common bean (*Phaseolus vulgaris* L.). *Funct. Integr. Genom.* **2022**, *22*, 743–756. [CrossRef]
30. Huang, Y.; Li, M.-Y.; Wang, F.; Xu, Z.-S.; Huang, W.; Wang, G.-L.; Ma, J.; Xiong, A.-S. Heat shock factors in carrot: Genome-wide identification, classification, and expression profiles response to abiotic stress. *Mol. Biol. Rep.* **2015**, *42*, 893–905. [CrossRef]
31. Liu, Z.-W.; Wu, Z.-J.; Li, X.-H.; Huang, Y.; Li, H.; Wang, Y.-X.; Zhuang, J. Identification, classification, and expression profiles of heat shock transcription factors in tea plant (*Camellia sinensis*) under temperature stress. *Gene* **2016**, *576*, 52–59. [CrossRef] [PubMed]
32. Chidambaranathan, P.; Jagannadham, P.T.K.; Satheesh, V.; Kohli, D.; Basavarajappa, S.H.; Chellapilla, B.; Kumar, J.; Jain, P.K.; Srinivasan, R. Genome-wide analysis identifies chickpea (*Cicer arietinum*) heat stress transcription factors (Hsfs) responsive to heat stress at the pod development stage. *J. Plant Res.* **2018**, *131*, 525–542. [CrossRef] [PubMed]
33. Kumar, R.R.; Goswami, S.; Singh, K.; Dubey, K.; Rai, G.K.; Singh, B.; Singh, S.; Grover, M.; Mishra, D.; Kumar, S.; et al. Characterization of novel heat-responsive transcription factor (TaHSFA6e) gene involved in regulation of heat shock proteins (HSPs)—A key member of heat stress-tolerance network of wheat. *J. Biotechnol.* **2018**, *279*, 1–12. [CrossRef] [PubMed]
34. Bin, J.; Zhu, M.; Ding, H.; Zai, Z.; Shi, T.; Wang, L.; Yang, X.; Yue, Y. New Insights into the Roles of Osmanthus Fragrans Heat-Shock Transcription Factors in Cold and Other Stress Responses. *Horticulturae* **2022**, *8*, 80. [CrossRef]
35. Xu, P.; Guo, Q.; Pang, X.; Zhang, P.; Kong, D.; Liu, J. New Insights into Evolution of Plant Heat Shock Factors (Hsfs) and Expression Analysis of Tea Genes in Response to Abiotic Stresses. *Plants* **2020**, *9*, 311. [CrossRef]

36. Shyamli, P.S.; Pradhan, S.; Panda, M.; Parida, A. De novo Whole-Genome Assembly of Moringa oleifera Helps Identify Genes Regulating Drought Stress Tolerance. *Fron. Plant Sci.* **2021**, *12*, 766999. [CrossRef]
37. Li, P.-S.; Yu, T.-F.; He, G.-H.; Chen, M.; Zhou, Y.-B.; Chai, S.-C.; Xu, Z.-S.; Ma, Y.-Z. Genome-wide analysis of the Hsf family in soybean and functional identification of GmHsf-34 involvement in drought and heat stresses. *BMC Genom.* **2014**, *15*, 1009. [CrossRef]
38. Ma, J.; Xu, Z.-S.; Wang, F.; Tan, G.-F.; Li, M.-Y.; Xiong, A.-S. Genome-wide analysis of HSF family transcription factors and their responses to abiotic stresses in two Chinese cabbage varieties. *Acta Physiol. Plant.* **2014**, *36*, 513–523. [CrossRef]
39. Ma, G.; Shen, J.; Yu, H.; Huang, X.; Deng, X.; Hu, Z.; Amee, M.; Chen, L.; Cao, L. Genome-wide identification and functional analyses of heat shock transcription factors involved in heat and drought stresses in ryegrass. *Environ. Exp. Bot.* **2022**, *201*, 104968. [CrossRef]
40. Liu, M.; Huang, Q.; Sun, W.; Ma, Z.; Huang, L.; Wu, Q.; Tang, Z.; Bu, T.; Li, C.; Chen, H. Genome-wide investigation of the heat shock transcription factor (Hsf) gene family in Tartary buckwheat (*Fagopyrum tataricum*). *BMC Genom.* **2019**, *20*, 871. [CrossRef]
41. Gong, C.; Pang, Q.; Li, Z.; Li, Z.; Chen, R.; Sun, G.; Sun, B. Genome-Wide Identification and Characterization of Hsf and Hsp Gene Families and Gene Expression Analysis under Heat Stress in Eggplant (*Solanum melongema* L.). *Horticulturae* **2021**, *7*, 149. [CrossRef]
42. Chen, C.; Chen, H.; Zhang, Y.; Thomas, H.R.; Xia, R. TBtools: An Integrative Toolkit Developed for Interactive Analyses of Big Biological Data. *Mol. Plant* **2020**, *13*, 1194–1202. [CrossRef] [PubMed]

horticulturae

Article

Transcriptome Analysis of *Rhododendron liliiflorum* H. Lév. Flower Colour Differences

Hang Zhang [1,†], Meifeng Chen [1,†], Xinglin Wang [1,†], Jin Dai [1], Xu Zhang [2], Zhengdong Zhang [3], Ximin Zhang [1], Ming Tang [1], Jing Tang [1], Jiyi Gong [1], Lunxian Liu [1,*] and Yin Yi [1,*]

1 Key Laboratory of National Forestry and Grassland Administration on Biodiversity Conservation in Karst Mountainous Areas of Southwestern China, Key Laboratory of Plant Physiology and Development Regulation, Key Laboratory of Environment Friendly Management on High Altitude Rhododendron Diseases and Pests, Institutions of Higher Learning in Guizhou Province, School of Life Science, Guizhou Normal University, Guiyang 550001, China
2 Guizhou Academy of Forestry Sciences, Guiyang 550001, China
3 College of Mathematics and Information Science, Guiyang University, Guiyang 550001, China
* Correspondence: llunxian@163.com (L.L.); gzklppdr@gznu.edu.cn (Y.Y.); Tel.: +86-15285161657 (L.L.); +86-51-83227345 (Y.Y.)
† These authors contributed equally to this work.

Abstract: *Rhododendron liliiflorum* H. Lév., with white outer edges and yellow inner edges of petals, is an ornamental flower that originated in China. In this study, we analysed the white (W) and yellow (Y) parts of *R. liliiflorum* flowers by RNA sequencing. Then, unigene assembly, unigene annotation, and classification of Eukaryotic Orthologous Groups (KOGs) were performed. Gene ontology (GO) classification and pathway enrichment analysis for unigenes were also conducted. A total of 219,221 transcripts and 180,677 unigenes of *R. liliiflorum* were obtained from 48.52 Gb of clean reads. Differentially expressed gene (DEG) analysis indicated that 2310 unigenes were upregulated and 3062 were downregulated in W vs. Y. Thirty-six of these DEGs were involved in the flavonoid biosynthesis pathway. Pathway enrichment analysis showed that DEGs were significantly enriched in phenylpropanoid, flavonoid, and isoflavone biosynthesis. The expression of dihydroflavonol-4-reductase (DFR) and chalcone synthase (CHS) may affect differences in *R. liliiflorum* flower colour. The findings on flavonoid biosynthesis and other related genes in this study will provide guidance for exploring the mechanism of flower colour formation in *Rhododendron*.

Keywords: coloration; unigene; flavonoids; gene expression; functional annotation

Citation: Zhang, H.; Chen, M.; Wang, X.; Dai, J.; Zhang, X.; Zhang, Z.; Zhang, X.; Tang, M.; Tang, J.; Gong, J.; et al. Transcriptome Analysis of *Rhododendron liliiflorum* H. Lév. Flower Colour Differences. *Horticulturae* **2023**, *9*, 82. https://doi.org/10.3390/horticulturae9010082

Academic Editor: Sergey Dolgov

Received: 16 November 2022
Revised: 24 December 2022
Accepted: 5 January 2023
Published: 9 January 2023

1. Introduction

Flower colour polymorphisms, considered to be one of the key adaptive characteristics chiefly correlated with pollinators (such as insects and birds) and animals for seed dispersal, have fascinated botanists, ecologists, and horticulturists [1–3]. The colouring of floral organs is predominantly caused by the accumulation of carotenoids, flavonoids, and alkaloids [4]. Flavonoids, a large group of secondary metabolites, have the structure of 2-phenylchromone. Flavonoids are one of the most important pigments in the petals of various ornamental plants, and they produce the widest range of colors from pale yellow to violet. They have been reported as the main pigment in *Chrysanthemum morifolium* Ramat. [5], *Dahlia pinnata* Cav. [6], *Rosa chinensis* Jacq. [2], *Matthiola incana* (L.) R. Br. [7], and *Paeonia lactiflora* Pall. [8].

Flavonoids, produced by a branch of the phenylpropanoid biosynthesis pathway, generally include chalcones, flavonoids, flavonols, and anthocyanins [9,10]. In the flavonoid biosynthesis pathway (Figure 1), the upstream substrate L-phenylalanine is first converted to chalcone by phenylalanine ammonia-lyase (PAL), cinnamate-4-hydroxylase (C4H), 4-coumaroyl-CoA synthetase (4CL), chalcone synthase (CHS), and chalcone isomerase

(CHI) [11]. Then, the synthesised chalcone is converted to flavanones by CHI, and the subsequent pathway is further divided into three branches, including flavonoid 3′,5′-hydroxylase (F3′5′H), flavanone 3-hydroxylase (F3H), and isoflavone synthetase (IFS), which catalyses the production of pentahydroxy flavanone, dihydrokaempferol (DHK) and isoflavones. DHK is further differentiated into two branches: the production of flavanol under the co-catalysis of flavonoid 3′-hydroxylase (F3′H), F3′5′H, and flavonol synthase (FLS) and the production of flavanol under the catalysis of FLS, followed by the production of leucoanthocyanidin under the catalysis of dihydroflavonol-4-reductase (DFR). The subsequent production of flavonol is catalysed by ANS, leucoanthocyanidin reductase (LAR), and anthocyanidin reductase (ANR). A reduction in the expression of CHI, the second key enzyme in this pathway, can induce an increase in chalcone, leading to colour changes [12]. There are basically two colour series, red and pure yellow, of flavonoid compounds [3]. The red series mainly includes anthocyanins, which can regulate the colour from pink to violet [13]. Other flavonoid compounds principally belong to the pure yellow series, for instance, dark yellow chalcone and aurora ketone, as well as pale yellow or almost colourless flavonoids and flavonols [14]. Flavonoid components may also vary widely in different coloured petals of the same species; for example, white *C. morifolium* contains only flavonoids and flavonols, while pink *C. morifolium* primarily contains anthocyanins, flavonoids, and flavonols [15].

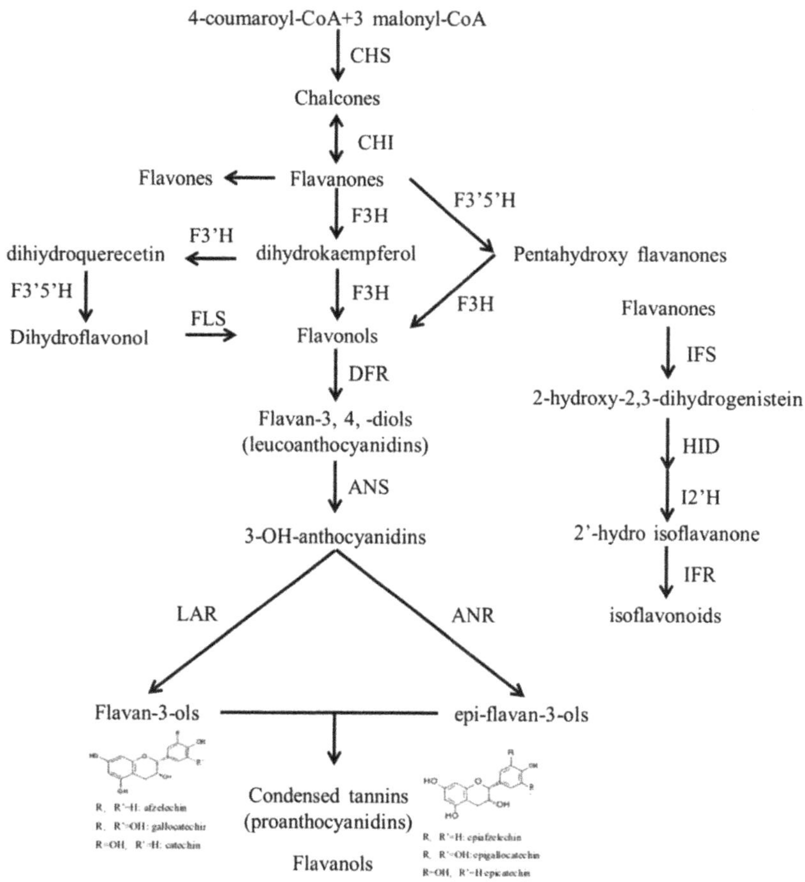

Figure 1. Flavonoid biosynthesis pathway.

The floral colours of the genus *Rhododendron* are varied, including red, pink, purple, blue, yellow, and white [16], indicating that rhododendrons are ideal materials for investigating differences in floral colour traits. Ye et al. [3] researched the flower colour divergence of three varieties of *Rhododendron sanguineum* Franch. Complex, namely *R. sanguineum* var. *sanguineum* (with bright crimson flowers), *R.* var. *haemaleum* (with deep blackish crimson flowers), and *R.* var. *didymoides* (with yellow flushed with pink flowers). It was found that FLS, ANS, acyltransferase (AT), CHI, and glutathione S-transferase (GST) were highly expressed in deep blackish crimson flowers, which was related to the production of anthocyanins [17]. In addition, β-glucosidase (BGLU) and peroxidase (PER) related to the degradation of anthocyanins were highly expressed in yellow flushed with pink flowers [18,19]. Du et al. [16] systematically studied the flower colours of 30 species of azalea (classified into red, purplish pink, purple, and white groups) via anthocyanin and flavonol identification, and found that the red group mainly contained cyanidin monoglycosides and had much higher total anthocyanin contents than the purplish pink and purple groups. White flowers did not accumulate anthocyanins [20], but the petals contained similar amounts of flavonols as the coloured flowers [16]. *Rhododendron liliiflorum* H. Lév. is a *Rhododendron* species native to Guangxi, Guizhou, Hunan, and southeast Yunnan, China, where it grows in sparse forests and shrubs on the hillside at altitudes of 800–1800 m [21]. The flower of *R. liliiflorum* is tubular and bell shaped, with white outer edges and yellow inner edges of petals. At present, the mechanism of flower colour formation in *R. liliiflorum* has rarely been studied. Therefore, we used transcriptomic analysis to investigate gene expression during the process of flower colour formation in *R. liliiflorum* in this study. We discuss the gene expression of enzymes related to chalcone, flavonoid, and anthocyanin biosynthesis in the flavonoid biosynthesis pathway, which can provide a basis for the genetic improvement of flower colour in *R. liliiflorum* and the genus *Rhododendron*.

2. Materials and Methods

2.1. Experimental Materials

The experimental materials (*R. liliiflorum*) transplanted from the Baili Rhododendron Nature Reserve (N 27°12′54″, E 105°55′5″) (Figure 2a), which is located in northwest Guizhou, China, were collected in the Key Laboratory of Plant Physiology and Development Regulation, Guizhou Normal University, in May 2021. *R. liliiflorum* flowers that had just bloomed for 2–3 days were picked. Afterwards, the yellow and white parts of the petals were quickly divided into five parts from the top to the base at −18 °C to maintain the activity of the petals and labelled from A to E. The samples were subsequently rapidly frozen in liquid nitrogen and kept at −80 °C.

2.2. RNA Extraction, Library Construction and Sequencing

The experiments were performed using three biological replicates of W (white part of petals) and Y (yellow part of petals) samples (Figure 2b) for library construction and sequencing. Total RNA was isolated from petal samples using an RNA Prep Pure Plant Kit (Tiangengjin, Beijing). RNA quality, purity, and integrity were determined using agarose gel electrophoresis, a Nano Photometer (OD260/280 and OD260/230), and an Agilent 2100 biological analyser. The RNA concentration was measured using a Qubit 2.0 fluorometer. After passing quality control, the mRNA purified from the total RNA of each of the three replicates using poly-T oligo-attached magnetic beads was cut into short fragments by adding a fragmentation buffer and then used as a template to synthesize the first-strand cDNA with random hexamer primers. The double-stranded cDNA was generated by adding buffer, dNTPs (dUTP, dATP, dGTP, and dCTP), and DNA polymerase I to the first-strand cDNA and purified with AMPure XP beads. The purified double-stranded cDNA was end-repaired and elongated by A-tail addition, and sequencing adapters were ligated. The cDNA fragment was size selected with AMPure XP beads and enriched by PCR to construct the final cDNA library. Qubit 2.0 was used for the preliminary quantitation of the cDNA library, Agilent 2100 was used to detect the insertion size of the library, and qPCR was used to accurately quantify the effective concentration of the library (the effective content of the library > 2 nM) to complete the quality inspection of the library. After evaluating and qualifying the cDNA library, it was sequenced on an Illumina HiSeq platform (Wuhan Metware Biotechnology Co., Ltd., Wuhan, China).

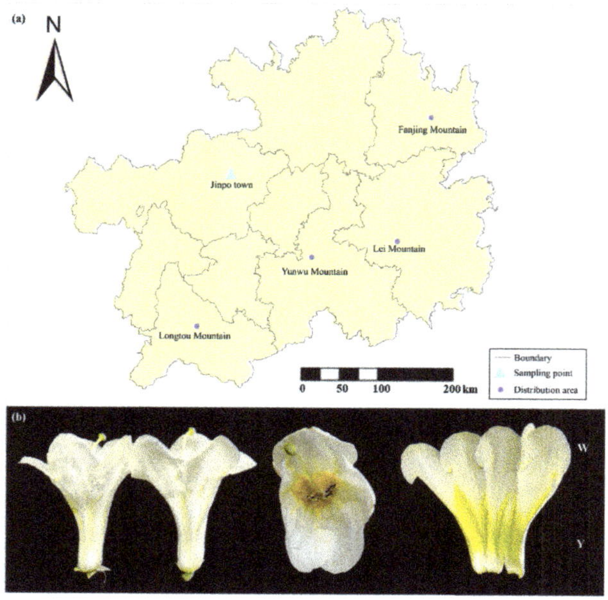

Figure 2. (**a**) Distribution of *R. liliiflorum* in Guizhou Province, China; (**b**) Pictures of *R. liliiflorum* flowers and white (W) vs. yellow (Y) parts.

2.3. Sequence Filtering, Splicing and Assembly

The raw data were first filtered by removing reads with adapter sequences, containing poly-Ns, with ambiguous nucleotides, and those of low quality. Then, sequencing error check was conducted and the phred scores (Q20, Q30) were calculated. The quality of the remaining reads was evaluated with FastQC, including GC content, sequence length distribution, and sequence duplication level of the clean data [3]. The obtained clean reads were assembled using Trinity v. 2.6.5 software [22]. The longest cluster sequence obtained after Corset hierarchical clustering was used as the unigene for subsequent analysis.

2.4. Unigene Functional Annotation and Expression Level Analysis

Using BLASTx (v.2.2.26) [23], the unigene sequences were compared against the NCBI non-redundant protein database (Nr), the Kyoto Encyclopedia of Genes and Genomes (KEGG) database (http://www.genome.jp/kegg/) accessed on 1 June 2021 [24], Eukaryotic Orthologous Groups database (KOG) (http://ftp.ncbi.nih.gov/pub/COG/KOG) accessed on 1 June 2021 [25], the Swiss-Prot database (http://www.uniprot.org/) accessed on 1 June 2021 [26], and the TrEMBL database. Gene ontology (GO) annotations were analysed using the Blast2GO (V.2.5) program (http://www.geneontology.org/) accessed on 2 June 2021 [27]. After predicting the amino acid sequences of the unigenes, Hmmer 3.0 (http://hmmer.janelia.org/) accessed on 2 June 2021 was used to compare the results with the Pfam database to obtain the annotation information of the unigenes. Taking the transcriptome spliced by Trinity as the reference sequence (Ref), the clean reads of each sample were mapped to Ref using RESM v. 1.3.1 [28]. The number of fragments per transcript was related to the amount of sequenced data (or mapped data), the length of the transcript, and the expression level of the transcript. To truly reflect the expression level of transcripts, the number of mapped reads and the length of transcripts in the sample must be normalised. Based on the alignments, the read counts of each gene were calculated and normalised to fragments per kilobase of transcript per million mapped fragments (FPKM) values. The FPKM method was used as an index to measure the level of transcript or gene expression.

2.5. Screening and Enrichment Analysis of DEGs

Setting $|\text{log2fold change}| \geq 1$ and false discovery rate (FDR) ≤ 0.05 as the threshold, the differentially expressed genes between W (white part of petals) and Y (yellow part of petals) were calculated using R 4.1.0 package DESeq2 [29,30]. The screened differentially expressed genes were annotated using KEGG pathway enrichment (among which pathways with a p value < 0.05 were considered significantly enriched) and Gene Ontology (GO). KEGG pathway annotation was performed online using DEG query sequences mapped to the KEGG Automatic Annotation Server (KAAS) (http://www.genome.jp/tools/kaas/) accessed on 3 June 2021. The gene ontology (GO) annotations of DEGs were analysed using the Blast2GO (V.2.5) program (http://www.geneontology.org/) accessed on 3 June 2021 [27].

3. Results

3.1. Transcriptome Sequencing and Assembly

Through transcriptome sequencing of the white and yellow parts of *R. liliiflorum*, a total of 6 sample transcripts of W vs. Y samples were obtained. The total number of raw reads was 335,805,834. After filtering, the number of clean reads was 323,525,900, and the clean read ratio reached 96.34% for a total of 48.52 Gb of data. The GC contents of the six samples were between 46.14% and 46.76%, and the ranges of Q20 and Q30 were 97.91–98.13% and 94.03–94.55%, respectively (Table 1). The base quality values Q20 and Q30 of the clean reads in each sample were higher than 97% and 94%, respectively, and the GC content was more than 46%, which indicated that the base recognition accuracy was high, and the sequencing result quality was good. It could be seen that the transcriptome sequencing results met the quality requirements for subsequent assembly. To improve the sequencing depth and integrity of transcript assembly, Trinity software was used to assemble and splice the obtained clean reads. A total of 219,221 transcripts were obtained, with an average assembly length of 1135 bp and an N50 length of 1928 bp. Among them, 82,737 transcripts were more than 1000 bp long, accounting for 37.74% of the total transcripts. Subsequently, the longest sequence in the transcript was selected as the unigene sequence. A total of 180,677 unigenes were obtained, with an average assembly length of 1317 bp and an N50 length of 2016 bp (Table 2). Among them, 82,719 unigenes were more than 1000 bp long, accounting for 45.78% of the total transcripts. The length distributions of the assembled transcripts and unigenes are shown in Table 3.

Table 1. Clean reads data quality statistics.

Sample	Raw Reads	Clean Reads	Clean Bases (G)	Q20 (%)	Q30 (%)	GC Content (%)
W1	60,163,298	58,377,468	8.76	98.02	94.29	46.69
W2	63,100,240	61,086,422	9.16	98.05	94.35	46.59
W3	58,213,406	56,532,014	8.48	97.91	94.03	46.76
Y1	51,367,344	49,409,492	7.41	98.11	94.51	46.36
Y2	48,489,068	46,025,466	6.9	98.11	94.53	46.14
Y3	54,472,478	52,095,038	7.81	98.13	94.55	46.16

Q20: The percentage of alkali base with Qphred value no less than 20 in total alkali base; Q30: Percentage of alkali base with Qphred value no less than 30 in total alkali base; GC content: the percentage of the sum of G and C in high-quality reads to the base of total alkali.

Table 2. Assembly results of transcriptome using Trinity software.

Item	Total Sequence Number	Average Length (bp)	N50 Length (bp)	N90 Length (bp)
Transcripts	219,221	1135	1928	470
Unigenes	180,677	1317	2016	590

N50/N90: Sorting the spliced transcripts from long to short, and accumulating the length of transcripts to spliced transcripts no less than 50%/90% of the total length.

Table 3. Length statistics of transcripts and unigenes from the transcriptome assembly.

Sequence Number	200–500 bp	500–1000 bp	1000–2000 bp	≥2000 bp
Transcripts	85,520	50,964	45,446	37,291
Unigenes	47,886	50,072	45,429	37,290

3.2. Functional Annotation Results

BLAST was used to compare the unigene sequences with the KEGG, Nr, Swiss-Prot, GO, COG/KOG, Trembl, and protein family (Pfam) databases. After predicting the amino acid sequence of the unigenes, HMMER was used for comparison with the Pfam database to obtain the annotation information for the unigenes. A total of 171,948 unigenes were annotated (Table 4). Among the seven databases, 63.25% of unigenes were annotated by Nr, 62.9% of unigenes were annotated by Trembl, and the fewest genes (38.47%) were annotated by the KOG database. Comparison with the Nr database reported the percentage of total homologous sequences with other species (Figure 3). It was found that the sequences annotated with *Vitis vinifera* L.-related genes were the most abundant, up to 16,589, which accounted for 15.25% of the genes annotated by the Nr database. The related species with sequence homology greater than 2% included *Quercus suber* L., *Juglans regia* L., *Olea europaea* var. *sylvestris*, *Coffea canephora* Pierre ex Froehn., *Sesamum indicum* L., *Nelumbo nucifera* Gaertn., *Theobroma cacao* L., and *Daucus carota* subsp. *Sativus* (Hoffm.). We also found that 616 unigenes (0.6%) were annotated as *Rhododendron*.

Table 4. Statistics related to unigenes.

Database	Number of Genes	Percentage (%)
KEGG	81,784	47.56%
Nr	108,753	63.25%
Swiss-Prot	78,209	45.48%
Trembl	108,159	62.90%
KOG	66,156	38.47%
GO	90,560	52.67%
Pfam	81,639	47.48%

3.3. GO Terms and Classifications

GO annotations were performed for the unigenes, and 708,175 GO annotations were obtained for 90,560 (52.67%) unigenes. The GO annotations were divided into three ontologies: biological processes, cell components, and molecular functions. Among them, annotations for biological processes were the most abundant (accounting for 47.46% of unigenes), whereas annotations for molecular functions were the least numerous (accounting for 22.03% of unigenes) (Figure 4). The three ontologies were divided into 59 more detailed terms. Biological processes were divided into 28 terms, of which cellular process was associated with the most unigenes, accounting for 24.22% of this subcategory, followed by metabolic process (18.01%) and biological regulation (9.30%). Sulphur utilisation was associated with the lowest number of unigenes, accounting for only 0.0011% of all biological processes. Cell components were divided into 18 terms according to their positions in the cell, of which cell, cell part, and organelle accounted for 22.46%, 22.43%, and 17.01%

of unigenes in this ontology, respectively. Other organism and organism parts were less abundant, accounting for only 0.004% of unigenes. Among the 13 terms of molecular functions, catalytic activity and binding were the most abundant, accounting for 46.25% and 40.43% of unigenes in this ontology, respectively, while toxin activity and protein tag were the least abundant, accounting for 0.0008% and 0.011%, respectively. From these results, we can see that the expression of genes associated with cell activity and metabolic activity terms in *R. liliiflorum* was high, indicating strong metabolic capacity.

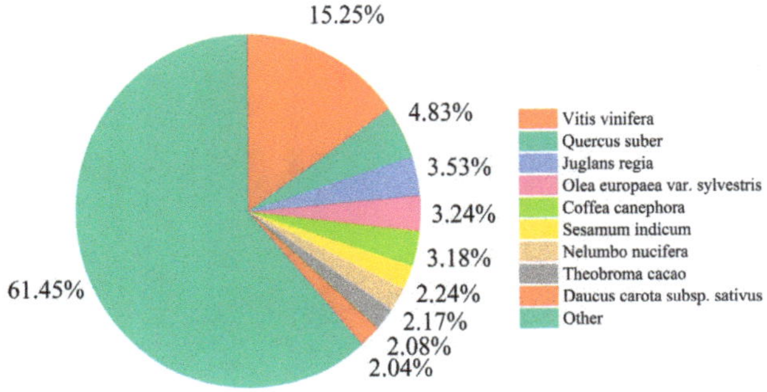

Figure 3. Species distribution of the *R. liliiflorum* unigenes in the Nr database.

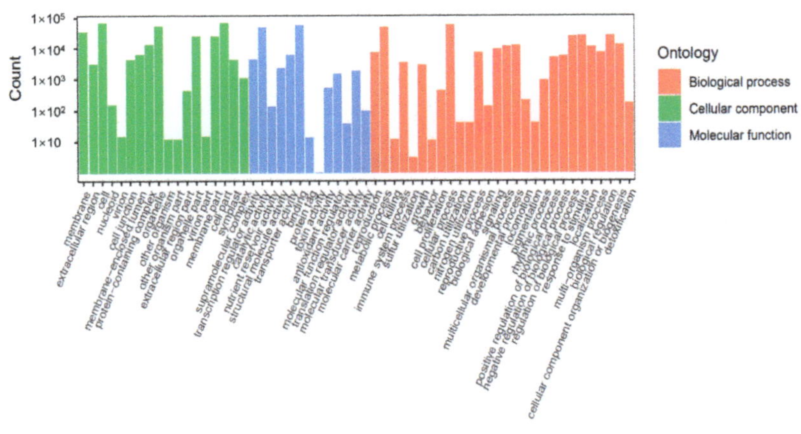

Figure 4. Gene ontology (GO) classification of unigenes.

3.4. KOG Terms and Classifications

When the assembled unigenes were compared to the Eukaryotic Orthologous Groups (KOG), the information obtained from the KOG functional annotation of *R. liliiflorum* showed that the terms could be subdivided into 25 categories and a total of 66,156 unigenes were annotated (accounting for 38.47% of the total number of unigenes). The gene expression abundance of each category varied greatly, including 76,226 functional annotations, which basically covered most of the life activities of *R. liliiflorum* (Figure 5). Among them, general function prediction only, signal transmission mechanisms, and posttranslational modification, protein turnover, and chaperones were the most common, with 16,605 (21.78%), 7768 (10.19%), and 7144 (9.37%) unigenes, respectively. There were fewer unigenes associated with cell motility and extracellular structures, only 22 (0.03%) and 207 (0.27%) unigenes, respectively.

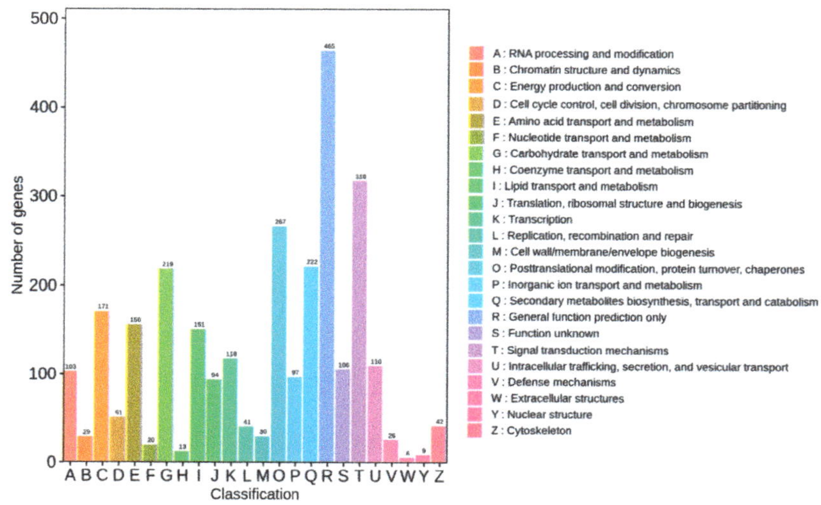

Figure 5. Eukaryotic Orthologous Groups (KOG) annotations of unigenes.

3.5. Screening and Enrichment Analysis of DEGs in Yellow and White Parts of *R. liliiflorum* Petals

3.5.1. Screening of DEGs

By comparing the gene expression levels of the yellow and white parts of the petals of *R. liliiflorum*, taking |log2fold Change| ≥ 1 and FDR < 0.05 as the screening conditions, 5372 DEGs were screened from the comparison of W vs. Y. There were 2310 upregulated unigenes and 3062 downregulated unigenes, and the total number of downregulated unigenes was higher than the total number of upregulated unigenes (Figure 6).

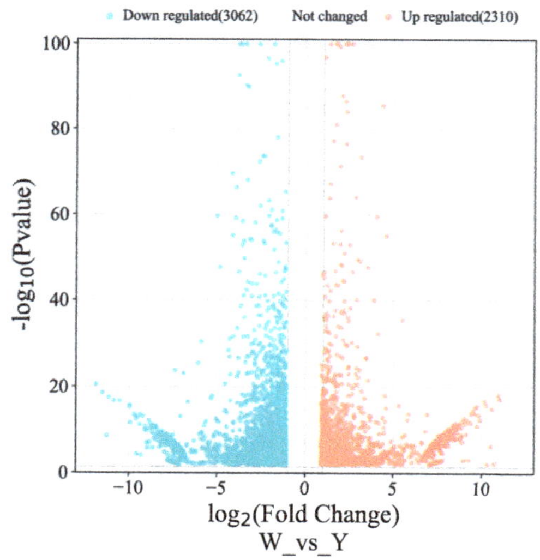

Figure 6. DEG volcano map. The red dots represent upregulated differentially expressed genes, the blue dots represent downregulated differentially expressed genes.

3.5.2. GO Enrichment of DEGs

The top 50 GO terms with significant enrichment were determined. As shown in the histogram of GO enrichment entries (Figure 7), the significantly enriched GO terms included photosystem and oxidoreductase activity, acting on peroxide as acceptor, peroxidase activity, response to chitin, and photosystem I.

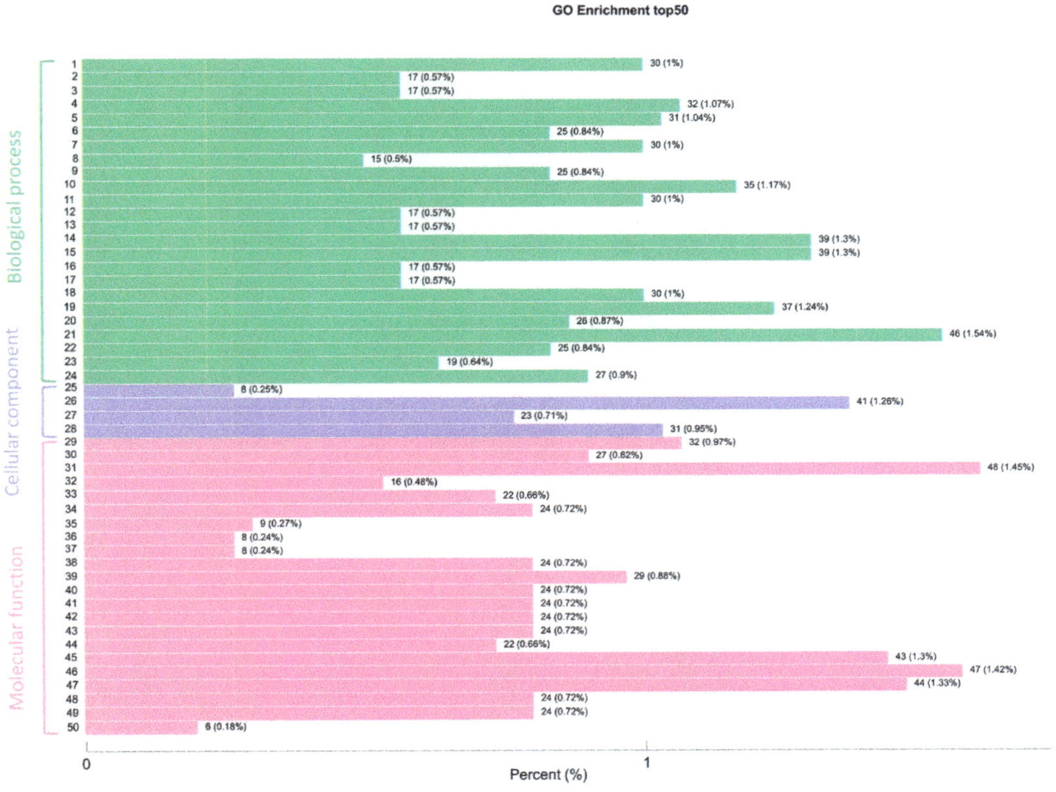

Figure 7. Gene Ontology analysis of top 50 differential expressed genes.

Among the 50 GO terms with the most significant enrichment, 24 terms were enriched in biological processes, of which the number of differentially expressed proteins enriched in response to chitin was the largest (46), followed by divalent metal ion transport and divalent inorganic cation transport (both 39).

There were 22 GO terms enriched in molecular functions, among which the differentially expressed genes related to calcium ion transmembrane transporter activity were the most abundant (48), followed by oxidoreductase activity, acting on peroxide as acceptor (47), and peroxidase activity. Four GO terms were enriched in cell components, all related to photosynthesis, among which photosystem (41) was the most abundant, followed by photosystem II (31) and photosystem I (22). GO terms related to flavonoid biosynthesis and pigments were not in the top 50, but they added up to 20 GO terms related to the biosynthesis of flavonol, flavanone, flavonoid, 2-isoflavonoid, and riboflavin.

The abscissa represents the ratio of genes annotated to the total number of genes annotated and the ordinate represents the name of the GO entry. The labels on the left side of the graph represent the categories to which the GO term belongs. The GO term definitions were: 1: amino acid transport; 2: amino sugar catabolic process; 3: aminoglycan catabolic process; 4: antibiotic catabolic process; 5: calcium ion transport; 6: calcium-mediated signaling; 7: carbohydrate derivative catabolic process; 8: cell wall macromolecule catabolic process; 9: cellular response to amino acid stimulus; 10: cellular response to nitrogen compound; 11: cellular response to organonitrogen compound; 12: chitin catabolic process; 13: chitin metabolic process; 14: divalent inorganic cation transport; 15: divalent metal ion transport;

16: glucosamine-containing compound catabolic process; 17: glucosamine-containing compound metabolic process; 18: hydrogen peroxide catabolic process; 19: hydrogen peroxide metabolic process; 20: response to amino acid; 21: response to chitin; 22: second-messenger-mediated signaling; 23: toxin catabolic process; 24: toxin metabolic process; 25: chloroplast thylakoid membrane protein complex; 26: photosystem; 27: photosystem I; 28: photosystem II; 29: amino acid transmembrane transporter activity; 30: calcium channel activity; 31: calcium ion transmembrane transporter activity; 32:chitinase activity; 33: chlorophyll binding; 34: extracellular ligand-gated ion channel activity; 35: gamma-aminobutyric acid transmembrane transporter activity; 36: germacradienol synthase activity; 37: germacrene-D synthase activity; 38: glutamate receptor activity; 39: glutathione transferase activity; 40: ionotropic glutamate receptor activity; 41: ligand-gated channel activity; 42: ligand-gated ion channel activity; 43: neurotransmitter receptor activity; 44: oxidoreductase activity, acting on diphenols and related substances as donors, oxygen as acceptor; 45: oxidoreductase activity, acting on paired donors, with incorporation or reduction of molecular oxygen, NAD(P)H as one donor, and incorporation of one atom of oxygen; 46: oxidoreductase activity, acting on peroxide as acceptor; 47: peroxidase activity; 48: transmitter-gated channel activity; 49: transmitter-gated ion channel activity; and 50: UDP-arabinose 4-epimerase activity.

3.5.3. KEGG Enrichment Analysis of DEGs

DEGs were annotated into 135 pathways, of which 28 metabolic pathways were significantly enriched ($p < 0.05$), and the top 20 pathways with significant enrichment were screened. As shown by the KEGG enrichment scatter diagram (Figure 8), KEGG significant enrichment was reflected in phenylpropanoid biosynthesis, photosynthesis, metabolic pathways, MAPK signalling pathway-plant, and biosynthesis of secondary metabolites. Among them, several metabolic pathways were related to flavonoid biosynthesis: the phenylpropanoid biosynthesis pathway (map00940) contained 110 unigenes, the flavonoid biosynthesis pathway contained 27 unigenes, and the isoflavone biosynthesis pathway contained 14 unigenes. According to the above annotation results, we found that 11 enzyme-related genes are involved in the flavonoid generation pathway (Table 5). Through the flavonoid biosynthesis pathway diagram (Figure 9), we found that the DEGs from the yellow part of the petals of *R. liliiflorum* were downregulated in pathways related to the biosynthesis of isoflavones, chalcone, flavanone, dihydrobrassinol, and flavonol, whereas two DEGs related to dihydroflavonol-4-reductase (DFR) were enriched in the anthocyanin biosynthesis pathway, of which one was upregulated and the other was downregulated. In addition, DEGs associated with LAR were upregulated in the anthocyanin biosynthesis pathway of flavanols.

Table 5. Genes associated with flavonoid biosynthesis were identified in the transcriptome of *R. liliiflorum*.

Gene Name	EC No.	Number of DEGs
CYP73A	E1.14.13.11	1
4CL	E6.2.1.12	3
CHS	E2.3.1.74	2
DFR	E1.1.12.34	2
F3′5′H	E1.14.14.81	1
FLS	E1.14.11.23	1
LAR	E1.14.11.19	2
CCoAOMT	E2.1.1.104	2
HCT	E2.3.1.133	16
I2′H	E1.14.13.89	2
HID	E4.2.1.105	4

Enzyme abbreviations: CYP73A: trans-cinnamate 4-monooxygenase; 4CL: 4-coumarate-CoA ligase; CHS: chalcone synthase; DFR: dihydroflavonol-4-reductase; F3′5′H: flavonoid 3′,5′-hydroxylase; FLS: flavonol synthase; ANR: anthocyanidin reductase; LAR: leucoanthocyanidin reductase; CCoAOMT: caffeoyl-CoA 0-methyltransferase; HCT: shikimate O-hydroxycinnamoyltransferase; I2′H: isoflavone 2′-hydroxylase; HID: 2-hydroxyisoflavanone dehydratase.

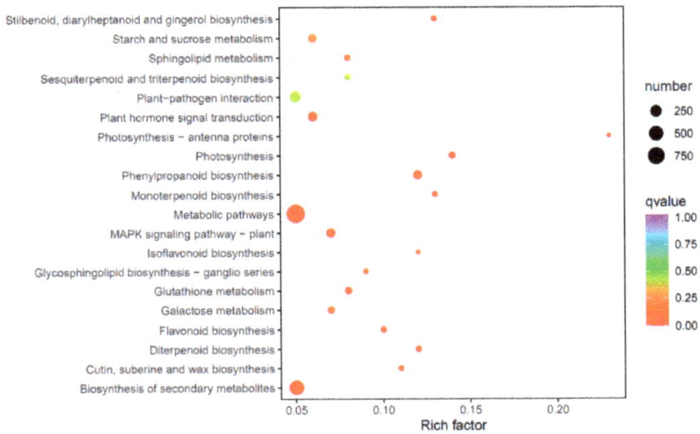

Figure 8. Kyoto Encyclopedia of Genes and Genomes (KEGG) pathway enrichment analysis of differentially expressed unigenes. The ordinate represents the KEGG pathway and the abscissa represents the Rich factor. The larger the Rich factor, the greater the degree of enrichment. The larger the point, the greater the number of differentially expressed genes enriched in the pathway. The redder the colour of the dot, the more pronounced the enrichment.

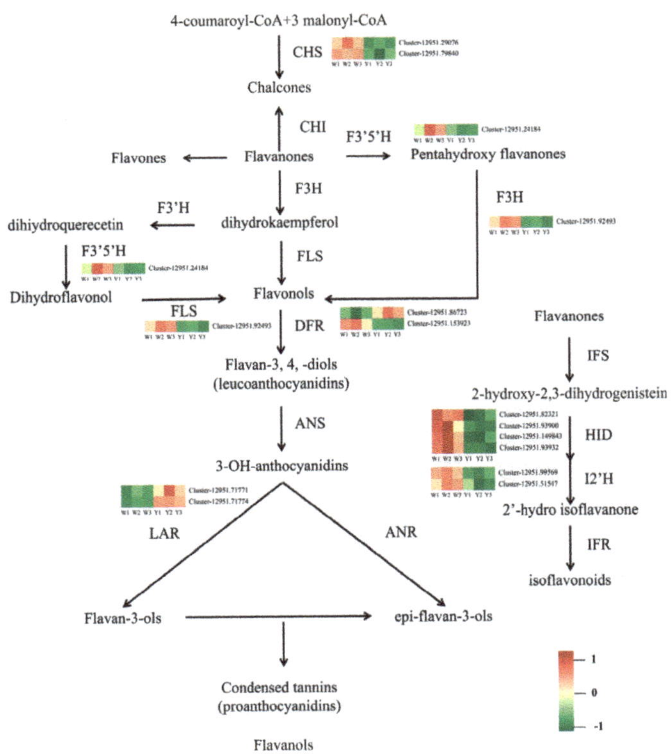

Figure 9. Schematic diagram of the flavonoid biosynthesis pathway related to flower pigment in the *R. liliiflorum*. Enzyme acronyms and expression patterns are shown beside each metabolic step and the direction of synthesis is marked with black arrows. The RNA-seq expression pattern of each gene is shown in heatmaps. The colour scale represents log2-transformed FPKM (fragments per kilobase of exon per million mapped reads) values. W vs. Y represent white and yellow parts of petals in *R. liliiflorum*, respectively.

4. Discussion

R. liliiflorum is a very important ornamental plant, and the colours of its petals are mainly yellow and white. To understand the reasons for this phenomenon, this study used the Illumina HiSeq sequencing platform to sequence the transcriptome of the yellow and white parts of the petals of *R. liliiflorum*. The results showed that each group of samples produced more than 6 GB of clean data, and a total of 180,677 unigenes were obtained, of which 171,948 were annotated. Through the differential expression comparison of W vs. Y, 5372 DEGs were obtained and further annotated through the KEGG pathway. The results showed that the DEGs of *R. liliiflorum* petals in different coloured parts were significantly enriched in phenylpropanoid biosynthesis, photosynthesis, metabolic pathways, MAPK signalling pathway-plant, biosynthesis of secondary metabolites, and flavonoid and isoflavone biosynthesis (Figure 8). In addition, GO classification also showed that a large number of DEGs in W vs. Y were involved in the primary metabolism of active substances in *R. liliiflorum*. This showed that the biosynthesis of flavonoids in the petals of *R. liliiflorum* was significantly differentially regulated during the development process. In this study, 36 genes related to the biosynthesis of flavonoids in the petals of *R. liliiflorum* were identified (Table 5), including CYP73A (1), 4CL (3), CHS (2), DFR (2), F3′5′H (1), FLS (1), LAR (2), CCoAOMT (2), HCT (16), I2′H (2), and HID (4).

Flavonoid metabolism is considered to be one of the most important ways to promote floral colour formation and is catalysed by multiple enzyme complexes [3]. In the flavonoid biosynthesis pathway (Figure 9), this study identified six key enzyme genes, including CHS, F3′5′H, FLS, DFR, I2′H, and LAR, involved in the regulation of flavonoid biosynthesis. The gene expression of CHS, F3′5′H, FLS, and I2′H was downregulated, and these genes were involved in the regulation of chalcone, dihydroflavonol, flavonol, and isoflavone biosynthesis, respectively. In contrast, LAR was upregulated in the anthocyanin to flavonoid biosynthesis pathway. The expression levels of two DEGs related to DFR were upregulated and downregulated. CHS is a key rate-limiting enzyme in the flavonoid biosynthesis pathway [31]. If it is inhibited, the total flavonoid level will be reduced [32]. The mutation of F3′5′H leads to a decrease in delphinid levels and an increase in anthocyanin levels [33]. Ectopic expression of FLS promotes the accumulation of kaempferol and a decrease in the anthocyanin content in flowers [34]. At the same time, FLS overexpression enhances quercetin signalling in roots [35]. HID can further convert the generated 2-hydroxy-2,3-dihydrogenistein and 2,7,40-trihydroxyisoflavone into the isoflavones genistein and daidzein [36], and then these two substances will lead to the accumulation of isoflavones under the continuous catalysis of a batch of enzymes, such as I2′H [37,38]. As a key regulatory enzyme for anthocyanin biosynthesis, DFR can catalyse the conversion of three colourless dihydroflavonol alcohols into light anthocyanins, playing an important role in the accumulation of anthocyanin aglycones [39]. Massive expression of upstream genes and low expression of DFR cause the biosynthesis of a great deal of anthoxanthin and a small amount of colourless leucopelargonidin and leucocyanidin [8]. Many yellow varieties of *Dahlia variabilis* Hort. do not show DFR activity [40]. LAR expression in *R. liliiflorum* was significantly upregulated. LAR is a member of the reductase–epimerase–dehydrogenase enzyme family and closely related to isoflavone reductase in the isoflavonoid biosynthesis pathway [25]. LAR affects the accumulation of anthocyanins in plants by participating in the biosynthesis of plant-derived proanthocyanidins (PAs), while higher levels of PAs inhibit the production of anthocyanins and quercetin [41]. It has been proven that LAR can convert white anthocyanin into (+)—catechin [42]. Wang et al. [43] studied the flavonoid biosynthesis of *Narcissus tazetta* L. var. *chinensis* Roem. and speculated that the lack of anthocyanins was due to the high expression of LAR and FLS and insufficient expression of ANS [43].

A previous study of the metabolites of some yellow and white flowers found that the flavonoid content of yellow flowers in medical *C. morifolium* [44] and *Freesia hybrida* Klatt. [45] was higher than that of white flowers, but the results were opposite in *P. lactiflora* [46] and *Pericallis hybrida* B. Nord. [47]. Specific to the type of flavonoids, it was found that the quercetin content in yellow flowers of *Gossypium arboreum* L. was higher than that of white flowers, while the kaempferol content was lower than that of white flowers [48]. Quercetin content in yellow flowers of *Michelia maudiae* Dunn was higher than that in white flowers [49]. These results proved that the flavonoid content affects the colour development of most yellow flowers. The transcriptome study on the corresponding parts of the above species found that the expression levels of FLS, F3H, DFR, and CHI were relatively high in yellow flowers. Wang et al. [50] found that the expression levels of isoflavonoids and dihydroflavonols in yellow flowers were higher than those in white flower of *Carthamus tinctorius* L., but the regulated gene expression was higher in white flowers than in yellow flowers [50]. Generally, in most flowers, the flavonoid content of yellow flowers is higher than that of white flowers, which may be regulated by FLS, F3H, DFR, etc., but the colour differentiation of the same flower may not be closely related to the flavonoid content. For example, the inner petal was yellow and the outer petal was white, but the flavonoid content of the white part was higher than that of the yellow part [46]. Bao et al. [51] found the pH of the yellow inner petals was higher than that of the white outer petals, and the moisture content was lower than that of white part in *P. lactiflora*. Similar to *R. liliiflorum*, DFR was upregulated and CHS was downregulated in *P. lactiflora*, indicating that these two genes may play a regulatory role in flower coloured differentiation of the same flower. Of course, apart from anthocyanins, flavonoids, and peptide factors, the pH of flower issues, epidermal cell surfaces, and anthocyanic vacuolar inclusion might also play roles in the formation of floral colour [16].

5. Conclusions

Using Illumina Hiseq transcriptome sequencing technology, this work analysed gene expression in the petals of *R. liliiflorum*. This study provides a preliminary understanding of the mechanism of gene regulation in the colour formation of *R. liliiflorum* petals. Our results show that the colour change of *R. liliiflorum* is related to differential changes in gene expression levels, and to a certain extent, it is related to genes associated with the flavonoid biosynthesis pathway. In the flavonoid biosynthesis pathway, we found that the expression of DEGs related to CHS, F3′5′H, FLS, I2′H, and HID in the chalcone, dihydroflavonol, and pentahydroxyflavanol flavonol biosynthesis pathways was downregulated. In the flavonol biosynthesis colourless anthocyanin pathway, the expression levels of two DEGs associated with DFR were upregulated and downregulated, while in the flavanol biosynthesis pathway, the expression of DEGs associated with LAR was upregulated. These results show that CHS, F3′5′H, FLS, I2′H, HID, DFR, LAR, and other enzyme-related genes play a key role in the formation of *R. liliiflorum* petal colour.

Author Contributions: Conceptualization, L.L. and Y.Y.; formal analysis, M.C., X.W. and J.D.; investigation, X.Z. (Xu Zhang) and Z.Z.; writing—original draft preparation, H.Z., M.C., X.W. and J.D.; writing—review and editing, X.Z. (Ximin Zhang), J.G., M.T. and J.T. All authors have read and agreed to the published version of the manuscript.

Funding: This research was supported by the Natural Science Foundation of China and the Karst Science Research Center of Guizhou Province, China (U1812401); the Science and Technology Support Plant Project, China ([2021] No. 224); Higher Education Science and Research Youth Project of Guizhou Education Department (Qianjiaoji [2022]130); QIANKEHEPINGTAIRENCAI ([2017] No. 5726-15); Science and Technology Fund Project of Guizhou Province ([2020] No. 4Y028); Innovation and entrepreneurship training plan for national and provincial college students, Guizhou Normal University, China (S202110663037); the Natural Science Foundation of China (NSFC) (32260393); and Guizhou forestry scientific research project, Qianlinkehe ([2022] No. 28).

Data Availability Statement: All sequencing data are available through the NCBI Sequence Read Archive under the accession number PRJNA898493.

Acknowledgments: The authors would like to thank the Natural Science Foundation of China and the Karst Science Research Center of Guizhou Province of China (U1812401) and the Science and Technology Support Plant Project of China ([2021]224) for their support in this research.

Conflicts of Interest: The authors declare no conflict of interest.

References

1. Bradshaw, H.D.; Schemske, D.W. Allele substitution at a flower colour locus produces a pollinator shift in monkeyflowers. *Nature* **2003**, *426*, 176–178. [CrossRef] [PubMed]
2. Schmitzer, V.; Veberic, R.; Osterc, G.; Stampar, F. Color and Phenolic Content Changes during Flower Development in Groundcover Rose. *J. Am. Soc. Hortic. Sci.* **2010**, *135*, 195–202. [CrossRef]
3. Ye, L.-J.; Möller, M.; Luo, Y.-H.; Zou, J.-Y.; Zheng, W.; Wang, Y.-H.; Liu, J.; Zhu, A.-D.; Hu, J.-Y.; Li, D.-Z. Differential expressions of anthocyanin synthesis genes underlie flower color divergence in a sympatric *Rhododendron sanguineum* complex. *BMC Plant Biol.* **2021**, *21*, 204. [CrossRef] [PubMed]
4. Mol, J.; Grotewold, E.; Koes, R. How genes paint flowers and seeds. *Trends Plant Sci.* **1998**, *3*, 212–217. [CrossRef]
5. Chen, S.; Li, C.; Zhu, X.; Deng, Y.M.; Sun, W.; Wang, L.; Chen, F.; Zhang, Z. The identification of flavonoids and the expression of genes of anthocyanin biosynthesis in the chrysanthemum flowers. *Biol. Plant.* **2012**, *56*, 458–464. [CrossRef]
6. Thill, J.; Miosic, S.; Ahmed, R.; Schlangen, K.; Muster, G.; Stich, K.; Halbwirth, H. 'Le Rouge et le Noir': A decline in flavone formation correlates with the rare color of black dahlia (*Dahlia variabilis* hort.) flowers. *BMC Plant Biol.* **2012**, *12*, 225. [CrossRef]
7. Tatsuzawa, F.; Saito, N.; Toki, K.; Shinoda, K.; Honda, T. Flower Colors and their Anthocyanins in *Matthiola incana* Cultivars (Brassicaceae). *J. Jpn. Soc. Hortic. Sci.* **2012**, *81*, 91. [CrossRef]
8. Zhao, D.; Tao, J.; Han, C.; Ge, J. Flower color diversity revealed by differential expression of flavonoid biosynthetic genes and flavonoid accumulation in herbaceous peony (*Paeonia lactiflora* Pall.). *Mol. Biol. Rep.* **2012**, *39*, 11263–11275. [CrossRef]
9. Winkel-Shirley, B. Flavonoid biosynthesis. A colorful model for genetics, biochemistry, cell biology, and biotechnology. *Plant Physiol.* **2001**, *126*, 485–493. [CrossRef]
10. Zhou, Z.; Gao, H.; Ming, J.; Ding, Z.; Lin, X.e.; Zhan, R. Combined Transcriptome and Metabolome analysis of Pitaya fruit unveiled the mechanisms underlying Peel and pulp color formation. *BMC Genom.* **2020**, *21*, 734. [CrossRef]
11. Liu, W.; Feng, Y.; Yu, S.; Fan, Z.; Li, X.; Li, J.; Yin, H. The Flavonoid Biosynthesis Network in Plants. *Int. J. Mol. Sci.* **2021**, *22*, 12824. [CrossRef] [PubMed]
12. Nishihara, M.; Nakatsuka, T.; Yamamura, S. Flavonoid components and flower color change in transgenic tobacco plants by suppression of chalcone isomerase gene. *FEBS Lett.* **2005**, *579*, 6074–6078. [CrossRef] [PubMed]
13. Dong, W.; Li, M.; Li, Z.; Li, S.; Zhu, Y.; Ding, H.; Wang, Z. Transcriptome analysis of the molecular mechanism of *Chrysanthemum* flower color change under short-day photoperiods. *Plant Physiol. Biochem.* **2020**, *146*, 315–328. [CrossRef] [PubMed]
14. Malik, W.; Khan, A.; Masooma, H.; Aslam, U.; Iqbal, M.; Qayyum, A.; Yasmeen, A.; Bibi, N. Transcriptome Analysis of Pigment Related Genes in Colored Cotton. *Int. J. Agric. Biol.* **2015**, *17*, 1560–8530.
15. Zheng, X.; Zhao, B.; Zeng, H.; Shen, H.; Xu, J. Comparative Analysis of Composition and Content of Pigments in Petals of Three Different Colors of *Rhododendron calophytum* in Qinling Mountains. *J. Northwest For. Univ.* **2017**, *32*, 62–68. (In Chinese)
16. Du, H.; Lai, L.; Wang, F.; Sun, W.; Zhang, L.; Li, X.; Wang, L.; Jiang, L.; Zheng, Y. Characterisation of flower colouration in 30 *Rhododendron* species via anthocyanin and flavonol identification and quantitative traits. *Plant Biol.* **2018**, *20*, 121–129. [CrossRef]
17. Jaakola, L.; Määttä, K.; Pirttilä, A.M.; Törrönen, R.; Kärenlampi, S.; Hohtola, A. Expression of genes involved in anthocyanin biosynthesis in relation to anthocyanin, proanthocyanidin, and flavonol levels during bilberry fruit development. *Plant Physiol.* **2002**, *130*, 729–739. [CrossRef]
18. Zipor, G.; Duarte, P.; Carqueijeiro, I.; Shahar, L.; Ovadia, R.; Teper-Bamnolker, P.; Eshel, D.; Levin, Y.; Doron-Faigenboim, A.; Sottomayor, M.; et al. *In planta* anthocyanin degradation by a vacuolar class III peroxidase in *Brunfelsia calycina* flowers. *New Phytol.* **2015**, *205*, 653–665. [CrossRef]
19. Liu, Y.; Tikunov, Y.; Schouten, R.E.; Marcelis, L.F.M.; Visser, R.G.F.; Bovy, A. Anthocyanin Biosynthesis and Degradation Mechanisms in *Solanaceous* Vegetables: A Review. *Front. Chem.* **2018**, *6*, 52. [CrossRef]
20. Mizuta, D.; Ban, T.; Miyajima, I.; Nakatsuka, A.; Kobayashi, N. Comparison of flower color with anthocyanin composition patterns in evergreen azalea. *Sci. Hortic.* **2009**, *122*, 594–602. [CrossRef]
21. Wang, S.; Zhang, Y.; Zou, Q.; Shan, W.; Li, X.; Zhang, L. Effects of IBA concentration and cutting time on rooting of *Rhododendron kiangsiense* and *Rh. liliiflorum* cuttings. *Guihaia* **2016**, *36*, 1468–1475. (In Chinese)
22. Haas, B.J.; Papanicolaou, A.; Yassour, M.; Grabherr, M.; Blood, P.D.; Bowden, J.; Couger, M.B.; Eccles, D.; Li, B.; Lieber, M.; et al. De novo transcript sequence reconstruction from RNA-seq using the Trinity platform for reference generation and analysis. *Nat. Protoc.* **2013**, *8*, 1494–1512. [CrossRef] [PubMed]
23. Altschul, S.F.; Madden, T.L.; Schäffer, A.A.; Zhang, J.; Zhang, Z.; Miller, W.; Lipman, D.J. Gapped BLAST and PSI-BLAST: A new generation of protein database search programs. *Nucleic Acids Res.* **1997**, *25*, 3389–3402. [CrossRef] [PubMed]
24. Kanehisa, M.; Goto, S.; Hattori, M.; Aoki-Kinoshita, K.F.; Itoh, M.; Kawashima, S.; Katayama, T.; Araki, M.; Hirakawa, M. From genomics to chemical genomics: New developments in KEGG. *Nucleic Acids Res.* **2006**, *34*, D354–D357. [CrossRef]
25. Tatusov, R.L.; Fedorova, N.D.; Jackson, J.D.; Jacobs, A.R.; Kiryutin, B.; Koonin, E.V.; Krylov, D.M.; Mazumder, R.; Mekhedov, S.L.; Nikolskaya, A.N.; et al. The COG database: An updated version includes eukaryotes. *BMC Bioinform.* **2003**, *4*, 41. [CrossRef]
26. UniProt Consortium. UniProt: A hub for protein information. *Nucleic Acids Res.* **2015**, *43*, D204–D212. [CrossRef]
27. Conesa, A.; Götz, S.; García-Gómez, J.M.; Terol, J.; Talón, M.; Robles, M. Blast2GO: A universal tool for annotation, visualization and analysis in functional genomics research. *Bioinformation* **2005**, *21*, 3674–3676. [CrossRef]

28. Li, B.; Dewey, C.N. RSEM: Accurate transcript quantification from RNA-Seq data with or without a reference genome. *BMC Bioinform.* **2011**, *12*, 323. [CrossRef]
29. Love, M.I.; Huber, W.; Anders, S. Moderated estimation of fold change and dispersion for RNA-seq data with DESeq2. *Genome Biol.* **2014**, *15*, 550. [CrossRef]
30. Varet, H.; Brillet-Guéguen, L.; Coppée, J.Y.; Dillies, M.A. SARTools: A DESeq2- and EdgeR-Based R Pipeline for Comprehensive Differential Analysis of RNA-Seq Data. *PLoS ONE* **2016**, *11*, e0157022. [CrossRef]
31. Deng, X.; Bashandy, H.; Ainasoja, M.; Kontturi, J.; Pietiäinen, M.; Laitinen, R.A.E.; Albert, V.A.; Valkonen, J.P.T.; Elomaa, P.; Teeri, T.H. Functional diversification of duplicated chalcone synthase genes in anthocyanin biosynthesis of *Gerbera hybrida*. *New Phytol.* **2014**, *201*, 1469–1483. [CrossRef] [PubMed]
32. Schijlen, E.; De Vos, R.; Martens, S.; Jonker, H.; Rosin, F.; Molthoff, J.; Tikunov, Y.; Angenent, G.; van tunen, A.; Bovy, A. RNA Interference Silencing of Chalcone Synthase, the First Step in the Flavonoid Biosynthesis Pathway, Leads to Parthenocarpic Tomato Fruits. *Plant Physiol.* **2007**, *144*, 1520–1530. [CrossRef] [PubMed]
33. Takahashi, R.; Dubouzet, J.G.; Matsumura, H.; Yasuda, K.; Iwashina, T. A new allele of flower color gene *W1* encoding flavonoid 3′5′-hydroxylase is responsible for light purple flowers in wild soybean *Glycine soja*. *BMC Plant Biol.* **2010**, *10*, 155. [CrossRef] [PubMed]
34. Jiang, X.; Shi, Y.; Fu, Z.; Li, W.-W.; Lai, S.; Wu, Y.; Wang, Y.; Liu, Y.; Gao, L.; Xia, T. Functional characterization of three flavonol synthase genes from *Camellia sinensis*: Roles in flavonol accumulation. *Plant Sci.* **2020**, *300*, 110632. [CrossRef] [PubMed]
35. Park, S.; Kim, D.-H.; Yang, J.-H.; Lee, J.-Y.; Lim, S. Increased Flavonol Levels in Tobacco Expressing *AcFLS* Affect Flower Color and Root Growth. *Int. J. Mol. Sci.* **2020**, *21*, 1011. [CrossRef]
36. Veremeichik, G.N.; Grigorchuk, V.P.; Butovets, E.S.; Lukyanchuk, L.M.; Brodovskaya, E.V.; Bulgakov, D.V.; Bulgakov, V.P. Isoflavonoid biosynthesis in cultivated and wild soybeans grown in the field under adverse climate conditions. *Food Chem.* **2021**, *342*, 128292. [CrossRef]
37. Uchida, K.; Akashi, T.; Aoki, T. The Missing Link in Leguminous Pterocarpan Biosynthesis is a Dirigent Domain-Containing Protein with Isoflavanol Dehydratase Activity. *Plant Cell Physiol.* **2017**, *58*, 398–408. [CrossRef]
38. Yao, S.; Lan, Z.; Huang, R.; Tan, Y.; Huang, D.; Gu, J.; Pan, C. Hormonal and transcriptional analyses provides new insights into the molecular mechanisms underlying root thickening and isoflavonoid biosynthesis in *Callerya speciosa* (Champ. ex Benth.) Schot. *Sci. Rep.* **2021**, *11*, 9. [CrossRef]
39. Sun, w.; Zhou, N.; Feng, C.; Sun, S.; Tang, M.; Tang, X.; Ju, Z.; Yi, Y. Functional analysis of a dihydroflavonol 4-reductase gene in *Ophiorrhiza japonica* (OjDFR1) reveals its role in the regulation of anthocyanin. *PeerJ* **2021**, *9*, e12323. [CrossRef]
40. Halbwirth, H.; Muster, G.; Stich, K. Unraveling the Biochemical Base of Dahlia Flower Coloration. *Nat. Prod. Commun.* **2008**, *3*, 1934578X0800300807. [CrossRef]
41. Li, H.; Tian, J.; Yao, Y.-y.; Zhang, J.; Song, T.-t.; Li, K.-t.; Yao, Y.-c. Identification of leucoanthocyanidin reductase and anthocyanidin reductase genes involved in proanthocyanidin biosynthesis in *Malus* crabapple plants. *Plant Physiol. Biochem.* **2019**, *139*, 141–151. [CrossRef] [PubMed]
42. Liu, C.; Wang, X.; Shulaev, V.; Dixon, R.A. A role for leucoanthocyanidin reductase in the extension of proanthocyanidins. *Nat. Plants* **2016**, *2*, 16182. [CrossRef] [PubMed]
43. Wang, G.; Yang, B.; Wu, J.; Luo, P.; Anwar, M.; Allan, A.C.; Lin-Wang, K.; Espley, R.V.; Zeng, L. Identification of Genes Involved in Flavonoid Biosynthesis of Chinese Narcissus (*Narcissus tazetta* L. var. *chinensis*). *Plant Mol. Biol. Report.* **2018**, *36*, 812–821. [CrossRef]
44. Zou, Q.; Wang, T.; Guo, Q.; Yang, F.; Chen, J.; Zhang, W. Combined metabolomic and transcriptomic analysis reveals redirection of the phenylpropanoid metabolic flux in different colored medicinal *Chrysanthemum morifolium*. *Ind. Crops Prod.* **2021**, *164*, 113343. [CrossRef]
45. Zhu, J.; Guo, X.; Li, X.; Tang, D. Composition of Flavonoids in the Petals of *Freesia* and Prediction of Four Novel Transcription Factors Involving in *Freesia* Flavonoid Pathway. *Front. Plant Sci.* **2021**, *12*, 756300. [CrossRef]
46. Wu, Y.-q.; Wei, M.-r.; Zhao, D.-q.; Tao, J. Flavonoid content and expression analysis of flavonoid biosynthetic genes in herbaceous peony (*Paeonia lactiflora* Pall.) with double colors. *J. Integr. Agric.* **2016**, *15*, 2023–2031. [CrossRef]
47. Jin, X.; Huang, H.; Wang, L.; Sun, Y.; Dai, S. Transcriptomics and Metabolite Analysis Reveals the Molecular Mechanism of Anthocyanin Biosynthesis Branch Pathway in Different *Senecio cruentus* Cultivars. *Front. Plant Sci.* **2016**, *7*. [CrossRef]
48. Xing, A.S.; Wang, X.Y.; Nazir, M.F.; Zhang, X.M.; Wang, X.X.; Yang, R.; Chen, B.J.; Fu, G.Y.; Wang, J.J.; Ge, H.; et al. Transcriptomic and metabolomic profiling of flavonoid biosynthesis provides novel insights into petals coloration in Asian cotton (*Gossypium arboreum* L.). *BMC Plant Biol.* **2022**, *22*, 416. [CrossRef]
49. Lang, X.; Li, N.; Li, L.; Zhang, S. Integrated Metabolome and Transcriptome Analysis Uncovers the Role of Anthocyanin Metabolism in *Michelia maudiae*. *Int. J. Genom.* **2019**, *2019*, 4393905. [CrossRef]
50. Wang, R.; Ren, C.; Dong, S.; Chen, C.; Xian, B.; Wu, Q.; Wang, J.; Pei, J.; Chen, J. Integrated Metabolomics and Transcriptome Analysis of Flavonoid Biosynthesis in Safflower (*Carthamus tinctorius* L.) with Different Colors. *Front. Plant Sci.* **2021**, *12*. [CrossRef]
51. Bao, M.; Liu, M.; Zhang, Q.; Wang, T.; Sun, X.; Xu, J. Factors Affecting the Color of Herbaceous Peony. *J. Am. Soc. Hortic. Sci. J. Amer. Soc. Hort. Sci.* **2020**, *145*, 257–266. [CrossRef]

horticulturae

Review

Mechanisms Underlying the C_3–CAM Photosynthetic Shift in Facultative CAM Plants

Shuo Qiu, Ke Xia, Yanni Yang, Qiaofen Wu and Zhiguo Zhao *

Guangxi Key Laboratory of Plant Functional Phytochemicals and Sustainable Utilization, Guangxi Institute of Botany, Guangxi Zhuang Autonomous Region and Chinese Academy of Sciences, Guilin 541006, China
* Correspondence: zwskfc@139.com; Tel.: +86-773-3550103; Fax: +86-773-3550067

Abstract: Crassulacean acid metabolism (CAM), one of three kinds of photosynthesis, is a water-use efficient adaptation to an arid environment. CAM is characterized by CO_2 uptake via open stomata during the nighttime and refixation CO_2 via the Calvin cycle during the daytime. Facultative CAM plants can shift the photosynthesis from C_3 to CAM and exhibit greater plasticity in CAM expression under different environments. Though leaf thickness is an important anatomical feature of CAM plants, there may be no anatomical feature changes during the C_3–CAM transition for all facultative CAM plants. The shift from C_3 photosynthesis to CAM in facultative CAM plants is accompanied by significant changes in physiology including stomata opening, CO_2 gas exchange and organic acid fluxes; the activities of many decarboxylating enzymes increase during the shift from C_3 to CAM; the molecular changes occur during the photosynthesis C_3–CAM shift involved DNA hypermethylation, transcriptional regulation, post-transcriptional regulation and protein level regulation. Recently, omics approaches were used to discover more proceedings underling the C_3–CAM transition. However, there are few reviews on the mechanisms involved in this photosynthetic shift in facultative CAM plants. In this paper, we summarize the progress in the comparative analysis of anatomical, physiological, metabolic and molecular properties of facultative CAM plants between C_3 and CAM photosynthesis. Facultative CAM plants also show the potential for sustainable food crop and biomass production. We also discuss the implications of the photosynthesis transition from C_3 to CAM on horticultural crops and address future directions for research.

Keywords: C_3 photosynthesis; crassulacean acid metabolism (CAM); shift mechanisms; facultative crassulacean acid metabolism (CAM) plants; environments

Citation: Qiu, S.; Xia, K.; Yang, Y.; Wu, Q.; Zhao, Z. Mechanisms Underlying the C_3–CAM Photosynthetic Shift in Facultative CAM Plants. *Horticulturae* **2023**, *9*, 398. https://doi.org/10.3390/horticulturae9030398

Academic Editor: Riccardo Lo Bianco

Received: 6 February 2023
Revised: 11 March 2023
Accepted: 15 March 2023
Published: 19 March 2023

1. Introduction

In the plant kingdom, there are three kinds of photosynthetic pathways: C_3, C_4 and crassulacean acid metabolism (CAM). CAM is characterized by CO_2 uptake during the nighttime via open stomata, when CO_2 is combined with phosphoenolpyruvate (PEP) and stored as organic acids (mainly malic acid). Then, organic acids are decarboxylated in the vacuoles during daytime and CO_2 is refixed via the Calvin cycle [1,2]. Some plants can switch their photosynthesis between C_3 and CAM, which are referred to as facultative (inducible or C_3/CAM intermediate) CAM plants. The first discovered facultative CAM plant was *Mesembryanthemum crystallinum*, the common iceplant [3]. Other well-known facultative CAM plants include *Sedum album* [4], *Clusia minor* [5], *Talinum triangulare* [6], etc. These species are sampled as models to study other facultative CAM plants, as they can shift their photosynthetic mode in response to water deficit and other environmental stressors [7,8]. Furthermore, recent evidence suggests that facultative CAM plants may be more widespread among vascular plants than previously thought [9].

As we all know, photosynthesis is one of the most important chemical reactions on the earth [10]. Crassulacean acid metabolism (CAM), with higher photosynthetic, water-use and possibly nutrient-use efficiency, represents higher carbon-concentrating mechanisms (CCMs) than C_3 photosynthesis in response to a warmer and drier world [11]. The phenomena of CAM is a typical ecophysiological adaptation to arid conditions [12–14]. Expression of CAM modes includes obligate CAM, facultative CAM, CAM-idling and CAM-cycling. Facultative CAM mode is one of the four plastic expressions of CAM modes. Facultative CAM plants can struggle with variable environments through the facultative CAM mode [14]. Recently, engineering CAM-related genes to C_3 crops to improve water-use efficiency (WUE) has caused extensive attention [15–17]. In order to engineer CAM into C_3 crops, a deep understanding of CAM-related genes and metabolic pathways is urgently needed [18].

However, the shift mechanism underlying C_3 photosynthesis to CAM in facultative CAM plants is complicated, concerning the genetic changes required for the progression and reversion of this shift [19]. In order to engineer CAM into C_3 crops to increase the water-use efficiency (WUE), a few facultative CAM species (such as *M. crystallinum*) were regarded as key tools to identify the genes involved in the CAM pathway and their respective regulation mechanisms [11,15,20,21]. In this paper, we review the signaling stress factors inducing C_3-photosynthesis to CAM in facultative CAM plants and assess progress in the analysis of anatomical, physiological, metabolic and molecular differences for facultative CAM plants between the C_3 and CAM mode. We also review their implications on horticultural crops and address directions for future research.

2. Signaling Stress Factors

CAM is a plastic photosynthetic adaptation found in plants in abiotic stress environments (such as drought, salinity, extreme temperature, etc.) [13]. Environmental, hormonal and circadian changes can regulate the CAM expression in facultative CAM species [22]. In facultative CAM species, photosynthesis can switch from C_3 to CAM modes after induction by abiotic stress, such as atmospheric CO_2 concentration, drought, salinity, photoperiod and light [12,23–25]. For instance, *M. crystallinum* switches its photosynthetic mode from C_3 to CAM under water or salinity stress [7]. Light intensity and quality also play a crucial role for the C_3–CAM transition [12,26]. All these signaling factors connect via a closed network and directly or indirectly affect each other [2].

The exogenous application of ethylene or abscisic acid (ABA) could induce the C_3–CAM transition in a few facultative CAM species [27]. The degree of CAM expression was positively correlated with ABA and trans-zeatin, but negatively correlated with cytokinins and jasmonic acid (JA) [28,29]. Exogenous hydrogen peroxide (H_2O_2) and root signaling also could induce the C_3–CAM transition in *M. crystallinum*, respectively [30–32].

3. Anatomical Variations during the C_3–CAM Shift in Facultative CAM Plants

Leaf thickness is an important anatomical feature for CAM plants. CAM is often associated with succulent leaves; indeed, the tissue succulence of CAM species has been observed in many plant families, such as Crassulaceae, Orchidaceae and Clusiaceae [33–36]. Many arid CAM plants with succulent nature are beneficial as they store more water than C_3 and C_4 species [37]. A study reported, by analyzing the leaf thickness and leaf $\delta^{13}C$ values in 173 tropical orchids, that the leaf was the thickest in the strong CAM species [36]. However, some plants with thinner leaves can also fix CO_2 through the CAM pathway; for example, *Dendrobium bigibbum* (a CAM orchid) can yield $\delta^{13}C$ with $-11.9‰$, despite the leaf thickness being only 0.79 mm [34], which proved that the assumption about obligate CAM species possessing more succulent leaves than facultative CAM species was not accurate [38]. In *M. crystallinum*, leaf succulence increased during the C_3 to CAM transition after 5 days of the salt treatment [39], but some CAM species (e.g., bromeliads) do not have succulent photosynthetic organs [40,41]. Recently, it was also hypothesized that the evolution of facultative CAM plants did not require major changes in anatomy [42,43]. Winter thought that strong CAM plants needed significant anatomical modifications, whereas facultative or weak CAM plants may not require them, suggesting there may be no anatomical feature changes during the C_3–CAM transition in facultative CAM plants [9]. Investigations showed that leaf anatomy was not correlated to CAM function in *Yucca gloriosa* (facultative CAM species, a C_3+CAM hybrid species) [44], and the relationships between leaf anatomy and degree of CAM expression were not very close [45]. Herrera [39] reported that less succulence is not a typical feature for facultative CAM plants. Thus, the leaf thickness, as an indicator, cannot completely distinguish between plant species, suggesting that the relationship between anatomical leaf features and CAM expression requires further investigation.

That being said, there may be some changes in vacuole and chloroplast anatomy during the photosynthesis shift from C_3 to CAM. Malic acid accumulates and releases in the vacuole. More particularly, the fluidity of the tonoplast will reduce after the photosynthesis shift from C_3 to CAM, in turn decreasing the vacuolar mobilization of malic acid [46], indicating that vacuole size may increase before the shift to CAM in the leaves [47]. Chloroplasts, the main sites of photosynthesis in plants, can regulate the facultative CAM plants to acclimate to high salinity environments [48], and show a severe thylakoid swelling at midday in CAM plants [49]. However, whether such changes in the ultrastructure level also occur in other facultative CAM species remains unclear. Young leaves in obligate CAM plants take up CO_2 by C_3 photosynthesis, while mature leaves take up CO_2 by CAM. It is worth exploring the changes in ultrastructure level (i.e., vacuole and chloroplast) that occur throughout the development process of obligate CAM plants.

4. Physiological Mechanisms during the C_3–CAM Shift in Facultative CAM Plants

Compared to C_3 and C_4 metabolism, CAM is characterized by CO_2 uptake at night and stomatal closure during the day. For facultative CAM species, the C_3 photosynthesis pathway is used to fix CO_2 when the plants are under well-conditions. However, the photosynthesis can shift to the CAM mode once induced by abiotic stress; stomata open at night and atmospheric CO_2 is absorbed by phosphoenolpyruvate carboxylase (EC 4.1.1.31; PEPC) via oxaloacetate into malic acid. On the following day, stomata close and malic acid is released from the vacuoles and decarboxylated, while CO_2 is refixed by the Calvin cycle [1]. Therefore, in facultative CAM plants, the CO_2 gas exchange exhibits a difference between daytime and nighttime, as well as the changes in the activities of PEPC and Rubisco, which are important features to distinguish the C_3 from the CAM mode. Figure 1 shows the physiological and metabolic shift of the facultative CAM plants from C_3 to CAM photosynthesis.

Since this C_3 to CAM transition was first recognized, *M. crystallinum* was used as a model plant to examine the associated changes in enzyme activities [50]. PEPC, one of the key enzymes, is involved in primary carboxylation during both CAM and C_4 photosynthesis. PEPC is widely distributed across plants, algae and bacterial species; it catalyzes the irreversible β-carboxylation of phosphoenolpyruvate (PEP) in the presence of HCO_3^- to yield oxaloacetate (OOA) and Pi. In CAM and C_4 photosynthesis, this enzyme is responsible for the primary fixation of inorganic carbon. PEPC also is a major anaplerotic enzyme in most non-photosynthetic organs and the leaves of C_3 plants [51,52]. Furthermore, the activities of many decarboxylating enzymes increase during the shift from C_3 to CAM, such as the cytosolic NADP-malic enzyme (NADPME, EC 1.1.1.40), the mitochondrial NAD-malic enzyme (NADME, EC 1.1.1.38) and PEP carboxykinase (PEPCK). This increased enzymatic activity is thought to be an indicator of the start of CAM photosynthesis [22,50].

In the plant, reactive oxygen species (ROS), including the superoxide radicals (O_2^-), the hydroxyl radical (OH) and hydrogen peroxide (H_2O_2), are always formed in response to environmental stress [53,54]. Oxidative stress could lead to the photosynthesis switch from C_3 to CAM in facultative CAM plants [55,56]. Enzymatic antioxidants and non-enzymatic antioxidants are involved to protect the plants from oxidative damage by the scavenging of ROS [57]. During the C_3–CAM shift in facultative CAM plants (e.g., *Sedum album*) induced by water stress, antioxidative enzymes, such as superoxide dismutase (SOD), peroxidase (POD), ascorbate peroxidase (APX), catalase (CAT), etc. [4,58,59]. In *M. crystallinum*, FeSOD activity increases more rapidly during the first few days before CAM appearance; then, MnSOD and Cu/ZnSOD activity increases after CAM occurs, induced by salt [48,60]. CAT is not only responsible for the removal of H_2O_2, but presents diel fluctuations [58]. Non-enzymatic antioxidants constituted by low molecular metabolites include ascorbic acid (AsA), glutathione (GSH), carotenoids, γ-tocopherol, etc. [57]. In *M. crystallinum*, a transition from C_3 to CAM induced by H_2O_2 or salinity, α-tocopherol, polyamines and proline showed accumulation and performed a crucial role in preventing oxidative damage [31,61,62]. Likewise, during the C_3–CAM shift in *Guzmania monostachia* (a facultative CAM plant) induced by water stress under high light PFD, carotenoids were proven to play an important role in the ROS scavenging system [63].

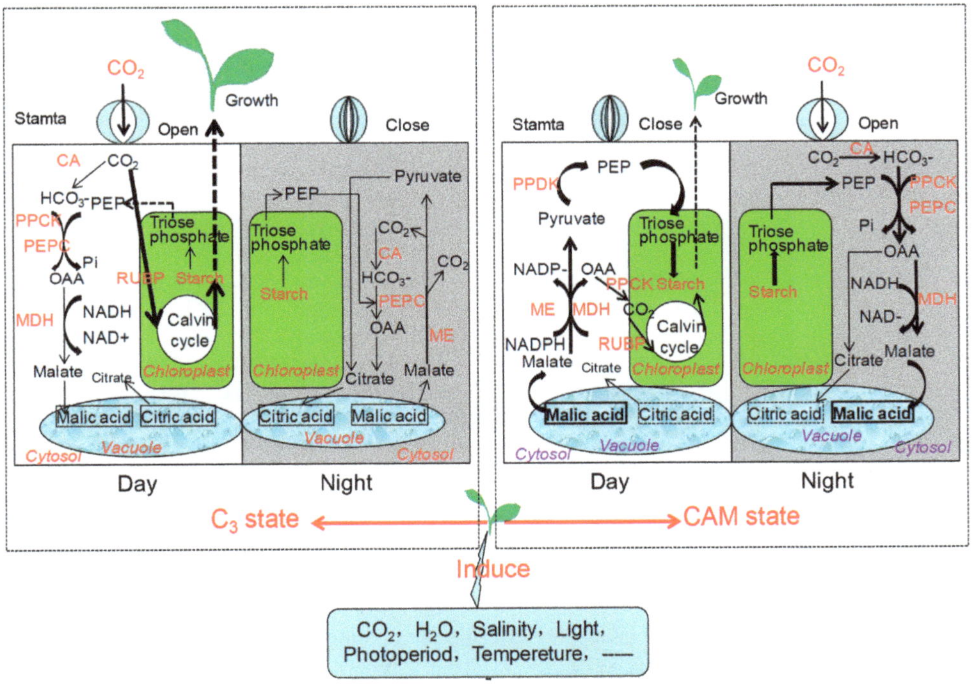

Figure 1. Physiological and Metabolic shift of the facultative CAM plants from C_3 to CAM photosynthesis. For the facultative CAM plants, C_3 state of photosynthesis is used to fix CO_2 under well conditions, when stomata open during the day and close at night. Once the facultative CAM plants stressed by abiotic stress (such as atmospheric CO_2 concentration, drought, salinity, photoperiod, light, etc.), the carbon assimilation pathway will be induced to the CAM mode, when CO_2 uptakes during the nighttime via open stomata and stores as organic acids (mainly malic acid). Then, organic acids are decarboxylated in the vacuoles during the daytime and CO_2 is refixed via Calvin cycle. CA, Carbonic anhydrase; PEP, Phosphoenol pyruvate; PEPC, Phosphoenolpyruvate carboxylase; PPCK, PEPC kinase; PPDK, Pyruvate phosphate kinase; ME, Malic enzyme; MDH, Malic dehydrogenase; OAA, Oxalacetic acid; RUBP, Ribulose-1,5-bisphosphate carboxylase; NAD, Nicotinamide adenine denucleotide; NADP, Nicotinamide adenine denucleotide phosphate.

5. Metabolic Mechanism during the C_3–CAM Shift in Facultative CAM Plants

CAM was controlled by the circadian clock and metabolites. Organic acids and carbohydrates are two kinds of major metabolites in facultative CAM plants [64]. Large diel changes in malic acid and transitory starch are important characteristics of CAM photosynthetic activity [65–67]. Plants, once stressed by adverse environmental factors, trigger their ABA- and Ca^{2+}-dependent signaling pathways, which leads to metabolite changes [28]. These diurnal changes in malic acid and/or citric acid regulate the expression of CAM photosynthesis [68], but the acid accumulation requires metabolic reprogramming [69]. Metabolic fluxes modeling of a starch/sugar-malate cycle was established to test whether a C_3–CAM continuum exists in CAM evolution [70].

Carbohydrates, including starch and sucrose or hexose, are produced by the CAM pathway at a high energetic cost [65]. They are converted into PEP and provide substrates for the nocturnal CO_2 fixation [71]. In *M. crystallinum*, this transitory starch reserve is critical for CAM photosynthesis and the oscillation in starch levels between day and night (when the plant is in CAM mode) is 20% greater than in the C_3 mode [72,73]. Starch-deficient mutants of this plant are characterized by their inability to photosynthesize in the CAM mode, as these mutant plants are deficient in leaf starch. It was reported that there was a transitory starch breakdown by the hydrolytic pathway when plants were in the C_3 mode, but that a shift to the phosphorolytic pathway occurred during CAM photosynthesis [66]. In *M. crystallinum*, about half of the starch degraded during the night was used to supply PEP to ensure CAM photosynthesis, and the others were exported as soluble sugars used during plant respiration [74]. In another example, the Agave plant (*Yucca aloifolia*), a CAM plant, uses soluble sugars as a carbohydrates during CAM photosynthesis, while the C_3 plant *Y. filamentosa* is thought to rely on starch pools. However, the diploid hybrid species *Y. gloriosa*, a facultative CAM species, relies on starch for carbohydrates (like its C_3 parent *Y. filamentosa*), though many features are more similar to that of its CAM parent *Y. aloifolia* [75,76]. In facultative CAM plants, the proportion of CO_2 taken up between nighttime and daytime is influenced by developmental and environmental factors (such as drought, salinity and extreme temperature) [72,77]. Various studies have revealed that many species express a low degree of CAM photosynthesis and a peak in C_3-type isotope signatures. Generally, the $\delta^{13}C$ value has more C_3-type isotope signatures than CAM in facultative CAM plants. In *S. nuttallianum*, the $\delta^{13}C$ value was −30.0‰ but $\triangle H+$ was significant under controlled conditions [36]. *C. minor* can shift from C_3 to CAM photosynthesis easily and reversibly, and its $\delta^{13}C$ value was about −21‰ [78]. These studies showed that carbon was mostly fixed through C_3 photosynthesis in facultative CAM plants, and dark CO_2 fixation made up, at most, 30% [39].

6. Molecular Mechanisms during C_3–CAM Shift in Facultative CAM Plants

The photosynthetic shift from C_3–CAM is regulated by some enzymes and metabolite transporters, making this process a part of complex metabolic adaptations to environmental stress [22]. However, the relative expression of genes for enzymes and metabolites is strictly regulated by environmental factors. Genes involved in the shift of C_3–CAM may be controlled by a co-expressed circadian master regulator [79]. Therefore, describing the molecular mechanism behind the C_3–CAM shift is important to understand the evolution of CAM. Although every documented gene in the CAM also exists in the ancestral C_3 species, the timing and magnitude are different between C_3 and CAM species [80].

The PEPC gene is one of the most important genes in CAM and C_4 photosynthesis. In the past decades, the presence and function of this gene have been investigated in many CAM and facultative CAM plants. PEPC is a ubiquitous enzyme in higher plants and belongs to a small multigene family with several PEPC isozymes [52,81]. Plant PEPC activity is regulated by PEPC kinase (PPCK) [82]. Boxall et al. found that silencing PEPC (*Ppc*1) in the obligate CAM species *Kalanchoë laxiflora* can prevent CO_2 uptake and malate accumulation at night [83]. Here, the main molecular mechanisms that occur during the photosynthesis transition from C_3 to CAM in facultative CAM plants are as follows.

7. DNA Level Regulation

The photosynthesis shift from C_3 to CAM involved DNA hypermethylation [84]. For instance, variations in cytosine methylation were found in the *Ppc*1 promoter during the transition from C_3 to CAM in *M. crystallinum* [85].

8. Transcriptional Regulation

Between obligate CAM and C_3 photosynthesis plants, all of the CAM-related genes exist in ancestral C_3 species, but transcriptional regulation cascades are very important for the C_3–CAM transition, especially the expression of CAM-specific genes [86,87]. Some enzymes and genes that are involved in some facultative CAM plants during the photosynthesis from C_3 to CAM are shown in Table 1. The expression of CAM-specific PEPC, NAD-GAPDH and PPDK is important for the onset during the CAM induction; for example, PEPC mRNA accumulation occurs within 2–3 h, stressed by salinity in *M. crystallinum* or *K. blossfeldiana* [27,88,89]. Additionally, there are discrete changes observed in protein sequences [9,75,90]; the promoter regions of these CAM-specific genes contain GT motifs, which may function in light-responsive or ABA-mediated gene expression events [91,92]. The promoter sequences were different in the CAM-specific *PEPC* (*Ppcl*) and the C_3 "housekeeping" *PEPC*, that is the former containing TATA and CAAT box motifs, but there was absence in the latter [91]. For CAM-specific genes, there may be the same common cis-acting regulatory elements for regulating the stress-induced expression patterns in different facultative CAM plants. During CAM induction, the distal regions between −977 and −721 control the expression of *Ppcl* for salt-responsiveness, while the regions between −735 and −675 control the expression of NADGAPDH(*GapC*1) induced by salt [93]. Therefore, the *Ppcl* and *GapCl* promoters in the distal regions share multiple consensus binding sites of transcription factor which control the salt-inducible gene expression. In *Kalanchoë*, *PPC*1 is essential for the practice of CAM [83]. In facultative CAM plant *Talinum triangulare*, transcriptional regulation of the C_3–CAM transition revealed that the levels of the CAM-cycle enzyme transcripts are increased in response to drought stress [6]. In facultative CAM plants, during CAM induction induced by abiotic stress, a few transcription factors may control some transcriptional activation events to improve their tolerance [6,94].

Recently, several transcription factors (TF) families take part in regulating CAM induction by salinity or drought stress, such as AP2/ERF, MYB, WRKY, NAC, NF-Y, bZIP and McHB7 [93,95–98]. In *T. triangulare*, during CAM induction stressed by ABA, transcription factors such as HSFA2, NF-YA9 and JMJ27 were identified as regulators for the CAM induction [21]. Cushman and Bohnert [99] demonstrated that one factor (designated *PCAT*-1) binds in the *Ppcl* promoter, and the *PCAT*-1 expressed is abundant and may play an important role in the assembly of active transcription complexes during the photosynthesis shift from C_3 to CAM.

Table 1. Some enzymes and genes involved in some facultative CAM plants during the photosynthesis shift from C_3 to CAM.

Enzyme	Gene	Source/Species	Inducer	References
Phosphenolpyruvate carboxylase	*Ppc*1	*M. crystallinum*	salt, ABA, drought, cytokinin	[25,100]
	Kb-1, *Kb*-2	*K. blossfeldiana*	short-day, drought	[101]
	Ppc 3	*T. triangulare*	ABA	[21]
	C_3-*type PEPCs*	*C. minor*	drought	[102]
Alpha Carbonic Anhydrase 1	*ACA*1	*T. triangulare*	ABA	[21]
Beta Carbonic Anhydrase 5	*BCA*5	*T. triangulare*	ABA	[21]
Malic Enzymes	*MEs*	*T. triangulare*	ABA	[21]
PEPC Kinase	*PPCK*1	*M. crystallinum* *T. triangulare*	Salt ABA	[21,103,104]
Pyruvate orthophosphate dikinase(PPDK)	*Ppdk*1	*M. crystallinum* *T. triangulare*	salt, ABA	[21,105]
Enolase	*Pgh*1	*M. crystallinum*	salt, drought, cold, hypoxia, ABA, 6-BA	[106]

Table 1. *Cont.*

Enzyme	Gene	Source/Species	Inducer	References
phosphoglyceromutase (PGM)	*Pgm*1	*M. crystallinum*	salt, drought, ABA, 6-BA	[107]
GAD-Glyceraldehyde 3-phosphate dehydrogenase (GAPDH)	*GapC*1	*M. crystallinum*	salt	[93,108]
NADP-Malic enzyme	*Mod*1	*M. crystallinum* *T. triangulare*	salt ABA	[21,109]
	*Mod*4	*T. triangulare*	ABA	[21]
NADP-Malate dehydrogenase	*MDH*1	*M. crystallinum*	salt	[80]
NAD-Malate dehydrogenase	*MDH*2	*M. crystallinum*	salt	[110]
H^+-ATPase. c subunit	*Atpvc*	*M. crystallinum*, *K. daigremontiana*	salt, ABA, light	[111–113]
H^+-ATPase, E subunit	*AtpvE*	*M. crystallinum*	salt	[114]
SNF1 kinase	*MK9*	*M. crystallinum*	salt	[115]
RNA-binding protein	*Rbp*1	*M. crystallinum*	salt	[116]
Ribosome inactivating proteins	*Rip*1	*M. crystallinum*	salt	[117]

9. Post-Transcriptional Regulation

Post-transcriptional regulations also take part in regulating CAM expression, such as that *Ppc*1 may facilitate long-term CAM build up by increasing mRNA stability in *M. crystallinum* during salt stress [118,119]. In *M. crystallinum,* many cDNA libraries were constructed with different tissues and stress treatments [120]. Large-scale steady-state mRNA abundance was found to change significantly during CAM induction in plants subjected to salinity-related stress [121]. There was at least one CAM-specific PEPC isoform (*Ppc*1) responsible for CO_2 fixation during the night, which is more abundant than during the day and is responsible for CAM expression [75,122]. During the photosynthesis shifts from C_3 to CAM in *T. triangulare*, PEPC isoform 1(*Ppc*1) and isoform 3 (*Ppc*3) transcript abundance increased and PPDK transcripts started to accumulate after 80 min, but the Alpha Carbonic Anhydrase 1 (*ACA*1) transcript decreased [21]. When *M. crystallinum* was treated for seven days with salt, *Ppc*1 and *PPCK*1 were up-regulated in guard cells [104].

Omics approaches were used by many scientists to explore the underlying molecular mechanism during the transition from C_3 to CAM [6,123]. Among monocot species, the partial transcriptomes or genomes in the genus *Phalaenopsis* have been characterized [124]. The rapid reversible C_3–CAM shift in the genus *Clusia* is based on the C_3 isoform of *PEPC* (a housekeeping gene) [102]. The post-transcriptional regulation of photosynthetic genes is a key driver of C_4 leaf ontogeny, determined by using exon–intron split analysis [125]. In order to explore the complex regulatory mechanisms, Heyduk et al. elaborated on a comparative analysis with closely related C_3 and CAM species [86]. Aside from PEPC, other enzymes are also important for CAM photosynthesis, which are the products of isogenes. Other studies showed that following CAM induction by salinity-related stress, several CAM-related glycolysis/gluconeogenesis genes showed increasing transcript abundance [106–108]. In order to identify the genes involved during the transition from C_3 to CAM in *M. crystallinum*, genomics and transcriptomics analyses were combined; twenty genes encoding six main enzymes were identified and one of four MDH genes presented a specific function in CAM photosynthesis [126]. Many CAM-related starch synthesis/degradation genes have been identified in *M. crystallinum*. During the photosynthesis shift from C_3 to CAM, ADP glucose pyrophosphorylase small (*Agp*1 and

*Agp*2) and large subunit (*Agp*3) catalyzed the starch biosynthesis and showed an increase in mRNA expression; additionally, there were three genes including *AmyA*1 (α-amylase isogene), *AmyB*1 and *AmyB*2 which exhibited increased remarkable mRNA abundance during nocturnal starch degradation [121].

Many microRNAs (miRNAs) also play a regulatory role in CAM photosynthesis in the leaves of obligate CAM species, such as *Ananas comosus* [127]. MiRNAs in *M. crystallinum* seedlings under salinity-related stress were analyzed by RNA sequencing and were found to be involved in the post-transcriptional regulation of salt tolerance [128]. Hu et al. identified some miRNAs that were involved in the regulation of CAM in *Kalanchoë* and found that the miR530-TZPs module regulates CAM-related gene expression [129].

10. Protein Level Regulation

Regulatory proteins exist in C_3, CAM and C_4 species and are essential to the C_3–CAM photosynthesis shift in facultative CAM plants. In *M. crystallinum*, salt-stressed, heat shock proteins and early light-inducible proteins were found increased in the cDNA libraries, and a few proteins were found increased in guard cells but decreased in mesophyll cells [130,131]. There were seven proteins with increased expression and four proteins with decreased expression in *M. crystallinum* induced by salt [98]. Additionally, new major phosphorylation events during the transition from C_3 to CAM stressed by salt were identified and characterized using proteomics and phosphoproteomics [132].

11. Implications in Horticultural Crops

CAM photosynthesis enables plants to assimilate carbon under environmental stress conditions. In contrast to C_3 and C_4 plants, obligate CAM plants have a higher transpiration efficiency and a lower photosynthetic rate; hence, obligate CAM plants often grow more slowly than their C_3 and C_4 counterparts [133]. Thus, CAM is not the best choice for highly productive plants [14]. However, facultative CAM plants can function in a C_3 mode to increase photosynthetic rates and growth when there are no physical limitations present, and shift to the CAM mode to decrease water loss and overcome environmental stressors. For facultative CAM plants, the photosynthesis switching between C_3 and CAM has important ecological implications. Therefore, engineering CAM into C_3 crops can improve their WUE and sustain crop productivity in hot and dry climates [15,16,134].

CAM plants are widely distributed within the plant kingdom, i.e., 343 genera in 34 families, approximately 6.5% of flowering plant species [135]. In fact, most crops practice either C_3 or C_4 photosynthesis, but not CAM photosynthesis, such as wheat, rice and maize, which have higher production. However, many horticultural plants belong to obligate CAM plants or facultative CAM plants; some of them (such as pineapple, an obligate CAM plant, practice a facultative C_3/CAM metabolism in the first 2 months of growth) can also be very productive when cultivated under well-conditions [55]. However, there was no exact number in facultative CAM plants. Winter (2019) reported that facultative CAM plants exist in at least 15 families, and he thought there may be over 1000 facultative CAM species in Aizoaceae alone [9]. Many orchids (*Dendrobium* spp., *Oncidium* spp. and *Phalaenopsis* spp.) [52,124,136,137], with higher ornamental values or medicinal values, were identified as facultative CAM plants since they could switch the pathway between C_3 and CAM according to the environmental condition. For example, *D. officinale*, an important traditional herb with higher commercial value in China, uses the facultative CAM pathway to increase its drought tolerance [138]. Many species of Portulacaceae belong to facultative CAM plants [9]. *Jatropha curcas*, an oil crop, could also practice CAM photosynthesis for survival in response to environmental stress [139]. Thus, this plastic photosynthetic adaptation results in important implications for many horticultural crops.

12. Future Perspectives

For the facultative CAM plants, the photosynthesis switches from C_3 to CAM have important ecological implications. Facultative CAM can prevent CO_2 loss and favors plant growth and reproduction in responses to environmental stress. During the C_3–CAM shift in facultative CAM plants, although there are some progresses in anatomy, physiological, metabolic and molecular properties of facultative CAM plants, in the near future, there are still some works that should be carried out. First, the argument on anatomical variations should be comprehensive and ongoing, and studied with more genera that include obligate C_3, facultative CAM and obligate CAM plants. For example, many characters and mechanisms can be explored in the genera *Clusia* (Clusiaceae), *Dendrobium* and *Oncidium* (Orchidaceae) and *Yucca* (Asparagaceae), which are well known for containing the obligate C_3, facultative CAM and obligate CAM plants in a single genus [19,76,102,136]. Second, molecular mechanisms underlying the transition from C_3 to CAM in plants is still limited. Recently, omics approaches including transcriptomic, genomic, proteomic, metabolomics and ionomics were used frequently to reveal the molecular changes during the C_3–CAM transition (See Table 2). For example, altered gene regulatory networks and expression profiles were found in the transition from C_3 to CAM in *Erycina* (Orchidaceae) and *Yucca* (Asparagaceae) [76,86], which will benefit the clarification of the key molecular switches underlying this transition of C_3 to CAM in facultative CAM plants. Third, understanding the functional genomics of CAM plants is important to elucidate the relationship between genotype and phenotype [9]. Synthetic biology toolboxes, such as the CRISPR/Cas 9 system which was confirmed to be effective for genome editing in *K. fedtschenkoi*, will accelerate the ongoing research about CAM and the C_3 to CAM transition mechanisms [17,140]. Assisted with all this prior knowledge, we should obtain more genetic transformation information on facultative CAM plants and transfer more CAM-related genes into C_3 types of horticultural crops in future.

Table 2. Omics approaches involved to reveal the molecular changes during the C_3–CAM transition in facultative CAM plants.

Omics Approaches	Source/Species	Photosynthesis Type	Year	Reference
Proteomics, Metabolomics	*M. crystallinum*	facultative CAM plants	2013	[141]
Transcriptomics	*M. crystallinum*	facultative CAM plants	2015	[142]
Metabolomics	*M. crystallinum*	facultative CAM plants	2015	[143]
Transcriptomics	*M. crystallinum*	facultative CAM plants	2015	[144]
Proteomics, Ionomics	*M. crystallinum*	facultative CAM plants	2016	[145]
Transcriptomics, Metabolomics	*T. triangulare*	facultative CAM plants	2016	[6]
Transcriptomics	*D. catenatum*	facultative CAM plants	2016	[124]
Transcriptomics	*Agave* (CAM), *Polianthes* (weak CAM), *Manfreda* (CAM), *Beschorneria* (weak CAM)	CAM plants	2018	[95]
Transcriptomics	*D. catenatum*	facultative CAM plants	2018	[137]
Transcriptomics	*Erycina pusilla* (CAM), *Erycina crista-galli* (C3),	CAM plants, C_3 plants	2019	[86]
Transcriptomics, Metabolomics	*T. triangulare*	facultative CAM plants	2019	[21]

Table 2. *Cont.*

Omics Approaches	Source/Species	Photosynthesis Type	Year	Reference
Metabolomics Transcriptomics	*Y. gloriosa* (C_3+ CAM), *Y. filamentosa* (C3), *Y. aloifolia* (CAM)	facultative CAM plants, C_3 plants, obligate CAM plants	2019	[76]
Genomics	*Sedum album*	facultative CAM plants	2019	[123]
Transcriptomics	*M. crystallinum*	facultative CAM plants	2020	[104]
Proteomics, Metabolomics	*M. crystallinum*	facultative CAM plants	2021	[131]
Proteomics	*M. crystallinum*	facultative CAM plants	2021	[98]
Proteomics, Phosphoproteomics	*M. crystallinum*	facultative CAM plants	2022	[132]
Transcriptomics	*Tamarix ramosissima*	facultative CAM plants	2022	[146]
Transcriptomics Genomics	*M. crystallinum*	facultative CAM plants	2022	[126]
Transcriptomics	11 species of Agavoideae	facultative CAM plants, C_3 plants, obligate CAM plants	2022	[147]

Author Contributions: Conceptualization, Z.Z. and S.Q.; writing and revision of the manuscript, S.Q.; revision and editing of the manuscript, K.X.; Y.Y. and Q.W. All authors have read and agreed to the published version of the manuscript.

Funding: This work was funded by the National Natural Science Foundation of China (31560567), National Natural Science Foundation of Guangxi (2020GXNSFAA297260), Start-up Fund of Innovation Team of Guangxi Academy of Sciences for Innovation and Utilization of Germplasm in Horticultural Crops (CQZ-E-1919), Fundamental Research Fund of Guangxi Institute of Botany (23011) and the fund of Guangxi Key Laboratory of Plant Functional Phytochemicals and Sustainable Utilization (ZRJJ2022-5 and ZRJJ2023-1).

Conflicts of Interest: The authors declare no conflict of interest.

Abbreviations

ABA	Abscisic acid
CAM	Crassulacean acid metabolism
APX	Ascorbate peroxidase
CAT	Catalase
H_2O_2	Hydrogen peroxide
MDH	Malate dehydrogenase
ME	Malic enzyme;
NO	Nitric oxide
PEPC	Phosphoenolpyruvate carboxylase
PEPCK	PEP carboxykinase
POD	Peroxidase
PFD	Photon flux density
Rubisco	Ribulosebisphosphate carboxylase/oxygenase
SOD	superoxide dismutase (SOD)
TF	Transcription factors
TZPs	Tandem zinc knuckle/PLU3 domain encoding genes
WUE	Water-use efficiency

References

1. Osmond, C.B. Crassulacean acid metabolism: A curiosity in context. *Annu. Rev. Plant Physiol.* **1978**, *29*, 379–414. [CrossRef]
2. Lüttge, U. Ecophysiology of Crassulacean acid metabolism (CAM). *Ann. Bot.* **2004**, *93*, 629–652. [CrossRef] [PubMed]
3. Winter, K.; von Willert, D.J. NaCl-induzierter Crassulaceen säurestoffwechsel bei *Mesembryanthemum crystallinum*. *Z. Pflanzenphysiol.* **1972**, *67*, 166–170. [CrossRef]
4. Castillo, F.J. Antioxidative protection in the inducible CAM plant *Sedum album* L. following the imposition of severe water stress and recovery. *Oecologia* **1996**, *107*, 469–477. [CrossRef] [PubMed]
5. Barrera Zambrano, V.A.; Lawson, T.; Olmos, E.; Fernández-García, N.; Borland, A.M. Leaf anatomical traits which accommodate the facultative engagement of Crassulacean acid metabolism in tropical trees of the genus *Clusia*. *J. Exp. Bot.* **2014**, *65*, 3513–3523. [CrossRef]
6. Brilhaus, D.; Bräutigam, A.; Mettler-Altmann, T.; Winter, K.; Weber, A.P. Reversible burst of transcriptional changes during induction of Crassulacean acid metabolism in *Talinum triangulare*. *Plant Physiology* **2016**, *170*, 102–122. [CrossRef]
7. Winter, K.; Holtum, J.A.M. Environment or development? Lifetime net CO_2 exchange and control of the expression of crassulacean acid metabolism in *Mesembryanthemum crystallinum*. *Plant Physiol.* **2007**, *143*, 98–107. [CrossRef]
8. Winter, K.; Garcia, M.; Holtum, J.A.M. On the nature of facultative and constitutive CAM: Environmental and developmental control of CAM expression during early growth of *Clusia*, *Kalanchoë*, and *Opuntia*. *J. Exp. Bot.* **2008**, *59*, 1829–1840. [CrossRef]
9. Winter, K. Ecophysiology of constitutive and facultative CAM photosynthesis. *J. Exp. Bot.* **2019**, *70*, 6495–6508. [CrossRef]
10. Johnson, G.N.; Lawson, T.; Murchie, E.H.; Raines, C. Photosynthesis in variable environments. *J. Exp. Bot.* **2015**, *66*, 2371–2372. [CrossRef]
11. DePaoli, H.C.; Borland, A.M.; Tuskan, G.A.; Cushman, J.C.; Yang, X.H. Synthetic biology as it relates to CAM photosynthesis: Challenges and opportunities. *J. Exp. Bot.* **2014**, *65*, 3381–3393. [CrossRef]
12. Grams, T.E.E.; Thiel, S. High light-induced switch from C_3-photosynthesis to crassulacean acid metabolism is mediated by UV-A/blue light. *J. Exp. Bot.* **2002**, *53*, 1475–1483. [CrossRef]
13. Cushman, J.C. Crassulacean acid metabolism. A plastic photosynthetic adaptation to arid environments. *Plant Physiology* **2001**, *127*, 1439–1448. [CrossRef] [PubMed]
14. Tay, S.; He, J.; Yam, T.W. CAM plasticity in epiphytic tropical orchid species responding to environmental stress. *Bot. Stud.* **2019**, *60*, 7. [CrossRef] [PubMed]
15. Borland, A.M.; Hartwell, J.; Weston, D.J.; Schlauch, K.A.; Tschaplinski, T.J.; Tuskan, G.A.; Yang, X.; Cushman, J.C. Engineering crassulacean acid metabolism to improve water-use efficiency. *Trends Plant Sci.* **2014**, *19*, 327–338. [CrossRef]
16. Yang, X.H.; Cushman, J.C.; Borland, A.M.; Edwards, E.J.; Wullschleger, S.D.; Tuskan, G.A.; Owen, N.A.; Griffiths, H.; Smith, J.A.C.; Paoli, H.C.D.; et al. A roadmap for research on crassulacean acid metabolism (CAM) to enhance sustainable food and bioenergy production in a hotter, drier world. *New Phytol.* **2015**, *207*, 491–504. [CrossRef]
17. Liu, D.G.; Chen, M.; Mendoza, B.; Cheng, H.; Hu, R.B.; Li, L.L.; Trinh, C.T.; Tuskan, G.A.; Yang, X.H. CRISPR/Cas9-mediated targeted mutagenesis for functional genomics research of crassulacean acid metabolism plants. *J. Exp. Bot.* **2019**, *70*, 6621–6629. [CrossRef] [PubMed]
18. Lim, S.; Lee, S.; Choi, W.; Yim, W.; Cushman, J. Laying the foundation for crassulacean acid metabolism (CAM) biodesign: Expression of the C4 metabolism cycle genes of CAM in *Arabidopsis*. *Front. Plant Sci.* **2019**, *10*, 101. [CrossRef] [PubMed]
19. Silvera, K.; Neubig, K.M.; Whitten, W.M.W.; Williams, N.H.; Winter, K.; Cushman, J.C. Evolution along the crassulacean acid metabolism continuum. *Funct. Plant Biol.* **2010**, *37*, 995–1010. [CrossRef]
20. Hartwell, J.; Dever, L.V.; Boxall, S.F. Emerging model systems for functional genomics analysis of crassulacean acid metabolism. *Curr. Opin. Plant Biol.* **2016**, *31*, 100–108. [CrossRef] [PubMed]
21. Maleckova, E.; Brilhaus, D.; Wrobel, T.J.; Weber, A.P.M. Transcript and metabolite changes during the early phase of abscisic acid-mediated induction of crassulacean acid metabolism in *Talinum triangulare*. *J. Exp. Bot.* **2019**, *70*, 6581–6596. [CrossRef] [PubMed]
22. Cushman, J.C.; Bohnert, H.J. Crassulacean acid metabolism: Molecular Genetics. *Annu. Rev. Plant Physiol. Plant Mol. Biol.* **1999**, *50*, 305–332. [CrossRef] [PubMed]
23. Lüttge, U. Carbon dioxide and water demand. Crassulacean acid metabolism (CAM), a versatile ecological adaptation exemplifying the need for integration in eco-physiological work. *New Phytol.* **1987**, *106*, 593–629. [CrossRef]
24. Borland, A.M.; Griffiths, H. The regulation of CAM and respiratory recycling by water supply and light regime in the C_3-CAM intermediate *Sedum telephium*. *Funct. Ecol.* **1990**, *4*, 33–39. [CrossRef]
25. Cushman, J.C.; Meyer, G.; Michalowski, C.B.; Schmitt, J.M.; Bohnert, H.J. Salt stress leads to differential expression of two isogenes of phosphoenolpyruvate carboxylase during crassulacean acid metabolism induction in the common ice plant. *Plant Cell* **1989**, *1*, 715–725.
26. Broetto, F.; Lüttge, U.; Ratajczak, R. Influence of light intensity and salt-treatment on mode of photosynthesis and enzymes of the antioxidative response system of *Mesembryanthemum crystallinum*. *Funct. Plant Biol.* **2002**, *29*, 13–23. [CrossRef] [PubMed]
27. Taybi, T.; Sotta, B.; Gehrig, H.; Giiclii, S.; Kluge, M.; Brulfert, J. Differential effects of abscisic acid on phosphoenolpyruvate carboxylase and CAM operation in *Kalanchoe blossfeldiana*. *Bot. Acta* **1995**, *108*, 240–246. [CrossRef]

28. Freschi, L.; Rodrigues, M.A.; Domingues, D.S.; Purgatto, E.; Van Sluys, M.A.V.; Magalhaes, J.R.; Kaiser, W.M.; Mercier, H. Nitric oxide mediates the hormonal control of crassulacean acid metabolism expression in young Pineapple plants. *Plant Physiology* **2010**, *152*, 1971–1985. [CrossRef]
29. Wakamatsu, A.; Mori, I.C.; Matsuura, T.; Taniwaki, Y.; Yoshida, R. Possible roles for phytohormones in controlling the stomatal behavior of *Mesembryanthemum crystallinum* during the salt-induced transition from C_3 to crassulacean acid metabolism. *J. Plant Physiol.* **2021**, *262*, 153448. [CrossRef]
30. Ślesak, I.; Karpinska, B.; Surówka, E.; Miszalski, Z.; Karpinski, S. Redox changes in the chloroplast and hydrogen peroxide are essential for regulation of C_3-CAM transition and photooxidative stress responses in the facultative CAM plant *Mesembryanthemum crystallinum* L. *Plant Cell Physiol.* **2003**, *44*, 573–581. [CrossRef]
31. Surówka, E.; Dziurka, M.; Kocurek, M.; Goraj, S.; Rapacz, M.; Miszalski, Z. Effects of exogenously applied hydrogen peroxide on antioxidant and osmoprotectant profiles and the C_3-CAM shift in the halophyte *Mesembryanthemum crystallinum* L. *J. Plant Physiol.* **2016**, *200*, 102–110. [CrossRef]
32. Eastmond, P.J.; Ross, J.D. Evidence that the induction of crassulacean acid metabolism by water stress in *Mesembryanthemum crystallinum* (L.) involves root signalling. *Plant Cell Environ.* **1997**, *20*, 1559–1565. [CrossRef]
33. Teeri, J.A.; Tonsor, S.J.; Turner, M. Leaf thickness and carbon isotope composition in the Crassulaceae. *Oecologia.* **1981**, *50*, 367–369. [CrossRef]
34. Winter, K.; Wallace, B.J.; Stocker, G.C.; Roksandic, Z. Crassulacean acid metabolism in Australian vascular epiphytes and some related species. *Oecologia* **1983**, *57*, 129–141. [CrossRef] [PubMed]
35. Borland, A.M.; Técsi, L.I.; Leegood, R.C.; Walker, R.P. Inducibility of crassulacean acid metabolism (CAM) in *Clusia* species; physiological/biochemical characterisation and intercellular localisation of carboxylation processes in three species which show different degrees of CAM. *Planta* **1998**, *205*, 342–351. [CrossRef]
36. Silvera, K.; Santiago, L.S.; Winter, K. Distribution of crassulacean acid metabolism in orchids of Panama: Evidence of selection of weak and strong modes. *Funct. Plant Biol.* **2005**, *32*, 397–407. [CrossRef]
37. Heyduk, K. Evolution of crassulacean acid metabolism in response to the environment: Past, present, and future. *Plant Physiol.* **2022**, *190*, 19–30. [CrossRef] [PubMed]
38. Veste, M.; Herppich, W.B.; von Willert, D.J. Variability of CAM in leaf deciduous succulents from the Succulent Karoo (South Africa). *Basic Appl. Biol.* **2001**, *2*, 283–288.
39. Herrera, A. Crassulacean acid metabolism and fitness under water deficit stress: If not for carbon gain, what is facultative CAM good for? *Ann. Bot.* **2009**, *103*, 645–653. [CrossRef] [PubMed]
40. Guan, Q.J.; Tan, B.; Kelley, T.M.; Tian, J.K.; Chen, S.X. Physiological changes in *Mesembryanthemum crystallinum* during the C_3 to CAM transition induced by salt stress. *Front. Plant Sci.* **2020**, *11*, 283. [CrossRef]
41. De-Santo, A.V.; Alfani, A.; Russo, G.; Fioretto, A. Relationship between CAM and succulence in some species of Vitaceae and Piperaceae. *Bot. Gaz.* **1983**, *144*, 342–346. [CrossRef]
42. Edwards, E.J. Evolutionary trajectories, accessibility and other metaphors: The case of C_4 and CAM photosynthesis. *New Phytol.* **2019**, *223*, 1742–1755. [CrossRef]
43. Heyduk, K.; Moreno-Villena, J.J.; Gilman, I.S.; Christin, P.A.; Edwards, E.J. The genetics of convergent evolution: Insights from plant photosynthesis. *Nat. Rev. Genet.* **2019**, *20*, 485–493. [CrossRef] [PubMed]
44. Heyduk, K.; Ray, J.N.; Leebens-Mack, J. Leaf anatomy is not correlated to CAM function in a C_3+CAM hybrid species, *Yucca gloriosa*. *Ann. Bot.* **2021**, *127*, 437–449. [CrossRef]
45. Herrera, A. Are thick leaves, large mesophyll cells and small intercellular air spaces requisites for CAM? *Ann. Bot.* **2020**, *125*, 859–868. [CrossRef] [PubMed]
46. Kliemchen, A.; Schomburg, M.; Galla, H.J.; Lüttge, U.; Kluge, M. Phenotypic changes in the fluidity of the tonoplast membrane of crassulacean acid metabolism plants in response to temperature and salinity stress. *Planta* **1993**, *189*, 403–409. [CrossRef]
47. Jones, M.B. Effect of leaf age on leaf resistance and CO_2 exchange of the CAM plant *Bryophyllum fedtschenkoi*. *Planta* **1975**, *123*, 91–96. [CrossRef] [PubMed]
48. Ślesak, I.; Miszalski, Z.; Karpinska, B.; Niewiadomska, E.; Ratajczak, R.; Karpinski, S. Redox control of oxidative stress responses in the C_3–CAM intermediate plant *Mesembryanthemum crystallinum*. *Plant Physiol. Biochem.* **2002**, *40*, 669–677. [CrossRef]
49. Niewiadomska, E.; Bilger, W.; Gruca, M.; Mulisch, M.; Miszalski, Z.; Krupinska, K. CAM-related changes in chloroplastic metabolism of *Mesembryanthemum crystallinum* L. *Planta* **2011**, *233*, 275–285. [CrossRef]
50. Holtum, J.A.M.; Winter, K. Activities of enzymes of carbon metabolism during the induction of crassulacean acid metabolism in *Mesembryanthemum crystallinum*. *Planta* **1982**, *155*, 8–16. [CrossRef]
51. O'Leary, B.; Park, J.; Plaxton, W.C. The remarkable diversity of plant PEPC (phosphoenolpyruvate carboxylase): Recent insights into the physiological functions and post-translational controls of non-photosynthetic PEPCs. *Biochem. J.* **2011**, *436*, 15–34. [CrossRef] [PubMed]
52. Silvera, K.; Winter, K.; Rodriguez, B.L.; Albion, R.L.; Cushman, J.C. Multiple isoforms of phosphoenolpyruvate carboxylase in the Orchidaceae (subtribe Oncidiinae): Implications for the evolution of Crassulacean acid metabolism. *J. Exp. Bot.* **2014**, *65*, 3623–3636. [CrossRef] [PubMed]
53. Mallick, N.; Mohn, F.H. Reactive oxygen species: Response of algal cells. *J. Plant Physiol* **2000**, *157*, 183–193. [CrossRef]

54. Reddy, A.R.; Chaitanya, K.V.; Vivekanandan, M. Drought-induced responses of photosynthesis and antioxidant metabolism in higher plants. *J. Plant Physiol.* **2004**, *161*, 1189–1202. [CrossRef] [PubMed]
55. Aragón, C.; Carvalho, L.; González, J.; Escalona, M.; Amâncio, S. The physiology of ex vitro pineapple (*Ananas comosus* L. Merr. var MD-2) as CAM or C_3 is regulated by the environmental conditions. *Plant Cell Rep.* **2012**, *31*, 757–769. [CrossRef]
56. Aragón, C.; Pascual, P.; González, J.; Escalona, M.; Carvalho, L.; Amancio, S. The physiology of ex vitro pineapple (*Ananas comosus* L. Merr. var MD-2) as CAM or C_3 is regulated by the environmental conditions: Proteomic and transcriptomic profiles. *Plant Cell Rep.* **2013**, *32*, 1807–1818. [CrossRef] [PubMed]
57. Gill, S.S.; Tuteja, N. Reactive oxygen species and antioxidant machinery in abiotic stress tolerance in crop plants. *Plant Physiol. Biochem.* **2010**, *48*, 909–930. [CrossRef]
58. Niewiadomska, E.; Miszalski, Z.; Ślesak, I.; Ratajczak, R. Catalase activity during C_3-CAM transition in *Mesembryanthemum crystallinum* L. leaves. *Free Radic. Res.* **1999**, *31*, 251–256. [CrossRef]
59. Nosek, M.; Gawrońska, K.; Rozpądek, P.; Szechyńska-Hebda, M.; Kornaś, A.; Miszalski, Z. Withdrawal from functional crassulacean acid metabolism (CAM) is accompanied by changes in both gene expression and activity of antioxidative enzymes. *J. Plant Physiol.* **2018**, *229*, 151–157. [CrossRef]
60. Miszalski, Z.; Slesak, I.; Niewiadomska, E.; Baczek-Kwinta, R.; Luttge, U.; Ratajczak, R. Subcellular localization and stress responses of superoxide dismutase isoforms from leaves in the C_3-CAM intermediate halophyte *Mesembryanthemum crystallinum* L. *Plant Cell Environ.* **1998**, *21*, 169–179. [CrossRef]
61. Sanada, Y.; Ueda, H.; Kuribayashi, K.; Andoh, T.; Hayashi, F.; Tamai, N.; Wada, K. Novel light-dark change of proline levels in halophyte (*Mesembryanthemum crystallinum* L.) and glycophytes (*Hordeum vulgare* L. and *Triticum aestivum* L.) leaves and roots under salt stress. *Plant Cell Physiol.* **1995**, *36*, 965–970. [CrossRef]
62. Shevyakova, N.I.; Shorina, M.V.; Rakitin, V.Y.; Kuznetsov, V. Stress-dependent accumulation of spermidine and spermine in the halophyte *Mesembryanthemum crystallinum* under salinity conditions. *Russ. J. Plant Physiol.* **2006**, *53*, 739–745. [CrossRef]
63. Abreu, M.E.; Carvalho, V.; Mercier, H. Antioxidant capacity along the leaf blade of the C_3-CAM facultative bromeliad *Guzmania monostachia* under water deficit conditions. *Funct. Plant Biol.* **2018**, *45*, 620–629. [CrossRef] [PubMed]
64. Borland, A.M.; Taybi, T. Synchronization of metabolic processes in plants with crassulacean acid metabolism. *J. Exp. Bot.* **2004**, *55*, 1255–1265. [CrossRef] [PubMed]
65. Black, C.C.; Osmond, C.B. Crassulacean acid metabolism photosynthesis: 'working the night shift'. *Photosynth. Res.* **2003**, *76*, 329–341. [CrossRef] [PubMed]
66. Weise, S.E.; van Wijk, K.J.; Sharkey, T.D. The role of transitory starch in C_3, CAM, and C_4 metabolism and opportunities for engineering leaf starch accumulation. *J. Exp. Bot.* **2011**, *62*, 3109–3118. [CrossRef] [PubMed]
67. Habibi, G.; Hajiboland, R. Comparison of photosynthesis and antioxidative protection in *Sedum album* and *Sedum stoloniferum* (Crassulaceae) under water stress. *Photosynthetica* **2012**, *50*, 508–518. [CrossRef]
68. Haider, M.S.; Barnes, J.D.; Cushman, J.C.; Borland, A.M. A CAM- and starch-deficient mutant of the facultative CAM species *Mesembryanthemum crystallinum* reconciles sink demands by repartitioning carbon during acclimation to salinity. *J. Exp. Bot.* **2012**, *63*, 1985–1996. [CrossRef]
69. Winter, K.; Smith, J.A.C. CAM photosynthesis: The acid test. *New Phytol.* **2022**, *233*, 599–609. [CrossRef]
70. Tay, I.Y.Y.; Odang, K.B.; Cheung, C.Y.M. Metabolic modeling of the C_3-CAM continuum revealed the establishment of a starch/sugar-malate cycle in CAM evolution. *Front. Plant Sci.* **2021**, *11*, 573197. [CrossRef]
71. Christopher, J.T.; Holtum, J.A.M. Patterns of carbon partitioning in leaves of crassulacean acid metabolism species during deacidification. *Plant Physiol.* **1996**, *112*, 393–399. [CrossRef]
72. Dodd, A.N.; Borland, A.M.; Haslam, R.P.; Griffiths, H.; Maxwell, K. Crassulacean acid metabolism: Plastic, fantastic. *J. Exp. Bot.* **2002**, *53*, 569–580. [CrossRef]
73. Cushman, J.C.; Agarie, S.; Albion, R.L.; Elliot, S.M.; Taybi, T.; Borland, A.M. Isolation and characterization of mutants of common ice plant deficient in crassulacean acid metabolism. *Plant Physiol.* **2008**, *147*, 228–238. [CrossRef] [PubMed]
74. Borland, A.M.; Dodd, A.N. Carbohydrate partitioning in crassulacean acid metabolism plants: Reconciling potential conflicts of interest. *Funct. Plant Biol.* **2002**, *29*, 707–716. [CrossRef] [PubMed]
75. Abraham, P.E.; Yin, H.; Borland, A.M.; Weighill, D.; Lim, S.D.; De-Paoli, H.C.; Engle, N.; Jones, P.C.; Agh, R.; Weston, D.J.; et al. Transcript, protein and metabolite temporal dynamics in the CAM plant Agave. *Nat. Plants* **2016**, *178*, 16178. [CrossRef] [PubMed]
76. Heyduk, K.; Ray, J.N.; Ayyampalayam, S.; Moledina, N.; Borland, A.; Harding, S.A.; Tsai, C.J.; Leebens-Mack, J. Shared expression of crassulacean acid metabolism (CAM) genes pre-dates the origin of CAM in the genus *Yucca*. *J. Exp. Bot.* **2019**, *70*, 6597–6609. [CrossRef]
77. Cushman, J.C.; Borland, A.M. Induction of Crassulacean acid metabolism by water limitation. *Plant Cell Environ.* **2002**, *25*, 295–310. [CrossRef] [PubMed]
78. Lüttge, U. *Clusia: A Woody Neotropical Genus of Remarkable Plasticity and Diversity*; Ecological Studies; Springer: Berlin, Germany, 2007; p. 194.
79. Borland, A.M.; Hartwell, J.; Jenkins, G.I.; Wilkins, M.B.; Nimmo, H.G. Metabolite control overrides circadian regulation of phosphoenolpyruvate carboxylase knase and CO_2 fixation in crassulacean acid metabolism. *Plant Physiol.* **1999**, *121*, 889–896. [CrossRef]

80. Cushman, J.C. Molecular cloning and expression of chloroplast NADP-malate dehydrogenase during crassulacean acid metabolism induction by salt stress. *Photosynth. Res.* **1993**, *35*, 15–27. [CrossRef]
81. Gehrig, H.H.; Wood, J.A.; Cushman, M.A.; Virgo, A.; Cushman, J.C.; Winter, K. Research note: Large gene family of phosphoenolpyruvate carboxylase in the crassulacean acid metabolism plant *Kalanchoe pinnata* (Crassulaceae) characterised by partial cDNA sequence analysis. *Funct. Plant Biol.* **2005**, *32*, 467–472. [CrossRef]
82. Taybi, T.; Patil, S.; Chollet, R.; Cushman, J.C. A minimal serine/threonine protein kinase circadianly regulates phosphoenolpyruvate carboxylase activity in crassulacean acid metabolism-induced leaves of the common ice plant. *Plant Physiol.* **2000**, *123*, 1471–1482. [CrossRef]
83. Boxall, S.F.; Kadu, N.; Dever, L.V.; Kneřová, J.; Waller, J.L.; Gould, P.J.D.; Hartwell, J. *Kalanchoë PPC1* is essential for crassulacean acid metabolism and the regulation of core circadian clock and guard cell signaling genes[CC-BY]. *Plant Cell* **2020**, *32*, 1136–1160. [CrossRef] [PubMed]
84. Dyachenko, O.V.; Zakharchenko, N.S.; Shevchuk, T.V.; Bohnert, H.J.; Cushman, J.C.; Buryanov, Y.I. Effect of hypermethylation of CCWGG sequences in DNA of *Mesembryanthemum crystallinum* plants on their adaptation to salt stress. *Biochem. Biokhimiia* **2006**, *71*, 461–465. [CrossRef] [PubMed]
85. Huang, N.C.; Li, C.H.; Lee, J.Y.; Yen, H.E. Cytosine methylation changes in the ice plant *Ppc*1 promoter during transition from C_3 to crassulacean acid metabolism. *Plant Sci.* **2010**, *178*, 41–46. [CrossRef]
86. Heyduk, K.; Hwang, M.; Albert, V.A.; Silvera, K.; Lan, T.Y.; Winter, K.; Leebens-Mack, J. Altered gene regulatory networks are associated with the transition from C_3 to crassulacean acid metabolism in *Erycina* (Oncidiinae: Orchidaceae). *Front. Plant Sci.* **2019**, *9*, 2000. [CrossRef] [PubMed]
87. Hu, R.B.; Zhang, J.; Jawdy, S.; Sreedasyam, A.; Sreedasyam, A.; Lipzen, A.; Wang, M.; Ng, V.; Daum, C.; Keymanesh, K.; et al. Comparative genomics analysis of drought response between obligate CAM and C_3 photosynthesis plants. *J. Plant Physiol.* **2022**, *277*, 153791. [CrossRef]
88. Schmitt, J.M. Rapid concentration changes of phosphoenolpyruvate carboxylase mRNA in detached leaves of *Mesembryanthemum crystallinum*. *Plant Cell Environ.* **1990**, *13*, 845–850. [CrossRef]
89. Brulfert, J.; Güclü, S.; Taybi, T.; Pierre, J.N. Enzymatic responses to water stress in detached leaves of the CAM plant *Kalanchoë blossfeldiana* Poelln. *Plant Physiol. Biochem.* **1993**, *31*, 491–497.
90. Yang, X.H.; Hu, R.B.; Yin, H.F.; Jenkins, J.; Shu, S.Q.; Tang, H.B.; Liu, D.G.; Weighill, D.A.; Yim, W.C.; Ha, J.M.; et al. The *Kalanchö* genome provides insights into convergent evolution and building blocks of crassulacean acid metabolism. *Nat. Commun.* **2017**, *8*, 1899. [CrossRef]
91. Cushman, J.C.; Vernon, D.M.; Bohnert, H.J. ABA and the transcriptional control of CAM induction during salt stress in the common ice plant. In *Control of Plant Gene Expression*; Verma, D.P.S., Ed.; CRC Press: Boca Raton, FL, USA, 1993; pp. 287–300.
92. Schmitt, J.M.; Fißlthaler, B.; Sheriff, A.; Lenz, B.; Bäßler, M.; Meyer, G. Environmental control of CAM induction in *Mesembryanthemum crystallinum*—A role for cytokinin, abscisic acid and jasmonate? Crassulacean Acid Metabolism. *Biochem. Ecophysiol. Evol.* **1996**, *114*, 159–175.
93. Schaeffer, H.J.; Forstheoefel, N.R.; Cushman, J.C. Identification of enhancer and silencer regions involved in salt-responsive expression of crassulacean acid metabolism (CAM) genes in the facultative halophyte *Mesembryanthemum crystallinum*. *Plant Mol. Biol.* **1995**, *28*, 205–218. [CrossRef]
94. Amin, A.B.; Rathnayake, K.N.; Yim, W.C.; Garcia, T.M.; Wone, B.; Cushman, J.C.; Wone, B.W.M. Crassulacean acid metabolism abiotic stress-responsive transcription factors: A potential genetic engineering approach for improving crop tolerance to abiotic stress. *Front. Plant Sci.* **2019**, *10*, 129. [CrossRef] [PubMed]
95. Heyduk, K.; Ray, J.N.; Ayyampalayam, S.; Leebens-Mack, J. Shifts in gene expression profiles are associated with weak and strong crassulacean acid metabolism. *Am. J. Bot.* **2018**, *105*, 587–601. [CrossRef] [PubMed]
96. Moseley, R.C.; Mewalal, R.; Motta, F.; Tuskan, G.A.; Haase, S.; Yang, X.H. Conservation and diversification of circadian rhythmicity between a model crassulacean acid metabolism plant *Kalanchoë fedtschenkoi* and a model C_3 photosynthesis plant *Arabidopsis thaliana*. *Front. Plant Sci.* **2018**, *9*, 1757. [CrossRef]
97. Yin, H.B.; Guo, H.B.; Weston, D.J.; Borland, A.M.; Ranjan, P.; Abraham, P.E.; Jawdy, S.S.; Wachira, J.; Tuskan, G.A.; Tschaplinski, T.J.; et al. Diel rewiring and positive selection of ancient plant proteins enabled evolution of CAM photosynthesis in Agave. *BMC Genom.* **2018**, *19*, 588. [CrossRef] [PubMed]
98. Zhang, X.M.; Tan, B.W.; Zhu, D.; Dufresne, D.; Jiang, T.B.; Chen, S.X. Proteomics of Homeobox7 enhanced salt tolerance in *Mesembryanthemum crystallinum*. *Int. J. Mol. Sci.* **2021**, *22*, 6390. [CrossRef]
99. Cushman, J.C.; Bohnert, H.J. Salt stress alters A/T-rich DNA-binding factor interactions within the phosphoenolpyruvate carboxylase promoter from *Mesembryanthemum crystallinum*. *Plant Mol. Biol.* **1999**, *20*, 411–424. [CrossRef]
100. McElwain, E.F.; Bohnert, H.J.; Thomas, J.C. Light moderates the induction of phosphoenolpyruvate carboxylase by NaCl and abscisic acid in *Mesembryanthemum crystallinum*. *Plant Physiol.* **1992**, *99*, 1261–1264. [CrossRef]
101. Gehrig, H.; Taybi, T.; Kluge, M.; Brulfert, J. Identification of multiple PEPC isogenes in leaves of the facultative crassulacean acid metabolism (CAM) plant *Kalanchoe blossfeldiana* Poelln. cv.Tom Thumb. *FEBS Lett.* **1995**, *377*, 399402.
102. Vaasen, A.; Begerow, D.; Hampp, R. Phosphoenolpyruvate carboxylase genes in C_3, crassulacean acid metabolism (CAM) and C_3/CAM intermediate species of the genus *Clusia*: Rapid reversible C_3/CAM switches are based on the C_3 housekeeping gene. *Plant Cell Environ.* **2006**, *29*, 2113–2123. [CrossRef]

103. Li, B.; Chollet, R. Salt induction and the partial purification/characterization of phosphoenolpyruvate carboxylaseprotein-serine kinase from an inducible crassulacean acid-metabolism (CAM) plant, *Mesembryanthemum crystallinum* L. *Arch. Biochem. Biophys.* **1994**, *314*, 247–254. [CrossRef] [PubMed]
104. Kong, W.W.; Yoo, M.J.; Zhu, D.; Noble, J.D.; Kelley, T.M.; Li, J.; Kirst, M.; Assmann, S.M.; Chen, S.X. Molecular changes in *Mesembryanthemum crystallinum* guard cells underlying the C_3 to CAM transition. *Plant. Mol. Biol.* **2020**, *103*, 653–667. [CrossRef] [PubMed]
105. Fissßthaler, B.; Meyer, G.; Bohnert, H.J.; Schmitt, J.M. Age-dependent induction of pyruvate, orthophosphate dikinase in *Mesembryanthemum crystallinum* L. *Planta* **1995**, *196*, 492–500.
106. Forsthoefel, N.R.; Cushman, M.A.F.; Cushman, J.C. Posttranscriptional and posttranslational control of enolase expression in the facultative crassulacean acid metabolism plant *Mesembryanthemum crystallinum* L. *Plant Physiol.* **1995**, *108*, 1185–1195. [CrossRef]
107. Forsthoefel, N.R.; Vernon, D.M.; Cushman, J.C. A salinity-induced gene from the halophyte *M. crystallinum* encodes a glycolytic enzyme, cofactor-independent phosphoglyceromutase. *Plant Mol. Biol.* **1995**, *29*, 213–226. [CrossRef]
108. Ostrem, J.A.; Vernon, D.M.; Bohnert, H.J. Increased expression of a gene coding for NAD: Glyceraldehyde-3-phosphate dehydrogenase during the transition from C_3 photosynthesis to crassulacean acid metabolism in *Mesembryanthemum crystallinum*. *J. Biol. Chem.* **1990**, *265*, 3497–3502. [CrossRef]
109. Cushman, J.C. Characterization and expression of a NADP-malic enzyme cDNA induced by salt stress from the facultative CAM plant, *Mesembryanthemum crystallinum*. *Eur. J. Biochem.* **1992**, *208*, 259–266. [CrossRef]
110. Ocheretina, O.; Scheibe, R. Cloning and sequence analysis of cDNAs encoding cytosolic malate dehydrogenase. *Gene* **1997**, *199*, 145–148. [CrossRef]
111. Löw, R.; Rockel, B.; Kirsch, M.; Ratajczak, R.; Hörntensteiner, S. Early salt stress effects on the differential expression of vacuolar H+-ATPase genes in roots and leaves of *Mesembryanthemum crystallinum*. *Plant Physiol.* **1996**, *110*, 259–265. [CrossRef]
112. Bartholomew, D.M.; Rees, D.J.G.; Rambaut, A.; Smith, J.A.C. Isolation and sequence analysis of a cDNA encoding the c subunit of a vacuolar-type H+-ATPase from the CAM plant *Kalanchoë diagremontiana*. *Plant Mol. Biol.* **1996**, *31*, 435–442. [CrossRef]
113. Tsiantis, M.S.; Bartholomew, D.M.; Smith, J.A.C. Salt regulation of transcript levels for the c subunit of a leaf vacuolar H+-ATPase in the halophyte *Mesembryanthemum crystallinum*. *Plant J.* **1996**, *9*, 729–736. [CrossRef]
114. Dietz, K.J.; Arbinger, B. cDNA sequence and expression of subunit E of the vacuolar H(+)-ATPase in the inducible crassulacean acid metabolism plant *Mesembryanthemum crystallinum*. *Biochem. Biophys. Acta Gene Struct. Funct.* **1996**, *1281*, 134–138. [CrossRef]
115. Baur, B.; Fisher, K.; Winter, K.; Dietz, K.J. cDNA sequences of a protein kinase from the halophyte *Mesembryanthemum crystallinum* L., encoding a SNF-1 homologue. *Plant Physiol.* **1994**, *106*, 1225–1226. [CrossRef] [PubMed]
116. Breiteneder, H.; Michalowski, C.B.; Bohnert, H.J. Environmental stress mediated differential 3′ end formation of chloroplast RNA-binding protein transcripts. *Plant Mol. Biol.* **1994**, *26*, 833–849. [CrossRef]
117. Rippmann, J.F.; Michalowski, C.B.; Nelson, D.E.; Bohnert, H.J. Induction of a ribosome-inactivating protein upon environmental stress. *Plant Mol. Biol.* **1997**, *35*, 701–709. [CrossRef]
118. Cushman, J.C.; Michalowski, C.B.; Bohnert, H.J. Developmental control of crassulacean acid metabolism inducibility by salt stress in the common ice plant. *Plant Physiol.* **1990**, *94*, 1137–1142. [CrossRef]
119. Cushman, J.C.; Bohnert, H.J. Transcriptional activation of CAM genes during development and environmental stress. In *Ecological Studies*; Smith, J.A.C., Winter, A.C., Eds.; Metabolism: Biochemistry, Ecophysiology and Evolution; Springer: Berlin, Germany, 1996; pp. 135–158.
120. Bohnert, H.J.; Ayoubi, P.; Borchert, C.; Bressan, R.A.; Burnap, R.L.; Cushman, J.C.; Cushman, M.A.; Deyholos, M.; Fischer, R.; Galbraith, D.W.; et al. A genomics approach towards salt stress tolerance. *Plant Physiol. Biochem.* **2001**, *39*, 295–311. [CrossRef]
121. Cushman, J.C.; Tillett, R.L.; Wood, J.A.; Branco, J.M.; Schlauch, K.A. Large-scale mRNA expression profiling in the common ice plant, *Mesembryanthemum crystallinum*, performing C_3 photosynthesis and crassulacean acid metabolism (CAM). *J. Exp. Bot.* **2008**, *59*, 1875–1894. [CrossRef] [PubMed]
122. Boxall, S.F.; Dever, L.V.; Kneřová, J.; Gould, P.D.; Hartwell, J. Phosphorylation of phosphoenolpyruvate ccarboxylase is essential for maximal and sustained dark CO_2 Fixation and core circadian clock operation in the obligate crassulacean acid metabolism species *Kalanchoë fedtschenkoi*. *Plant Cell* **2017**, *29*, 2519–2536. [CrossRef] [PubMed]
123. Wai, C.M.; Weise, S.E.; Ozersky, P.; Mockler, T.C.; Michael, T.P.; VanBuren, R. Time of day and network reprogramming during drought induced CAM photosynthesis in *Sedum album*. *PLoS Genet.* **2019**, *15*, e1008209. [CrossRef] [PubMed]
124. Deng, H.; Zhang, L.S.; Zhang, G.Q.; Zheng, B.Q.; Liu, Z.J.; Wang, Y. Evolutionary history of PEPC genes in green plants: Implications for the evolution of CAM in orchids. *Mol. Phylogenetics Evol.* **2016**, *94*, 559–564. [CrossRef]
125. Fankhauser, N.; Aubry, S. Post-transcriptional regulation of photosynthetic genes is a key driver of C_4 leaf ontogeny. *J. Exp. Bot.* **2017**, *68*, 137–146. [CrossRef]
126. Shen, S.Q.; Li, N.; Wang, Y.J.; Zhou, R.; Sun, P.C.; Lin, H.; Chen, W.; Yu, T.; Liu, Z.; Wang, Z.Y.; et al. High quality ice plant reference genome analysis provides insights into genome evolution and allows exploration of genes involved in the transition from C_3 to CAM pathways. *Plant Biotechnol. J.* **2022**, *20*, 2107–2122. [CrossRef] [PubMed]
127. Wai, C.M.; VanBuren, R.; Zhang, J.S.; Huang, L.X.; Miao, W.; Edger, P.P.; Yim, W.C.; Priest, H.D.; Meyers, B.C.; Mockler, T.; et al. Temporal and spatial transcriptomic and microRNA dynamics of CAM photosynthesis in pineapple. *Plant J. Cell Mol. Biol.* **2017**, *92*, 19–30. [CrossRef] [PubMed]

128. Chiang, C.P.; Yim, W.C.; Sun, Y.H.; Ohnishi, M.; Mimura, T.; Cushman, J.C.; Yen, H.E. Identification of ice plant (*Mesembryanthemum crystallinum* L.) microRNAs using RNA-seq and their putative roles in high salinity responses in seedlings. *Front. Plant Sci.* **2016**, *7*, 1143. [CrossRef]
129. Hu, Z.K.; Nie, Z.Y.; Yan, C.; Huang, H.; Ma, X.J.; Wang, Y.P.; Ye, N.; Tuskan, G.A.; Yang, X.H.; Yin, H.F. Transcriptome and degradome profiling reveals a role of miR530 in the circadian regulation of gene expression in *Kalanchoë marnieriana*. *Cells* **2021**, *10*, 1526. [CrossRef] [PubMed]
130. Kore-eda, S.; Cushman, M.A.; Akselrod, I.; Bufford, D.; Fredrickson, M.; Clark, E.; Cushman, J.C. Transcript profiling of salinity stress responses by large-scale expressed sequence tag analysis in *Mesembryanthemum crystallinum*. *Gene* **2004**, *341*, 83–92. [CrossRef]
131. Guan, Q.J.; Kong, W.W.; Zhu, D.; Zhu, W.; Dufresne, C.; Tian, J.K.; Chen, S.X. Comparative proteomics of *Mesembryanthemum crystallinum* guard cells and mesophyll cells in transition from C_3 to CAM. *J. Proteom.* **2021**, *231*, 104019. [CrossRef]
132. Perron, N.; Tan, B.; Dufresne, C.P.; Chen, S.X. Proteomics and phosphoproteomics of C_3 to CAM transition in the common ice plant. *Methods Enzymol.* **2022**, *676*, 347–368.
133. Nobe, P. High productivities of certain agronomic CAM species. Crassulacean Acid Metabolism. In *Biochemistry, Ecophysiology and Evolution*; Springer: Berlin/Heidelberg, Germany, 1996; pp. 255–265.
134. Töpfer, N.; Braam, T.; Shameer, S.; Ratcliffe, R.G.; Sweetlove, L.J. Alternative crassulacean acid metabolism modes provide environment-specific water-saving benefits in a leaf metabolic model. *Plant Cell* **2020**, *32*, 3689–3705. [CrossRef]
135. Winter, K.; Smith, J.A.C. (Eds.) Taxonomic distribution of crassulacean acid metabolism. In *Crassulacean Acid Metabolism. Biochemistry, Ecophysiology and Evolution*; Springer: Berlin/Heidelberg, Germany, 1996; pp. 427–436.
136. Qiu, S.; Sultana, S.; Liu, Z.D.; Yin, L.Y.; Wang, C.Y. Identification of obligate C_3 photosynthesis in *Dendrobium*. *Photosynthetica* **2015**, *53*, 168–176. [CrossRef]
137. Zou, L.H.; Wan, X.; Deng, H.; Zheng, B.Q.; Li, B.J.; Wang, Y. RNA-seq transcriptomic profiling of crassulacean acid metabolism pathway in *Dendrobium catenatum*. *Sci. Data* **2018**, *5*, 180252. [CrossRef]
138. Zhang, Z.J.; He, D.X.; Gao, R.F. Concomitant CAM and C_3 photosynthetic pathways in *Dendrobium officinale* plants. *J. Am. Soc. Hortic. Sci.* **2014**, *139*, 290–298. [CrossRef]
139. Winter, K.; Holtum, J.A.M. Cryptic crassulacean acid metabolism (CAM) in *Jatropha curcas*. *Funct Plant Biol.* **2015**, *42*, 711–717. [CrossRef]
140. Yuan, G.L.; Hassan, M.M.; Liu, D.G.; Lim, S.D.; Yim, W.C.; Cushman, J.C.; Markel, K.; Shih, P.M.; Lu, H.W.; Weston, D.J.; et al. Biosystems Design to Accelerate C_3-to-CAM Progression. *BioDesign Res.* **2020**, *2020*, 3686791. [CrossRef]
141. Cosentino, C.; Silvestre, D.D.; Fischer-Schliebs, E.; Homann, U.; Palma, A.D.; Comunian, C.; Mauri, P.L.; Thiel, G. Proteomic analysis of *Mesembryanthemum crystallinum* leaf microsomal fractions finds an imbalance in V-ATPase stoichiometry during the salt-induced transition from C_3 to CAM. *Biochem. J.* **2013**, *450*, 407–415. [CrossRef] [PubMed]
142. Oh, D.H.; Barkla, B.J.; Vera-Estrella, R.; Pantoja, O.; Lee, S.Y.; Bohnert, H.J.; Dassanayake, M. Cell type-specific responses to Psalinity–the epidermal bladder cell transcriptome of *Mesembryanthemum crystallinum*. *New Phytol.* **2015**, *207*, 627–644. [CrossRef] [PubMed]
143. Barkla, B.J.; Vera-Estrella, R. Single cell-type comparative metabolomics of epidermal bladder cells from the halophyte *Mesembryanthemum crystallinum*. *Front. Plant Sci.* **2015**, *6*, 435. [CrossRef] [PubMed]
144. Tsukagoshi, H.; Suzuki, T.; Nishikawa, K.; Agarie, S.; Ishiguro, S.; Higashiyama, T. RNA-seq analysis of the response of the halophyte, *Mesembryanthemum crystallinum* (ice plant) to high salinity. *PLoS ONE* **2015**, *10*, e0118339. [CrossRef]
145. Barkla, B.J.; Vera-Estrella, R.; Raymond, C. Single-cell-type quantitative proteomic and ionomic analysis of epidermal bladder cells from the halophyte model plant *Mesembryanthemum crystallinum* to identify salt-responsive proteins. *BMC Plant. Biol.* **2016**, *16*, 110. [CrossRef]
146. Yan, X.; Chang, Y.; Zhao, W.J.; Qian, C.J.; Yin, X.Y.; Fan, X.K.; Zhu, X.Y.; Zhao, X.Q.; Ma, X.F. Transcriptome profiling reveals that foliar water uptake occurs with C_3 and crassulacean acid metabolism facultative photosynthesis in *Tamarix ramosissima* under extreme drought. *AoB Plants* **2022**, *14*, plab060. [CrossRef] [PubMed]
147. Heyduk, K.; McAssey, E.V.; Leebens-Mack, J. Differential timing of gene expression and recruitment in independent origins of CAM in the Agavoideae (Asparagaceae). *New Phytol.* **2022**, *235*, 2111–2126. [CrossRef] [PubMed]

MDPI AG
Grosspeteranlage 5
4052 Basel
Switzerland
Tel.: +41 61 683 77 34

Horticulturae Editorial Office
E-mail: horticulturae@mdpi.com
www.mdpi.com/journal/horticulturae

www.ingramcontent.com/pod-product-compliance
Lightning Source LLC
Chambersburg PA
CBHW061605210926
51850CB00009B/2217
* 9 7 8 3 7 2 5 8 1 4 9 1 6 *